Two-Dimensional Nanostructures

Two-Dimensional Nanostructures

Mahmood Aliofkhazraei
Nasar Ali

CRC Press
Taylor & Francis Group
Boca Raton London New York

CRC Press is an imprint of the
Taylor & Francis Group, an **informa** business

CRC Press
Taylor & Francis Group
6000 Broken Sound Parkway NW, Suite 300
Boca Raton, FL 33487-2742

Printed in the United States of America on acid-free paper
Version Date: 20120224

International Standard Book Number: 978-1-4398-6665-8 (Hardback)

Library of Congress Cataloging-in-Publication Data

Aliofkhazraei, Mahmood.
Two-dimensional nanostructures / Mahmood Aliofkhazraei, Nasar Ali.
p. cm.
Summary: "Discussing different fabrication methods for developing 2-D nanostructures, this book is the first of its kind to focus on the "size effect" of 2-D nanostructures. Using accessible language and simple figures, it classifies different methods by their ability to control the sizes of 2-D nanostructures and thus the relative properties of the resulting materials. The book also presents applications in both nanotechnology and materials science and covers mechanical, electrochemical, and physical properties and usage, including thin films and nanostructured coatings"-- Provided by publisher.
Includes bibliographical references and index.
ISBN 978-1-4398-6665-8 (hardback)
1. Nanostructured materials. I. Ali, Nasar. II. Title.

TA418.9.N35A435 2012
620.1'15--dc23 2012005186

Visit the Taylor & Francis Web site at
http://www.taylorandfrancis.com

and the CRC Press Web site at
http://www.crcpress.com

Contents

Preface

The focus on two-dimensional nanostructures has intensified significantly during the past decade. In addition to conventional nanolayers, newly discovered materials in this form with their unique properties have attracted considerable attention from the research community. New aspects of size effect in this kind of nanostructure show great potential for future applications. A nice example for one of the most important materials in this form, which has been the focus of researchers in recent decades, is graphene. Graphene is the two-dimensional crystalline form of carbon. It is a single layer of carbon atoms arranged in hexagons. It seems that such two-dimensional crystals are impossible to create; however, physicists at the University of Manchester (Professor Geim's group) confirmed attractive properties of graphene for the first time in 2004. As a result of research and development on such class of materials, researchers have been able to identify unique properties of this element. One of the novel applications of graphene is to create adjustable optical modulator devices used in communication, as well as other electronic instruments, in nanoscale. Graphite is a well-known phase of carbon, which is also used in manufacturing other materials. It is composed of layers of carbon atoms that are densely packed and stacked together in a surface. Although prior to 2004 none of these layers were examined separately, each of the constituent layers of graphite is graphene. As a crystal, two-dimensional graphene is quite dissimilar from three-dimensional materials such as silicon. One of the noticeable properties of the structure of electrons in this material is that graphene has electrons that are free to move around. Unlike other materials, the electrons in graphene move without collision over great distances, even at room temperature. Subsequently, the ability of electrons to conduct electrical current is 10 to 100 times greater than that of semiconductors such as silicon.

Even before research pioneers working on graphene received the Noble Prize in physics in 2010, this matter was declared the next wonder material. Many believe that graphene will spell an end for silicon. Graphene that has been touted as the "miracle material" of the twenty-first century is said to be the strongest material ever studied and a replacement for silicon. Its exceptional properties, such as the most conductive material known to humans, have shocked the scientific world. Researchers noticed graphene as the strongest material ever known. It is 200 times stronger than steel, and it would take the weight of an elephant behind a pencil to pass through a thin sheet of graphene. Since its properties are uncovered, more and more scientists are keen to work on relevant projects. About 200 companies are now involved in conducting research on graphene, and just in 2010 more than 3600 articles were published about this subject.* The benefits for both companies and consumers are obvious: faster and cheaper devices that are thinner and more flexible. Imagine you can roll up your intelligent phone and stick it behind your ear like a carpenter's pencil. If graphene is compared to the way plastic is used today, we should expect a day when everything from crisp packets to clothing would be digitized. The future could see credit cards contain as much processing power as your current smart-phones. Graphene can open completely new applications in transparent electronics, in flexible electronics, and electronics that are much faster than today. Beyond its digital applications, just one example of its use would be graphene powder added to tires to make them stronger. Samsung Company has been one of the biggest investors in research on graphene. It has already demonstrated a 25-inch flexible touch-screen using this technology, aiming to manufacture a dozen commercial products using graphene in the next five years. Nevertheless, companies like IBM and Nokia also have hope for the future of graphene. IBM has created a 150 GHz transistor, while the quickest comparable silicon device runs at about 40 GHz.

All of the microelectronic advances including transistors, memories, sensors, and so on owe the thin layers that employ these kinds of nanostructures. To improve the properties related to surface, such as fatigue, corrosion, and erosion, nanostructured coatings are also expanded. Optical glasses also employ nanostructured coatings. This means that a thin nanostructured layer, like ZnO, is purposefully deposited on a glass, which alters its properties. Heat and ultraviolet (UV) reflective glasses are of this category. These are just a few indications of what two-dimensional nanostructures can do in our future. Based on this importance we decided to write a book about their fabrication methods and properties. Scientists alter the surface properties of materials by using two-dimensional nanostructures of different materials with thicknesses less than 100 nm. The substrate material and nanocoating create a system with different enhanced properties. Nanotechnology creates instruments for control of three key parameters for two-dimensional nanostructures: (a) chemical composition, (b) thickness, and (c) surface geometry. Currently different processes are applied for production of two-dimensional nanostructures. We classified them in three chemical, mechanical, and physical routes in different chapters. This book also contains some basic introduction and classification of nanostructures from their fabrication methods to their different properties.

Chapter 1 discusses the principal concepts of different kinds of nanostructures. Properties of nanostructured materials are described and the size effect has been considered. Different influenced properties of nanomaterials with some interesting examples have been described. This chapter also discusses future perspectives of nanotechnology and continues with some applications of nanostructures. Chapter 2 introduces some classification

* Based on the search of "graphene" in (abstract + title + keywords) in the Scopus database.

groups for two-dimensional nanostructures. The main groups of fabrication methods have been described briefly. The chapter also describes different forms of growth for two-dimensional nanostructures. Chapter 3 discusses both characterization and fabrication methods of two-dimensional nanostructures in detail.

Chapter 4 is about both mechanical fabrication methods and mechanical properties of two-dimensional nanostructures. Usual usage of multiple nanolayers is introduced in this chapter. A new mechanical method for fabrication of graphene layers is also mentioned. The effect of a supermodule in multilayer systems and some aspects of tribological behavior of multilayer coatings are discussed in this chapter. Chapter 5 discusses chemical fabrication methods and affected properties of two-dimensional nanostructures. A conventional chemical deposition method is presented with theory and a thermodynamic method of codeposition, and also nucleation and the growth process. Some aspects of phase transition for two-dimensional nanostructures are also described. Chapter 6 discusses physical and especially light-scattering properties of two-dimensional nanostructures. It also contains important physical fabrication methods of nanostructures. Specifications of nanostructured thin films are discussed with a focus on a model to relate features and structures.

Finally we would like to express our thanks to Professors Andre Geim and Konstantin Novoselov (Nobel Laureates [Nobel Prize in Physics, 2010], for groundbreaking experiments regarding the two-dimensional material graphene) and also the Condensed Matter Physics group at the Manchester University for allowing us to use some of their related images of two-dimensional nanostructures on the front cover of this book.

Mahmood Aliofkhazraei
Nasar Ali

1

Synthesis, Processing, and Application of Nanostructures

1.1 Introduction to Nanotechnology

During the past years scientists have achieved significant successes in nano science and technology. Nanotechnology is a branch of sciences that deals with fine structures and materials with very small dimensions—less than 100 nm. The measurement unit of "nano" has been extracted from the *nano-* prefix, which is a Greek word meaning "extremely fine." One nano (10^{-9} m) is the length equivalent to 5 silicon atoms or 10 hydrogen atoms aligned side by side. To introduce a perspective note the following examples: A hydrogen atom is about 0.1 nm; a virus is about 100 nm; the diameter of a red blood cell is 7000 nm; and the diameter of a human hair is 10000 nm.[1–3]

Nanotechnology is a field of applied sciences focused on design, production, detection, and employing the nanosize materials, pieces, and equipment. Advances in nanotechnology lead to improvement of tools and equipment as well as their application in human life. *Nanoscience* is the study of the phenomena emerged by atomic or molecular materials with the size of several nanometers to less than 100 nm. In chemistry this size involves a range of colloids, micelles, polymer molecules, and structures such as very large molecules or dense accumulation of the molecules. In the physics of electrical engineering, nanoscience is strongly related to quantum behavior or electron behavior in structures with nano sizes. In biology and biochemistry, also, interesting cellular components and molecular structures such as DNA, RNA, and intercellular components are considered as nanostructures.[4–7]

Nanotechnology is applying the sciences in control of the phenomena with molecular dimensions. In this range, materials properties are significantly different than those of its mass state. These differences are related to design, properties' detection analysis, structures application, as well as tools and systems for nano shape and size control. In other words, nanoscience and technology is a study field including

1. Advances in the methods for producing the materials and surface analysis equipment for producing the materials and structures
2. Understanding the changes in physical and chemical properties in order to materials fining
3. Using these properties for the development of new highly used materials in the equipment

Alive cells are the best examples of machines working at the nano level. The study field of the nanotechnology major is interconnected to physics, chemistry, materials science, microbiology, biochemistry, and molecular biology. Integration of nanotechnology, biotechnology, and medical engineering is demonstrated in the use of nano-size structures for disease prognosis, defining the gene sequence, and medicine transport.[8,9]

The main applications of nanotechnology are

- Medicine: drug delivery and tissue engineering
- Chemistry and environmental studies: catalysts, filtration
- Energy: decrease in energy uses, increasing of the efficiency of the products yielding batteries recycling energy, use of environmentally friendly energy systems
- Information and communications: new semiconductors, new electrical pieces, display monitors with low energy use, quantum computers
- Consumer goods: food production, safety, and packaging, self-cleaner system, fibers and textiles, optical equipment, sport facilities

1.2 History of Nanotechnology

Use of nanotechnology by humans, contrary to the dominant belief, has a long history. There is evidence implying the nanostructured nature of the blue color used by the Mayans. Since then, Romans used these materials for fabrication of cups with live colors; they applied gold for dying these cups. One of these cups was the Lycurgus cup that was made in 4 BC. This cup involves gold and silver particles and reveals various colors once exposed to light. This method was also used in the Middle Ages for making church windows.[10,11]

Among various scientific fields, biology was the first that entered this realm. The basic mechanism of biological machines was nanodimensional reactions. Mosquitoes, ants, and flies can be named as examples of natural nanomachines. However, the scientific account of nanotechnology varies from that explained above. Early investigations on nanoparticles stem from Faraday. In 1811 working on the red colloid of Ag, he found that the red colloid color depends on metallic particles sizes, and the gold and silver found 2000 years ago were actually nanoparticles.

The use of nanostructured catalysts began 70 years ago. In the early 1940s the evaporated and deposited nanoparticles of silicon were fabricated and used as an alternative for tiny particles of black carbon for rubber reinforcement in the United States and Germany. These products have a wide range of applications, from dry milk used with coffee to car tires and optical fibers. In the 1960s and 1970s, metallic nanopowders were made for information storing on film. In 1976, nanocrystals were invented by some researchers using the complete gas evaporation method.

The first breakthrough in the history of nanotechnology originates from the speech of Richard Feynman in a conference in 1959. In this conference, he discussed materials manipulation in nano dimensions in his paper entitled "There Are Plenty of Rooms in the Bottom." Today this paper is used as part of a nanotechnology committee's regulation. He extracted his notion from biological mechanisms, which are very tiny on one hand, and very active on the other. He stated that "There is nothing that I can see in the physical laws that says the computer elements cannot be made enormously smaller than they are now."*

The first impetus for materials and structures' fining arose from the electronics industry. Their goal was to industrialize their equipment to produce tinier electronic tools on silicon chips with sizes of 40 to 70 nm. Nowadays, nanoengineering is rapidly developing and introduces mechanical, catalyzer, electrical, magnetic, optical, and electronic potentials.[12–17]

1.3 What Is a Nanomaterial?

A nanomaterial belongs to a group of substances that have exterior dimensions or constituent phases (at least in one dimension) less than 100 nm; however, most of their unique properties occur in dimensions less than 50 nm. With respect to this definition, nanomaterials are divided into the following groups:[18]

1. Atomic clusters or nanoparticles
2. Nanolayers
3. Nanotubes and nanorods
4. Nanocrystals
5. Nanocomposites

* "There's Plenty of Room at the Bottom, An Invitation to Enter a New Field of Physics," Richard P. Feynman, transcript of the classic talk that Richard Feynman gave on December 29th 1959 at the annual meeting of the American Physical Society.

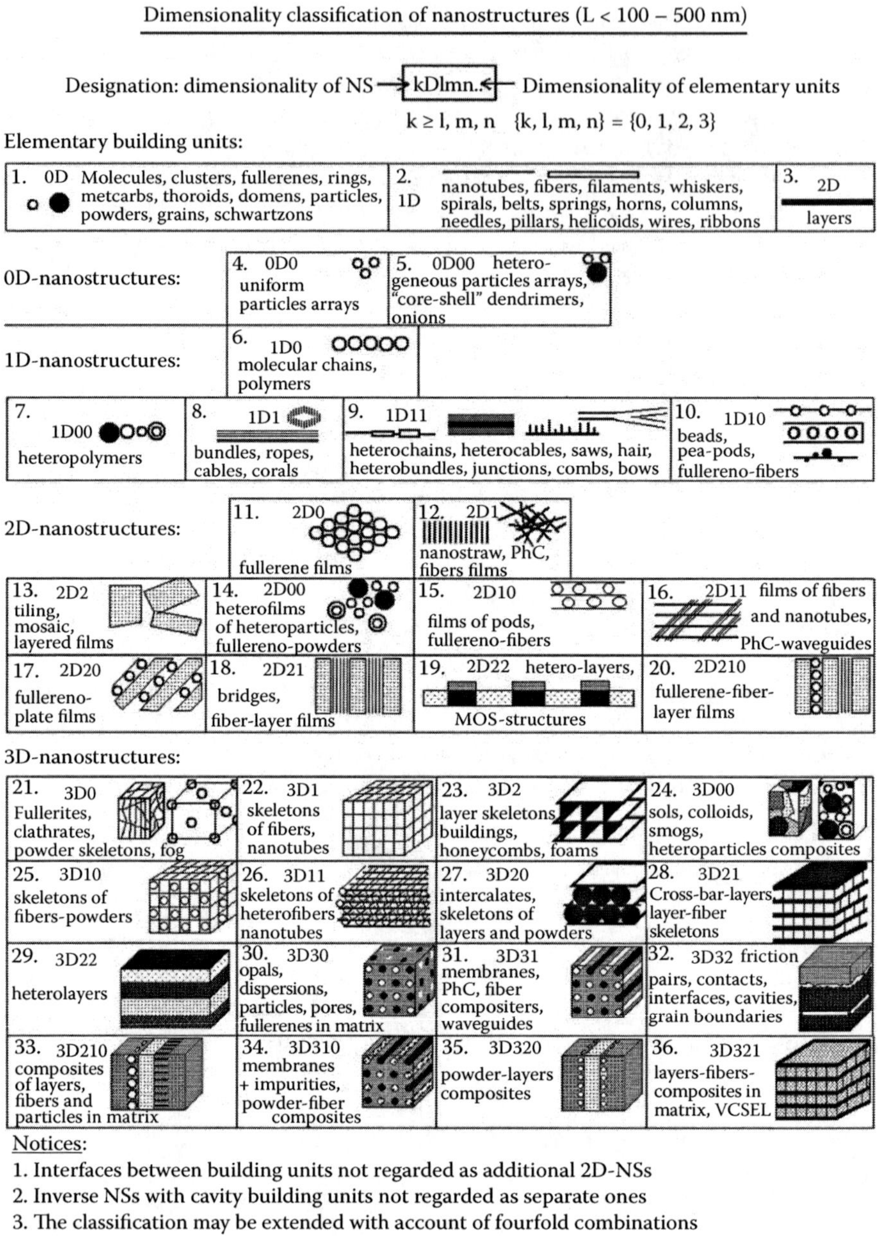

FIGURE 1.1
Dimensionality classification of nanostructures. (Reprinted from Pokropivny, V. V., and Skorokhod, V. V., Classification of nanostructures by dimensionality and concept of surface forms engineering in nanomaterial science, *Materials Science and Engineering C* 27 (5-8 SPEC. ISS.), 990–993, Copyright 2007, with permission from Elsevier.)

Categorization of nanomaterials by their nanometric dimensions is shown in Figure 1.1. Atomic clusters are known as nondimensional nanomaterials, while bulk nanocrystals and nanocomposites are called three-dimensional (3D) nanomaterials. Within the nanocrystals, the grains (crystallites) are co-axis and their size is in the nano range. The structure of nanocomposites is made of two or more phases, where one of the phases is presented in zero, one-, two-, or three-dimensional states. Nanocrystals and nanocomposites are generally known as nanostructured materials.

All nanomaterials share this feature that a significant portion of their atoms is located in interfaces such as free surfaces, grains boundaries, and interphase faces. Most nanomaterial properties come from this fact. There are several physical, chemical, and mechanical methods invented for production of nanomaterials. Nanomaterial is made of structural units that are laid side by side. In some methods for production of nanomaterials, first the structural units are made from atoms or molecules and then combining these structural units the given nanostructure is achieved. These methods are known as "bottom-up." However, there are some other methods known as "top-down," in which the large structural units of a material are gradually divided into smaller parts until we reach nano dimensions.[19–22]

The first point in production of nanomaterials can be started from a gas, liquid (or melted), and even solid phase. Table 1.1 shows some of the main processes in each group. Figure 1.2 presents an example for a method that has been used in order to fabricate reduced graphene oxide.

In each of these processes, it is possible to change the dimension, morphology, and texture of nanostructured material by controlling the involved parameters and achieving favorable characteristics. Nevertheless, it must be noted that in processes that involve a change of phase (e.g., changing the vapor phase into a solid or changing the liquid phase into a solid), creating the structure requires an increase in nucleating rate and a decrease of growth rate during the production process. Each of these processes involves advantages as well as disadvantages. Choosing a production method depends on multiple factors such as characteristics of the final structure (grain size, phase morphology, etc.) and purity of the final product.

Processes for producing nanomaterials in solid phases involve low production costs. The operation is performed in low temperatures (generally

TABLE 1.1

Classification of Nanomaterials Production Processes

Production from vapor phase	Physical and chemical deposition of the vapor
Production from liquid (or melted) phase	Sol-gel, electrochemical deposition, sudden freezing
Production from solid phase	Mechanical alloying, severe plastic deformation, mechanochemical process, crystallization of the amorphous phase

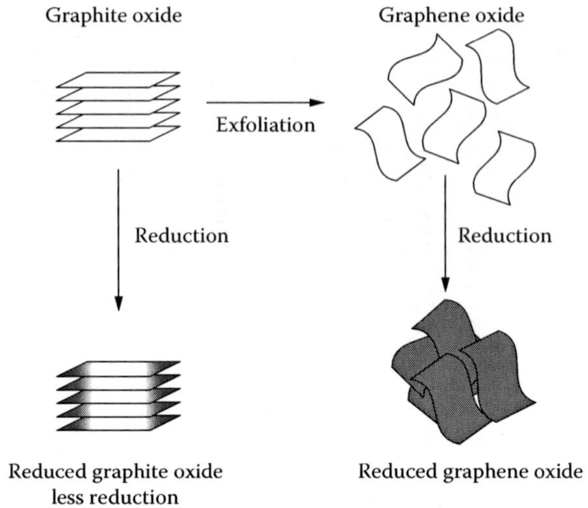

FIGURE 1.2
Representative scheme of a degree of reduction of graphite/graphene oxides. (Reprinted from Park, S., An, J., Potts, J. R., Velamakanni, A., Murali, S., and Ruoff, R. S., Hydrazine-reduction of graphite- and graphene oxide, *Carbon* 49 (9), 3019–3023, Copyright 2011, with permission from Elsevier.)

in room temperature). In addition, these processes can be applied for a variety of materials including pure metals, metallic alloys, intermetallic components, ceramics, and polymers. Moreover, it is possible to easily control production conditions to create ideal fine structures. This feature has caused wide use of solid-state production processes for creating 3D nanomaterials, particularly once the high rate of purity is not our objective. In the following sections of this chapter, the techniques applied for production of nanostructured materials will be discussed briefly.[24–27]

1.4 Properties of Nanostructured Materials

In general, a material's properties depend on atom arrangement and atom alignment in its structure. For instance, steel's properties differ from those of copper, because their atoms are different. Furthermore, properties of face-centered cubic (FCC) crystalline steel are different from body-centered cubic (BCC) crystalline steel because their atom arrangement in the crystal structure is not the same.

One distinct property of nanomaterials is their different behavior in macro- or microstructural states. All materials, irrespective of their composition, reveal new features (such as improvement of their optical features, decrease of melt point, increase of tensional strength, enhancement in catalytic properties, increase of semiconductors' band gap, increase of magnetic properties, and decrease of electrical resistance) once their size drops lower than 100 nm. As well as composition and structure of a substance, its dimension, once it goes below a specific limit, is another critical factor that has an effect on its properties. The following items can be mentioned as reasons for these behaviors of the materials:

1. Material size approaching molecular and atomic ranges
2. The high surface-to-volume ratio in nanomaterials; this means that the atom's distance from the surface is very short, so the interatomic forces and chemical bounds are of great importance and have a determining role
3. Increase in the volume of grains' boundaries with decrease of grains' sizes which, in turn, affects the material's physical features

It must be noted that with an increase in the surface ratio, the free energy of the material rises; this causes changes in the material's features. This can be explored in another way: According to the above-mentioned points, materials' characteristics also depend on their atoms arrangement, so as the grains sections increase, the larger number of atoms would be present in its exterior surface, implying that the material's atoms are in an environment different than that of atoms in conventional materials; hence, atoms placed at the surface experience a different environment than that of interior atoms.[28–30]

Materials properties considerably rely on their internal structures' dimensions. As the size of grains or constituent phases of a material decreases up to the nano range, common mechanisms of the mechanical and physical properties cannot be considered, and the material's properties dramatically change. For instance, pure microcrystalline metals indicate a low yield strength, which results from ease of dislocation movements and development across their grains. However, once grain size of pure metal drops to the nanometric range, the dislocations mechanism would not be active, and yield strength of the material will be significantly enhanced.

Properties of nanostructured materials depend on three main factors:

- Size of constituent phases or grains of the material
- Structure or nature of the existing interfaces such as grain contacts and interphase faces
- Chemical of the substance

In some cases it is possible that just one of the above-mentioned factors plays a main role in defining properties of the nanostructured material.

In different control methods, it may be possible to partially control these three factors and achieve the objective properties through control of process parameters. But the efficacy of controlling all these three factors is not equal in all methods.

One of the most important characteristics of nanostructured materials is their having an extensive interface surface with each other.[31–36] So, a considerable ratio of their atoms is located in their interfaces. For example, once we have a nanocrystalline material with spherical or cubic grains with a diameter of d, the surface-to-volume ratio will be $6/d$ taking the grain boundary's thickness (δ) into account and regarding this fact that each grain boundary is shared with two adjacent grains, the ratio of atoms located in grain boundaries (X_b) is defined by the following equation:

$$X_b = 3\delta/d \times 100 \tag{1.1}$$

Usually as the grain size drops, the number of atoms in grain boundaries greatly increases (e.g., when the grain size is 5 nm the atom number in the grain boundary of 0.5 nm will be 50%). If the grain boundary thickness is 1 nm, 80% of the atoms will be located in grain boundaries. Thus, it is predicted that in this condition the nanocrystalline material's properties is strongly influenced by grain boundaries' properties. The relation between the number of atoms per grain boundary (GB) interface and the calculated grain boundary interface area is calculated in Figure 1.3.

Although several studies have been performed to identify grain boundaries in nanocrystalline materials, it is not clear that weather atoms arrangement in grain boundaries of nanocrystalline and microcrystalline materials is the same, and weather structure and nature of grain sizes in nanocrystalline materials are influenced by their production technique. In any event, the average distance between atoms in grain boundaries is the maximum, and the number of closest neighbors for each atom is the minimum. In addition, interatomic bonds in grain boundaries are the weakest. In addition to an extensive interface surface, the presence of voids created by production steps and crystalline defects such as atomic vacancies, or residual dislocations, has a significant effect on the properties of nanostructured materials. Therefore, it is expected that nanostructured materials properties, even in grain size and chemical composition, be varied according to their production method.[38–43]

As previously mentioned, there is an extensive interface surface in nanostructured materials, where a considerable ratio of atoms can be accommodated. Such a structure results in different and superior chemical, physical, and mechanical properties of nanostructured materials compared with traditional substances (microcrystals). There is also a possibility to change different characteristics in a definite range by controlling the grain size and phases. Some of these changes are shown in Figure 1.4.

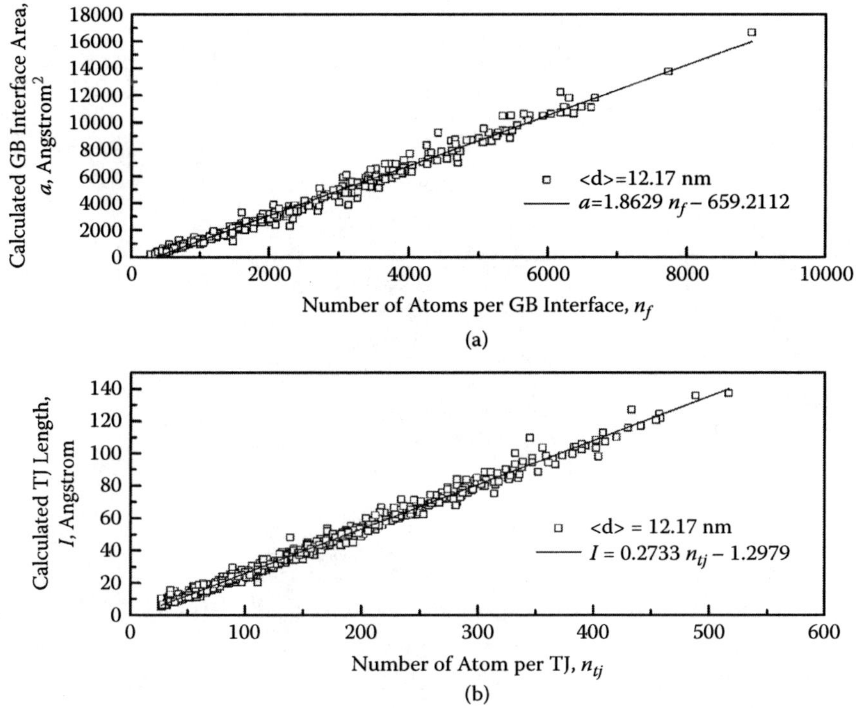

FIGURE 1.3
(a) The linear relation between the number of atoms per grain boundary (GB) interface, n_f, and the calculated GB interface area; (b) the linear relation between the number of atoms per triple junction (TJ), n_{tj}, and the calculated TJ length. The number of grains in the nanocrystalline Cu sample is 50, and the mean grain size is 12.17 nm. (Reprinted from Li, M., and Xu, T., Topological and atomic scale characterization of grain boundary networks in polycrystalline and nanocrystalline materials, *Progress in Materials Science* 56 (6), 864–899, Copyright 2011, with permission from Elsevier.)

1.4.1 Physical Properties

1.4.1.1 Diffusion Coefficient

For their wider atomic arrangement, grain boundaries and interphase faces are considered as quicker diffusion paths in polycrystalline materials, as activation energy of atoms' diffusion through grains boundary is about half of their activation energy for diffusion across the grains. The difference between the diffusion rate in a grain's boundary and the diffusion rate across the grains themselves is of a great importance in low temperatures.[44–56] Diffusion coefficient and temperature are correlated by the following exponential Arrhenius function:

$$D = D_0 \exp\left(\frac{-Q}{RT}\right) \tag{1.2}$$

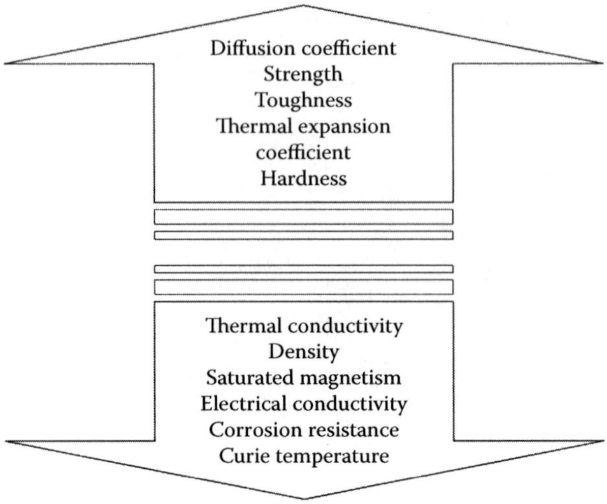

FIGURE 1.4
Usual mechanical and physical trends of changes for nanostructured materials.

where D is the diffusion coefficient, D_0 is the constant coefficient, Q is the diffusion activation energy, T is the temperature (K), and R is the universal gas constant. Table 1.2 presents grain boundary and lattice diffusion coefficients for self-diffusion of Au at two different temperatures.

In any case, in microcrystalline materials, the grain boundary share in whole diffusion is negligible, but for nanocrystalline materials, where there is an extensive grain boundary, it is predicted that the diffusion rate would be considerably high, in comparison with microcrystalline materials. Table 1.3 shows diffusion coefficients of silver atoms in copper at three different temperatures, which reveals that the diffusion coefficient for a nanocrystalline sample is higher than that of grain boundaries.

An increase in diffusion of nanostructures is also observed in some other metals. Here, it must be noted that the diffusion coefficient in a nanocrystalline sample (D_n) is even more than that of grain boundary (D_b) diffusion, suggesting that the structure of the grain boundary for nanocrystals might

TABLE 1.2
Diffusion Coefficient across the Grain Boundary (D_b) and Lattice (D_L) for Self-Diffusion of Au

Material	Temperature (°C ± 0.2°C)	$\alpha^{-5/3}t^{-1/2} \times 10^{-9}$	D_L (cm²/sec)	δD_b (cm³/sec)
(A) Bulk	444.1	1.23	1.85×10^{-14}	2.11×10^{-16}
	367.2	1.37	5.54×10^{-16}	4.19×10^{-17}
(B) Films (Au-Mo-SiO_2)	177.0	0.58	5.41×10^{-22}	1.78×10^{-20}
	117.0	0.33	4.17×10^{-25}	2.81×10^{-22}

TABLE 1.3

Diffusion Coefficients of Silver Atoms in Nanocrystalline Copper, in Copper Grain Boundaries, in the Lattice of Copper, and on Copper Surfaces

T (°K)	Nanocrystalline D_i (m²/s)	Grain boundaries[a] D_b (m²/s)	Lattice D_L (m²/s)	Surface[b] D_s (m²/s)
373	1.2×10^{-17}	22.5×10^{-20}	8.1×10^{-33}	9×10^{-14}
353	3.1×10^{-18}	6.9×10^{-21}	1.4×10^{-34}	5.3×10^{-14}
303	0.3×10^{-18}	1.3×10^{-22}	2.6×10^{-38}	1×10^{-14}

[a] Assuming a grain boundary thickness of 1 nm.
[b] Silver diffusion on Cu (110) and (331) surfaces.
Source: Reprinted from Schumacher et al., Diffusion of silver in nanocrystalline copper between 303 and 373k, *Acta Metallurgica* 37(9), 2485–2488, Copyright 1989, with permission from Elsevier.

be wider in microcrystalline materials though an increase of the diffusion coefficient can originate from the presence of fine voids in nanocrystalline samples. An effective diffusion coefficient in a polycrystalline material can be obtained by the following equation:

$$D_{eff} = (1-F)D_L + FD_b \tag{1.3}$$

where F is the grain boundary fraction. Once δ is the grain boundary thickness and d is the grain size, one can say

$$F \approx \frac{2\delta}{d} \tag{1.4}$$

Assuming that the thickness of the grain boundary is 7 nm, Equation (1.3) shows that decreasing grain size from 100 to 50 nm caused the effective diffusion coefficient to increase from 1×10^{-18} (m²/sec) to 5.7×10^{-15} (m²/sec).

Equation (1.3) also shows that supposing that diffusions are mainly performed through grain boundaries, the following condition must be achieved:

$$FD_b \gg (1-F)D \tag{1.5}$$

or

$$\frac{\delta}{d}D_b \gg \left(1-\frac{\delta}{d}\right)D_L \tag{1.6}$$

These conditions will be achieved once the grain size is very fine.

An increase of the diffusion rate in nanostructures is followed by acceleration of solid-state alloying and a considerable change in creep and superplastic behavior of the material. This also causes acceleration of the sintering process during preparation of pieces from nanostructured powders. This means that the sintering process of the powder metallurgy pieces can be performed in much lower temperatures. For instance, it was reported that the required temperature for sintering titanium dioxide powder (TiO_2) with grain sizes around 12 nm is about 400°C to 600°C lower than that of the same powder with grain size of 1.3 μm.

1.4.1.2 Thermal Expansion Coefficient

Metallic nanoparticles and semiconductors are of lower melt or phase transition temperature in contrast to their mass state. The lower melting temperature of the particles emerges once the particle size is less than 100 nm, which is due to a decrease of surface energy with grain size. A drop in phase transition temperature can be described with changes in surface energy to volume energy as a function of grain size.

It is supposed that the optimum diffusion in the grain boundaries of the nanomaterials is due to changes in their thermal features. In the metals, melt point and thermal conductivity decrease through use of nanotechnology (e.g., melt temperature of the nano-gold is 27°C less than that of conventional gold). Besides, the thermal expansion coefficient has been decreased in the nano-gold. Decrease in thermal conductivity of the nanostructured ceramics can develop their application as a thermal shield. Another application of these materials is in the coating blades of airplanes' turbines, which multiplies their life up to six times more than the common ones.[46,59–69]

Contradictory results were reported about the effect of grain size drop on the thermal coefficient of the materials. Basically, because a high fraction of atoms are located in grain boundaries, it is expected that the thermal expansion coefficient of nanostructures is a large figure. Some measurements on several metals and nanocrystalline alloys prove this prediction. For example, it was reported that the thermal expansion coefficient of copper with a grain size of 8 nm in the range of 110°K to 293°K is twice as much as that of a single crystal copper sample. On the other hand, some other results imply no change in the thermal expansion coefficient of nickel in the nanocrystalline state, in comparison with the microcrystalline state. In addition, some results have shown that the thermal expansion coefficient of nanocrystalline samples, which are made by consolidation of the initial powder, decreases with an increase of applied pressure. Based on these findings, one can conclude that there is basically no significant difference between the thermal expansion coefficient of common nanocrystalline and microcrystalline materials. An increase of the thermal expansion coefficient of some nanocrystalline materials, presented in some experiments, can be interpreted as a result of voids, fractures, and other structural defects in the final structure of the prepared samples. Figure 1.5 compares the thermal conductivity, the specific

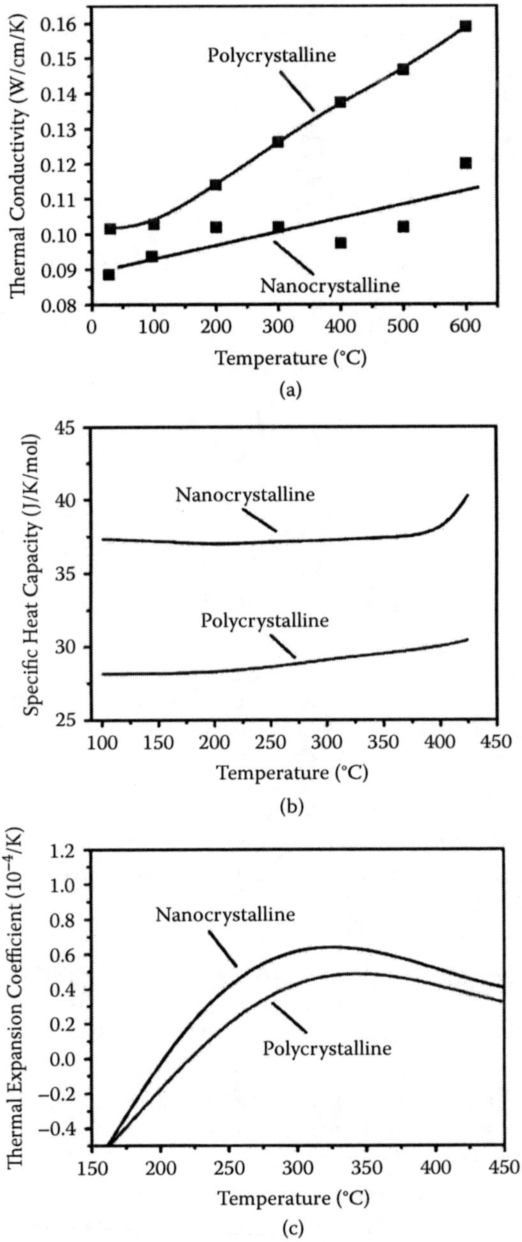

FIGURE 1.5
(a) The thermal conductivity, (b) the specific heat capacity, and (c) the thermal expansion coefficient of the prepared ultrafine nanocrystalline Gd bulk as a function of the temperature, together with a comparison with those of the polycrystalline Gd bulk. (Reprinted from Lu, N., Song, X., and Zhang, J., Microstructure and fundamental properties of nanostructured gadolinium (Gd), *Materials Letters* 63 (12), 1089–1092, Copyright 2009, with permission from Elsevier.)

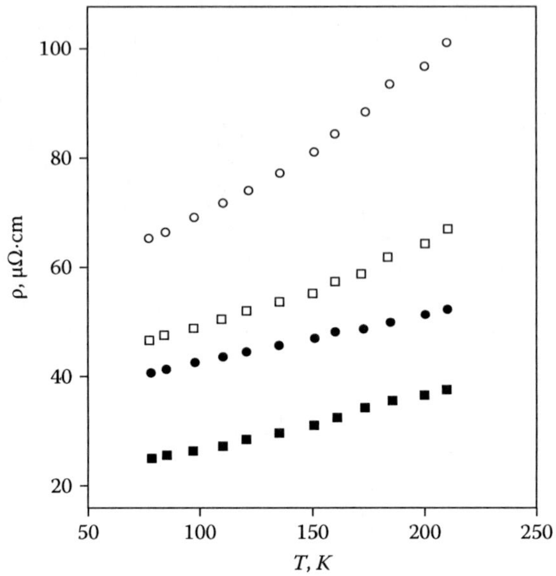

FIGURE 1.6
Dependence of electrical resistivity on temperature for nanocrystalline Fe-Cu-Si-B alloys with grain sizes of (○) 30 nm, (□) 40 nm, (•) 50 nm, and (■) 90 nm. (Reprinted from Wang, Y. Z., Qiao, G. W., Liu, X. D., Ding, B. Z., and Hu, Z. Q., Electrical resistivity of nanocrystalline Fe-Cu-Si-B alloys obtained by crystallization of the amorphous alloy, *Materials Letters* 17 (3–4), 152–154, Copyright 1993, with permission from Elsevier.)

heat capacity, and the thermal expansion coefficient of ultrafine nanocrystalline Gd bulk as a function of the temperature with those of the polycrystalline bulk Gd.

1.4.1.3 Electrical Resistance

Figure 1.6 indicates changes of specific electrical resistance with temperature for some nanocrystalline alloys of Fe-Cu-Si-B with different grain sizes.

For nanocrystalline materials, like microcrystals, specific electrical resistance increases with temperature. However, once the grain size rises, this increase grows with higher rates. Also, at a constant temperature, specific electrical resistance considerably intensifies with grain size increase.

Due to entropies in atoms' arrangement in grains' boundary, electrons are scattered during passing from the boundaries; this, in turn leads to enlargement of specific electrical resistance of the material. It is obvious that as grain size decreases, or in other words area of grain boundaries increases, electron dispersion and consequently electrical resistance, increases. Then, by controlling the grain size, it is possible to change the specific electrical resistance.[66,72–80]

1.4.1.4 Solubility of Alloy Elements

Solubility of elements in nanocrystalline materials is notably higher than common polycrystalline materials. Due to the presence of an expanded grain boundary in nanocrystalline materials and low concentration of atoms in grain boundaries, a large amount of alloy elements can be dissolved in nanocrystalline materials. For instance, solubility of mercury in polycrystalline copper is less than 1 atomic percent, while it is 17 atomic percent for mercury dissolved in nanocrystalline copper. This phenomenon even exists in alloy systems in which there is a solubility defect area (like alloy systems such as Ag-Fe, Ti-Mg, and Cu-Fe). For example, the solubility of Mg in Ti in the solid phase is less than 0.2 atomic percent, but decreasing the titanium grain size up to the nanometric range brings the number up to 30 times.[81–87] Thus, it is possible to produce new alloys with new properties via increasing solubility of alloy elements. Figure 1.7 indicates that lower temperatures are needed for nominal copper solubility in Al-Cu alloys with lower nanometric grain sizes.

1.4.1.5 Magnetic Properties

Because the magnetic properties increase with the increase of the surface-to-volume ratio, magnetic nanomaterials demonstrate unusual behaviors, which is due to particle size and their charge transmission properties. Magnetic properties in nanostructured materials vary with those in a mass

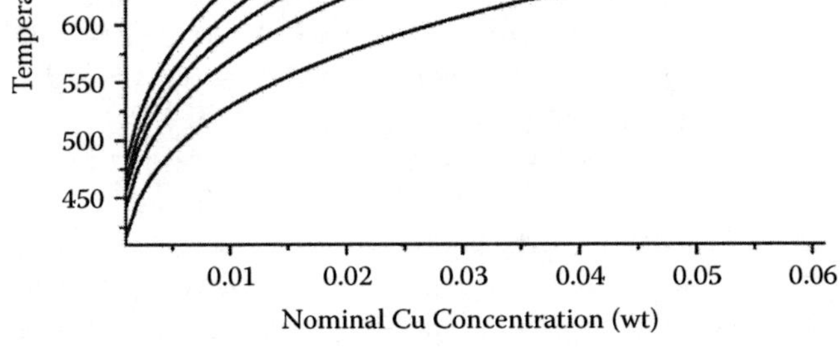

FIGURE 1.7
The nominal copper solubility (solvus) in Al-Cu alloys with different grain sizes. Grain size (a) $d = \infty$, (b) $d = 100$ nm, (c) $d = 50$ nm, (d) $d = 25$ nm, and (e) $d = 10$ nm. (Reprinted from Meng, Q. P., Rong, Y. H., and Hsu, T. Y., Distribution of solute atoms in nanocrystalline materials, *Materials Science and Engineering A* 471 (1–2), 22–27, Copyright 2007, with permission from Elsevier.)

state of the materials, because an increase of surface energy leads to development of enough energy for spontaneous changes of the magnetic fields in polar directions, which changes the material into a paramagnetic state. As the behavior of these modified paramagnetic differs from that of prevalent paramagnetic, it is called supermagnetic. In other words, due to high levels of the energy paramagnetic properties of the mass, materials are removed in nanodimensions and the material turns into a supermagnetic state.

Materials' ferromagnetic properties are controlled by their interatomic distances, grain size, and constituent phases. For this reason, saturated magnetic properties (M_s) and Curie temperature (T_C) in nanocrystalline materials are considerably low compared with microcrystalline materials. For instance, magnetism of microcrystalline Fe is 220 emu/g, while this declines to 130 emu/g once Fe grains size goes down to 6 nm. Or it is reported that Curie temperature of nanocrystalline Gadolinium (Gd) with grain size of 10 nm is about 10°C less than that of microcrystalline Gd. These changes all come from decreasing the grain size to the nanometric range, as each grain can act like an independent unit. In addition, a considerable fraction of the atoms in nanocrystalline structures are located in the grain boundary, where average atomic distances are more than those in grains.

Alloys of nanostructural Fe, such as $Fe_{73.5}Cu_1Nb_3Si_{13.5}B_9$ (commercially known as Finement) and $Fe_{91}Zr_7B_2$ (commercially known as Nanoperm) manifest an efficient complex of magnetic properties. These materials involve a low inhibition of 5 to 10 A/cm, high permeability of 10^5, and a low energy waste, as well as magnetic properties of about zero (less than 2×10^{-6}). Furthermore, due to a high electrical resistance, nanostructured materials display a low core waste of 200 KW/m^3. These characteristics all together have made nanocrystalline Fe alloys among the best magnetic materials and an alternative for matrix amorphous alloys of cobalt. Figure 1.8 presents a comparison of magnetic properties of nanocrystalline magnetic Fe alloy (Nanoperm) and conventional soft magnetic materials.[89–96]

Creating nanostructures in permanent magnets such as $Fe_{90}Nd_7B_3$ has significantly improved their magnetic properties. The magnetic properties of these alloys are induced from the $Nd_2Fe_{14}B$ phase, which has hard magnetic properties. This phase with the Fe phase (ferrite), which involves soft magnetic properties, is present in the structure of Fe-Nd-B alloy. Once this biphasic structure is created in nanodimensions, the residual magnetism (M_r) increases, and then M_r will be considerably larger than $1/2\ M_s$.

Development of nanostructures causes improvement of the high magnetic resistance phenomenon. This feature is defined as dramatic reduction of electrical resistance due to applying a magnetic field. This reduction is about 1% to 2% for microcrystalline materials, while for nanocrystalline materials this has been reported to be 50% or even more. The creation of nanostructures in magnetic coolers improves their performance and output.

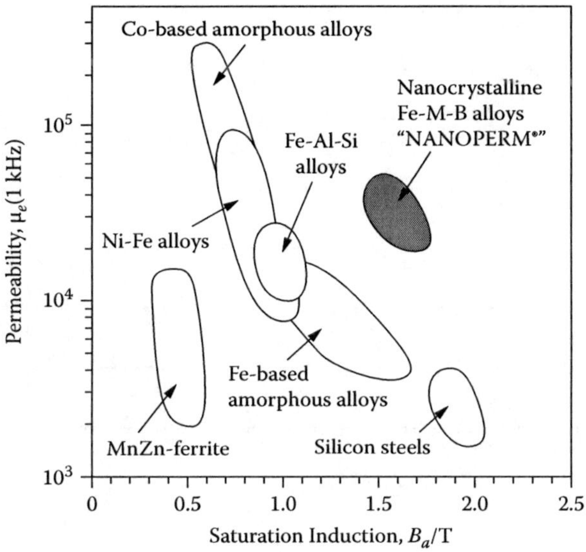

FIGURE 1.8
A comparison between magnetic properties of Nanoperm and conventional soft magnetic materials. (Reprinted from Makino, A., Inoue, A., and Masumoto, T., Nanocrystalline soft magnetic Fe-M-B (M = Zr, Hf, Nb), Fe-M-O (M = Zr, Hf, rare earth) alloys and their applications, *Nanostructured Materials* 12 (5), 825–828, Copyright 1999, with permission from Elsevier.)

1.4.1.6 Corrosion Resistance

A number of researchers have investigated corrosion resistance of nanocrystalline materials. A study on corrosion of Ni-P alloy shows that the corrosion rate of nanocrystalline alloys is higher than that of common microcrystalline alloys; however, corrosion is more general.

Higher rates of corrosion in nanocrystalline materials are due to their extensive grain boundary, where atoms are in balance and then have higher energy levels that make them able to participate in chemical corrosive reactions. More regular corrosion of nanostructured materials is due to their more homogenous structures. Experiments performed on nanocrystalline 304 stainless steel in HCl reveal that nanostructures have higher local resistance against corrosion than traditional structures.[98–103]

Decreasing the average size of nanocrystallites usually can increase the corrosion resistance, while some other factors such as the nature of the material and its porosity can affect this trend. Figure 1.9 shows the close relationship between corrosion resistance and average nanocrystallite size for a kind of hard coating fabricated by plasma electrolysis.

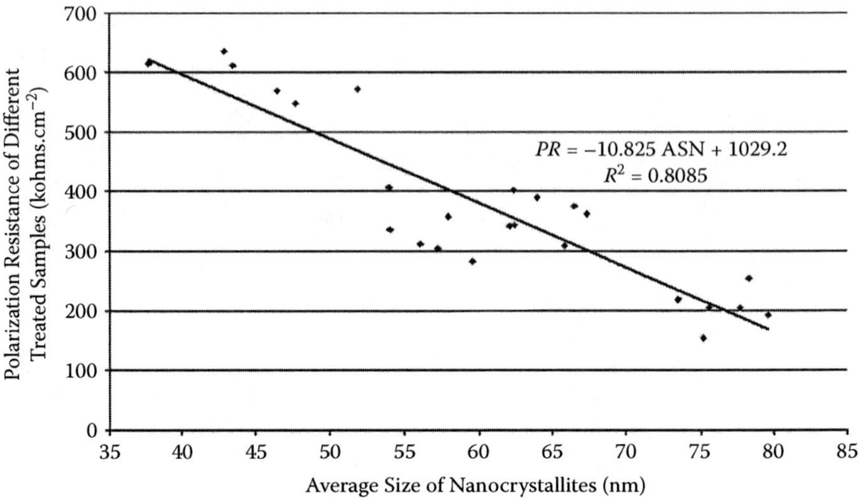

FIGURE 1.9
Polarization resistances of different coatings versus the average size of their nanocrystallites. (Reprinted from Aliofkhazraei, M., Sabour Rouhaghdam, A., and Heydarzadeh, A., Strong relation between corrosion resistance and nanostructure of compound layer of treated 316 austenitic stainless steel, *Materials Characterization* 60 (2), 83–89, Copyright 2009, with permission from Elsevier.)

1.4.1.7 Hydrogen Capacitance Properties

Hydrogen is among the best alternatives for fossil fuels such as oil and gas. Hydrogen combustion produces no pollution and is available from renewable energy resources. However, industrial use of hydrogen as a fuel requires the development of safe and economic capacitance systems. One of the best techniques to store the hydrogen is to use metallic hydrides. Compounds such as $LaNi_5$, ZrV_2, Mg_2Ni, and FeTi can store a considerable amount of hydrogen (about 8 weight percent) by the development of metallic hydrides and release it once there is a favorable condition of temperature and pressure. In this regard, required rate, temperature, and pressure for absorption or release of hydrogen by the material are of great importance in industry. These characteristics can be surprisingly improved by creating nanostructures.[105–113]

1.4.2 Mechanical Properties

Mechanical properties of the nanomaterials are improved as their size decrease. Most studies conducted in this area are focused on mechanical properties in one-dimensional structures such as nanowires. Enhanced mechanical strength in nanowires or nanorods is attributed to the high rate of internal defects in the nanowires. In general, defects, such as impurities, dislocations, and microtorsions, in crystals contain high energy content and must be removed from the completely crystalline structures. As the section area of the nanowires decreases, their defect rate drops because nanodimensions allow for removal of such defects.

Also, as the grain size increases, the volumetric ratio of the grain's size would increase in a single cell, as for the grains with a size of 5 nm about 50% of the volume would be empty space. One can explain the decrease of material strength due to the decrease of grain size as follows: The grain sizes start to slip because of high density of the defects in the stress field, so the atoms and dislocations would be widely expanded in the stress field.

During recent years, various studies have been conducted on mechanical properties of nanostructures. The results show that once the grain size is fine enough (up to nanometric range), many mechanical properties of materials—compared with common materials—dramatically change.

Mechanical properties of the nanostructured materials are mainly due to their extensive interfaces and the presence of a considerable fraction of atoms in these faces. Some factors including phase shape, distribution, size, impurities, and also density of crystalline defects influence mechanical behavior of the nanostructured materials. It must be noticed that the presence of voids or cracks in their structures, which mainly occur during the production processes, can also change their many other characteristics. Some discrepancy between results obtained by different researchers might be the result of these factors. Unfortunately, there is no comprehensive study about the role of voids, second-phase particles, alloy elements or impurities, and the nature of interfaces on mechanical properties of nanostructured materials. Hence, offering a complete summation is rather difficult.[114–123]

Table 1.4 shows mechanical properties of Ni in three different grain sizes. Some of these properties such as yield strength, final strength, hardness, and abrasive strength in the nanostructural state are several times higher than those in the microcrystalline state. But on the other hand, some other features such as Young elastic modulus (E) show no noticeable change and some, like flexibility, even show a reduction.

1.4.2.1 Elastic Properties

Initial measurements to determine elastic constants of nanocrystalline materials were done on samples produced with the powder metallurgy method. These research studies indicate that elastic properties such as E in nanocrystalline materials are greatly (up to 30%) more than traditional microcrystalline materials. However, laboratory test results and the following models clearly suggest that the presence of voids and cracks in samples can have a considerable effect on elastic modulus reduction of nanocrystalline materials, as it can decrease up to 80% regarding size, percentage, and shape of the voids.[36,121,125–130] The presented results in Table 1.4 show that for a nanocrystalline sample with insubstantial porosity, Young modulus basically does not change with

TABLE 1.4

Mechanical Properties of Ni in Nanocrystalline and Microcrystalline States in Ambient Temperature

Property	Grain Size		
	10 μm	100 nm	10 nm
Yield strength (MPa, 25°C)	103	690	>900
Ultimate tensile strength (MPa, 25°C)	403	1100	>2000
Tensile elongation (%, 25°C)	50	>15	1
Elongation in bending (%, 25°C)	—	>40	---
Modulus of elasticity (Gpa, 25°C)	207	214	204
Vickers hardness (kg/mm²)	140	300	650
Work hardening coefficient	0.4	0.15	0.0
Fatigue strength (MPa, 10⁸) cycles/air/25°C)	241	275	—
Wear rate (dry air pin on disc, μm³/μm)	1330	—	7.9
Coefficient of friction (dry air pin on disc)	0.9	—	0.5

Source: Reprinted from Robertson, A., Erb, U., and Palumbo, G., Practical applications for electrode-posited nanocrystalline materials, *Nanostructured Materials* 12(5), 1035–1040, Copyright 1999, with permission from Elsevier.

grain size decrease. It is predicted that this trend is true up to a grain size of 5 nm. But, once grain size is smaller than 5 nm the number of atoms located in the grain boundary will be significant[131] and can dramatically change the elastic properties of the material. Thus, most of the time nanocrystalline materials with a grain size of 10 nm have the similar elastic constants with common microcrystalline materials. The changes of different related properties for an ultrananocrystalline diamond can be seen in Figure 1.10.

1.4.2.2 Plastic Properties

Yield strength and hardness of common polycrystalline materials depend on their grain size. For polycrystalline materials the empirical Hall–Petch equation states the relationship between yield strength and grain size as follows:

$$\sigma_y = \sigma_0 + kd^{-1/2} \tag{1.7}$$

where σ_y is the yield stress, σ_0 is the innate yield stress for starting a single dislocation, d is the grain diameter, and k is a constant related to the grain boundaries' effect on dislocations. In the same way, Equation (1.8) shows the effect of grain size on the hardness of polycrystalline materials:

$$H = H_0 + k'd^{-1/2} \tag{1.8}$$

where H and H_0 are, respectively, hardness and innate hardness (hardness of a single crystal).

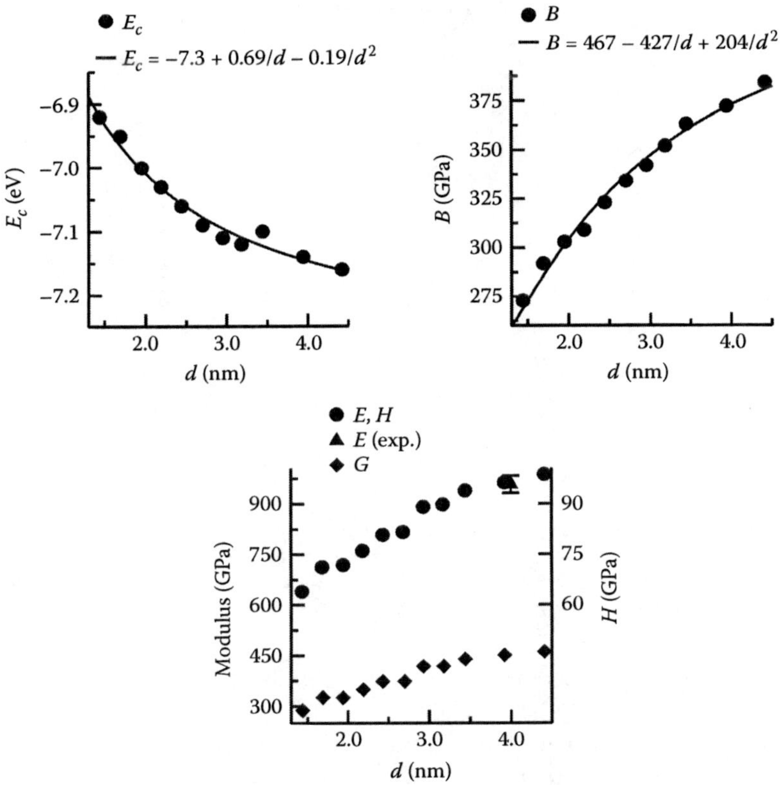

FIGURE 1.10
Cohesive energy (left panel), bulk modulus (central panel), Young's and shear moduli, and estimated hardness (right) of ultrananocrystalline diamond (UNCD) versus the average grain size. The solid lines in the left and central panels are fits to the data. The function used assumes different compressibilities for atoms in the bulk, grain boundaries, or edges. In the right panel, the experimental value from Espinosa and colleagues is also shown. (Reprinted from Remediakis, I. N., Kopidakis, G., and Kelires, P. C., Softening of ultra-nanocrystalline diamond at low grain sizes, *Acta Materialia* 56 (18), 5340–5344, Copyright 2008, with permission from Elsevier; also reprinted from Espinosa, H. D., Peng, B., Moldovan, N., Friedmann, T. A., Xiao, X., Mancini, D. C., Auciello, O., Carlisle, J., Zorman, C. A., and Merhegany, M., Elasticity, strength, and toughness of single crystal silicon carbide, ultrananocrystalline diamond, and hydrogen-free tetrahedral amorphous carbon, *Applied Physics Letters* 89 (7), Copyright 2006, with permission from the American Institute of Physics.)

Equations (1.7) and (1.8) show that drawing the σ_y (or H) in terms of $d^{-1/2}$ leads us to a line with gradient of k (or k'). To describe these empirical equations, several models have been suggested that are mainly based on production, movement, and concentration of dislocations behind the grains. If the Hall–Petch equation is true for nanocrystalline materials without changing the k (or k') coefficient, these materials' strength and hardness should increase up to 10 times.[134–136] For instance, if grain size reduces from 10 μm to 10 nm,

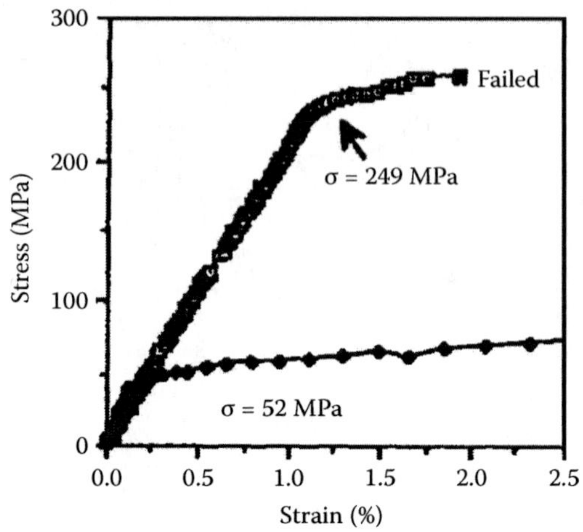

FIGURE 1.11
Stress–strain curve for nanocrystalline (grain size of 7 nm) and microcrystalline (grain size of 100 μm) Pd. (Reprinted from Nieman, G. W., Weertman, J. R., and Siegel, R. W., Tensile strength and creep properties of nanocrystalline palladium, *Scripta Metallurgica et Materiala* 24 (1), 145–150, Copyright 1990, with permission from Elsevier.)

the strength must increase up to 30 times, though laboratory results do not support this claim (Table 1.4).

Due to limitations in making laboratory samples with large scales, the performed studies (particularly mechanical properties of nanostructured materials) are limited for measuring fine hardness. However, some results about tensional test will also be available. Figure 1.11 shows the result of the tension test on Palladium (Pd) in both microcrystalline and nanocrystalline states. This obviously reveals that in nanostructured Pd there would be a dramatic rise in yield strength with a decrease of grain size.

In total, for nanocrystalline materials with a grain size of 10 nm, hardness and yield strength are about two to seven times more than common materials with grain sizes of more than 1 μm. As previously mentioned, the effect of grain size on hardness and yield strength of a material is typically described by the Hall–Petch equation. In Figures 1.12 and 1.13, some examples of this equation for nanocrystalline materials are presented.

Evaluating these results implies that in all cases, disregarding the production method, hardness and strength increase according to the Hall–Petch equation, but once the grain size is lower than a critical range (15 to 25 nm) it does not rise by the previous gradient. In other words, in grain sizes lower than a critical size k' (or k), the coefficient in the Hall–Petch equation will be reduced. In grain sizes lower than the critical size, nanocrystalline materials display three different behaviors.

In most of the cases, the Hall–Petch equation continues but its gradient declines. In a few cases Hall–Petch curve's gradient limits to zero. This occurs

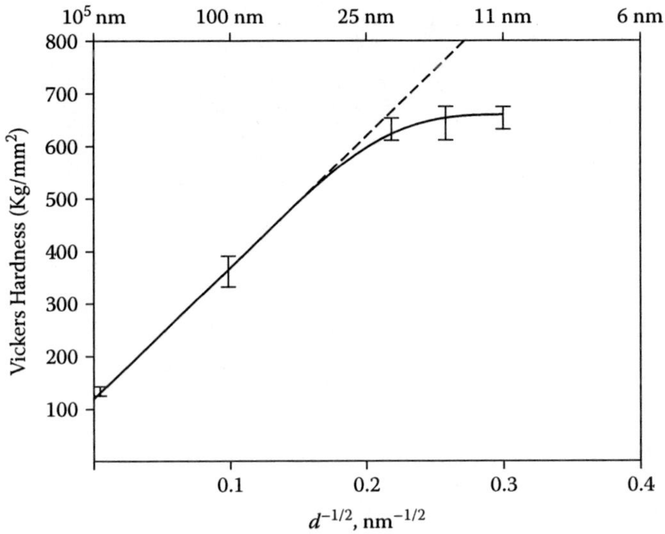

FIGURE 1.12
Hall–Petch diagram for nanocrystalline nickel (Ni). (Reprinted from El-Sherik, A. M., Erb, U., Palumbo, G., and Aust, K. T., Deviations from Hall–Petch behaviour in as-prepared nanocrystalline nickel, *Scripta Metallurgica et Materiala* 27 (9), 1185–1188, Copyright 1992, with permission from Elsevier.)

in a small group of materials in grain sizes lower than critical size where the curve's gradient is negative. This means that the smaller the grain size, the lower is the hardness. In this situation, we are faced with an inverse Hall–Petch equation, so instead of work hardening there will be work softening. Figure 1.13 shows these behaviors for Cu and Pd and alloys of γ-TiAl, respectively.

Why an inverse Hall–Petch curve behaves in this way is not yet fully answered. However, it is mainly believed that common mechanisms, about creation and movement of dislocations, offered for plastic deformation are not at play in nano ranges. Principally, in grain sizes less than 50 nm dislocations cannot be presented. The absence of dislocations in limited spaces such as whiskers was proven a long time ago. The required stress for activation of dislocations sources, such as the Frank–Read ones, are inversely correlated with distance between two dislocations' interlocking. Therefore, due to close grain boundaries in nanocrystalline materials, the distance between dislocations locking is negligible and the required stress for activation of dislocation sources can amount to theoretical shear stress in a sample with no dislocations. So, it is expected that in very small grain sizes some other phenomena control the plastic behavior of the materials. For example, it is suggested that processes such as grain boundary slips or grain disorientation (diffusive creep) are the main mechanism for plastic deformation of the nanocrystalline materials with critical grain sizes.

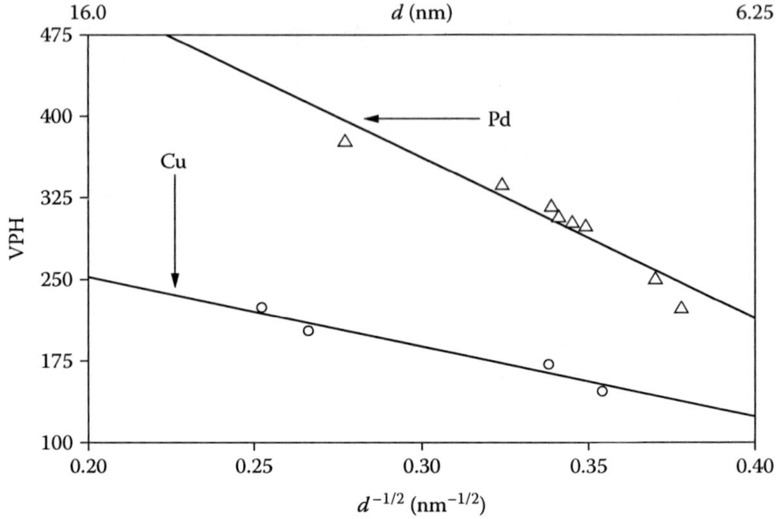

FIGURE 1.13
Inverse Hall–Petch curve for hardness of nanocrystalline Cu and Pd. (Reprinted from Chokshi, A. H., Rosen, A., Karch, J., and Gleiter, H., On the validity of the Hall–Petch relationship in nanocrystalline materials, *Scripta Metallurgica* 23 (10), 1679–1683, Copyright 1989, with permission from Elsevier.)

Although the results above are mainly for pure metals, the reported results for nanocrystalline intermetallic components or alloys bearing ceramic nanoparticles (oxides, carbides, etc.) are qualitatively similar to the results mentioned above. Observing inverse Hall–Petch behavior in ceramics and composites promises production of these materials in lower temperatures. Figure 1.14 schematically displays Hall–Petch behavior in two grain sizes for ZrN coating.

Like hardness and strength, it is predicted that flexibility of a substance would rise with the drop of grain sizes down to the nano range. This prediction is based on empirical experimental results on common microcrystalline materials, suggesting that smaller grain sizes lead to higher flexibility and fracture toughness. As a matter of fact, in these materials as the grain size decreases fracture strength intensifies more than yield strength, and the substance shows a soft behavior. Nevertheless, the empirical results do not prove this. As previously mentioned, nanostructured materials have lower flexibility (particularly in tension test) than the common materials. Polycrystalline Ni with a grain size of 10 μm would have a linear strain of up to 50%, while once the grain size is 100 nm this strain is 15% of its original length. If the grain size is higher than 100 nm, Ni displays an insignificant flexibility (about 1%). In this way, nanocrystalline Ni will basically behave like a brittle substance.

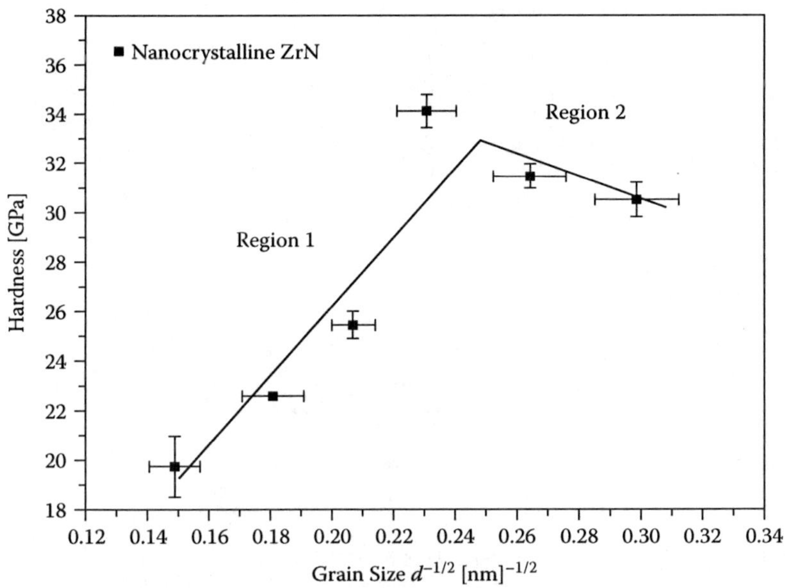

FIGURE 1.14
Hall–Petch plot of hardness of nanocrystalline ZrN coatings against the inverse square-root of grain size. (Reprinted from Qi, Z. B., Sun, P., Zhu, F. P., Wang, Z. C., Peng, D. L., and Wu, C. H., The inverse Hall–Petch effect in nanocrystalline ZrN coatings, *Surface and Coatings Technology* 205 (12), 3692–3697, Copyright 2011, with permission from Elsevier.)

It is worth mentioning that most of the tests for measuring mechanical properties of the nanostructured materials were performed on samples prepared by the powder metallurgy method. Under this condition, the samples will generally have a considerable amount of voids, and the measured flexibility will be influenced by the presence of voids and lack of complete consistency between particles of the powder. In addition, sample surface quality intensely affects the results of flexibility tests. To examine the role of these factors, samples with no porosity or cracks were prepared using some other production methods such as precipitation from the vapor phase or electrochemical deposition. Mechanical tests on these samples also prove this limited flexibility in the nanocrystalline state. Figure 1.15 shows flexibility of some metals and alloys in terms of grain size. Although the data are highly distributed, they all show that once the grain size decreases the flexibility also reduces, and if grain size is less than 25 nm the flexibility will be negligible. However, all these materials demonstrate a significant flexibility of 40% to 60% in the microcrystalline state. Figure 1.16 compares true stress–true strain curves for electrodeposited nanocrystalline nickel and coarse grained nickel.

The reported flexibility for nanostructured materials in a tensional test is less than a compressive test. For example, nanocrystalline copper samples (with a

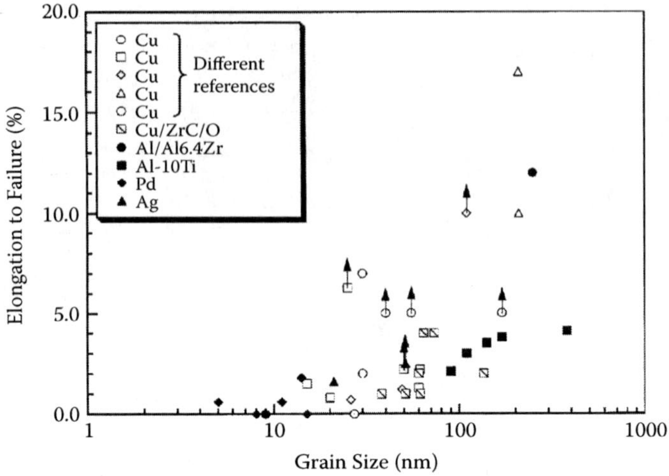

FIGURE 1.15
Flexibility (in tensional test) in terms of grain size for some alloys and metals. (Reprinted from Malow, T. R., and Koch, C. C., Mechanical properties in tension of mechanically attrited nanocrystalline iron by the use of the miniaturized disk bend test, *Acta Materialia* 46 (18), 6459–6473, Copyright 1998, with permission from Elsevier.)

grain size of 20 nm) show 12% to 18% of flexibility in a compressive test, while this sample's flexibility in a tensional test is about 2%. This difference is partially due to lack of sensitivity to the presence of voids and cracks of the sample in the compressive test, but in the tensional test these defects lead to stress concentration and drop of flexibility. One factor limiting the use of intermetallic

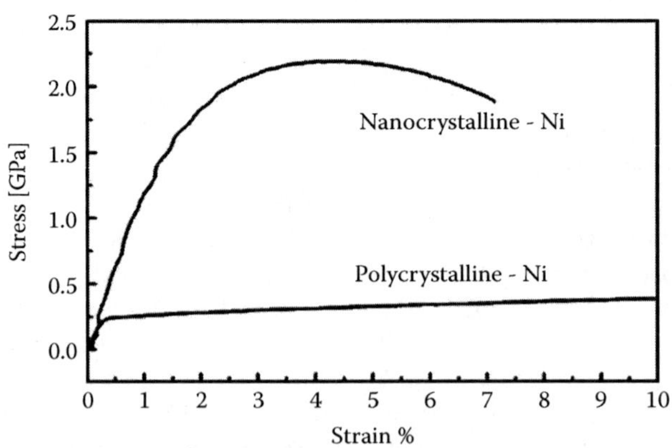

FIGURE 1.16
True stress–true strain curves for electrodeposited nanocrystalline nickel and coarse grained nickel. The coarse grained nickel reached a total elongation of 20%. (Reprinted from Van Swygenhoven, H., Footprints of plastic deformation in nanocrystalline metals, *Materials Science and Engineering A* 483–484 (1–2 C), 33–39, Copyright 2008, with permission from Elsevier.)

compounds and ceramics is their brittle behavior. It is expected that in the process of creating nanostructures, intermetallic compounds and ceramics show soft behavior in temperatures lower than half of their melting point (°K). This prediction is based on the fact that in common materials a grain size decrease leads to a quicker increase of fracture stress compared with yield stress, so brittle fracture changes into a soft fracture occur in lower temperatures.

On the other hand, based on creep theories, we have

$$R \approx D_b/d^3T \tag{1.9}$$

where *r* is the creep rate, D_b is the diffusion rate in grain boundary, *d* is the grain size, and *T* is temperature (in Kelvin).

According to Equation (1.9), if the grain size decreases from 10 μm to 10 nm (which is 1000 times) the creep rate will increase 10^9 times. In addition, the higher diffusion rate in the grain boundary for nanocrystalline materials leads to more growth in creep rate. It is expected that an increase in creep rate for nanostructured materials makes it possible to perform plastic deformation of ceramics and intermetallic compounds in lower temperatures. Although plastic behavior of ceramics and intermetallic compounds was observed in lower temperatures in some cases (including TiO_2, CaF_2, and NiAl), test repeatability has not yet been proved. Hence, it is not possible to offer a conclusion at this point.

Fatigue strength is another important mechanical characteristic of materials. To date, there have been few studies on fatigue behavior of nanostructured materials. The early results suggest that most nanocrystalline metals have a longer fatigue life in high stress cycles and vice versa. These results might originate from the fact that fatigue life in high stress cycles is closely related to strength of the materials, whereas short fatigue life in low stress cycles relates to flexibility of the material. Also, it was reported that surface hardening treatments, such as shot peening, have no effect on improvement of nanocrystalline materials' fatigue properties.

Because nanostructured materials have higher strength and hardness than common materials, it can be assumed that their abrasive strength is considerably higher than the customary materials (e.g., hardness of cobalt-tungsten carbide [WC-Co] nanocomposites is twice as much as their hardness in the microcrystalline state). Figure 1.17 illustrates the wear rate of WC-Co composites (summarized in Table 1.5) in both nanostructured and traditional states. This increase in wear rate can enhance the life span of cutting tools and piece exposed to abrasion up to five times. The variation of wear coefficient with the carbide grain size can also be seen in Figure 1.18.

1.4.2.3 Superplasticity

Some polycrystalline substances at a given temperature and strain rate demonstrate a high plastic deformation (100 to 1000 percent or even more)

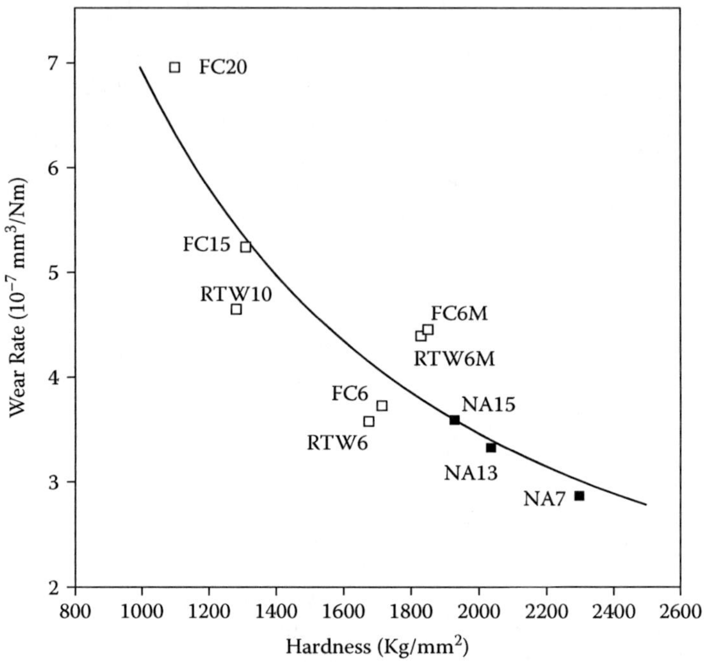

FIGURE 1.17
Comparison between hardness and wear rate in WC-Co composite in both nanostructured and conventional states. (Reprinted from Jia, K., and Fischer, T. E., Sliding wear of conventional and nanostructured cemented carbides, *Wear* 203–204, 310–318, Copyright 1997, with permission from Elsevier.)

without any fracture or necking phenomenon. This phenomenon is known as superplasticity that typically occurs in materials with fine grains (less than 10 μm) and temperatures higher than $T_m/2$ (where T_m is the melting point in Kelvin). As superplasticity behavior allows us to produce pieces with very complicated shapes form materials such as metallic matrix composites and intermetallic compounds (which cannot be machined), it is vital and useful in industrial issues.

To the moment, industrial application of superplastic deformation is limited due to its requirement for a low rate of deformation (about $10^{-3}s^{-1}$), which causes a longer deformation period (20 to 30 minutes). Still, the needed deformation rate for the occurrence of the superplasticity behavior is inversely related with square of grain sizes. Hence, one can claim that if grain size decreases up to 10 times, the optimum rate of plastic deformation for achieving superplastic behavior increases up to 100 times. To simplify, a decrease of grain size from 2 μm to 200 nm cut the superplasticity time from 20 to 30 minutes to 20 to 30 seconds. Moreover, with the decrease of grain size the temperature range necessary for superplasticity will shrink. Then, it is

TABLE 1.5

Nominal Composition and Structural Characteristics of WC-Co Composites in Figure 1.17

Sample	Structural Characteristics			Hardness and Toughness	
	WC Grain Size (μm)	Cobalt Content (vol%)	Measured Mean Free Path (μm)	Vickers Hardness (GPa)	K_{1C} (MPa.m$^{1/2}$)
FC20	1.2	30.5	1.06	11	NA
FC15	1.0	23.6	0.61	13	NA
FC6	1.0	10.1	0.39	17	9
FC6M	0.7	10.1	0.25	18.6	8.2
RTW10	2.5	16.3	1.0	12.8	16.4
RTW6	1.2	10.1	0.43	16.8	10
RTW6M	0.8	10.1	0.14	18.4	8.4
NA15	0.07	23.6	0.068	19.4	8
NA13	0.07	20.8	0.068	20.5	8.3
NA7	0.07	11.7	0.039	23.0	8.4

Note: The number in the sample identification indicates the cobalt content in wt%. For example, NA15 contains 15 wt% and 23.6 vol% Co. Hardness indicated in GPa. 1000 kg.mm^{-2} = 9.81 GPa.

Source: Reprinted from Jia, K. and Fischer, T. E., Sliding wear of conventional and nanocrystalline cemented carbides, *Wear* 203–204, 310–318, Copyright 1997, with permission from Elsevier.

predicted that industrial applications of superplastic shaping with nanostructures would have a huge development.[144,145]

Superplastic behaviors in low temperatures have been reported in some nanostructured metals or even intermetallic compounds. For example, nanocrystalline nickel in a temperature of 0.36 T_m shows a superplastic behavior.

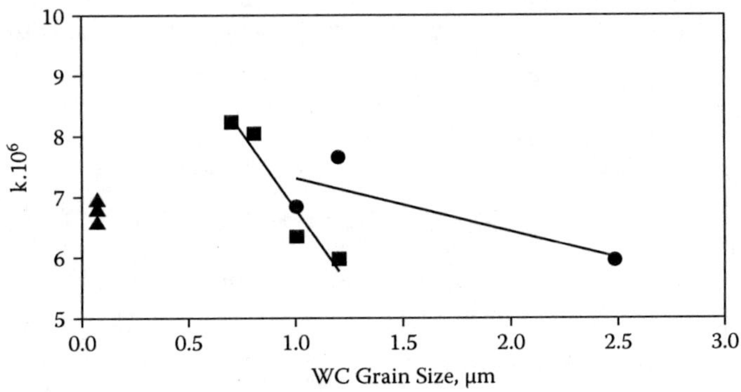

FIGURE 1.18

The variation of wear coefficient with the carbide grain size in WC-Co composites: Triangles, nanocomposites; squares, cermets with 6% Co; circles, RTW10, FC15, and FC20 from Table 1.5. (Reprinted from Jia, K., and Fischer, T. E., Sliding wear of conventional and nanostructured cemented carbides, *Wear* 203–204, 310–318, Copyright 1997, with permission from Elsevier.)

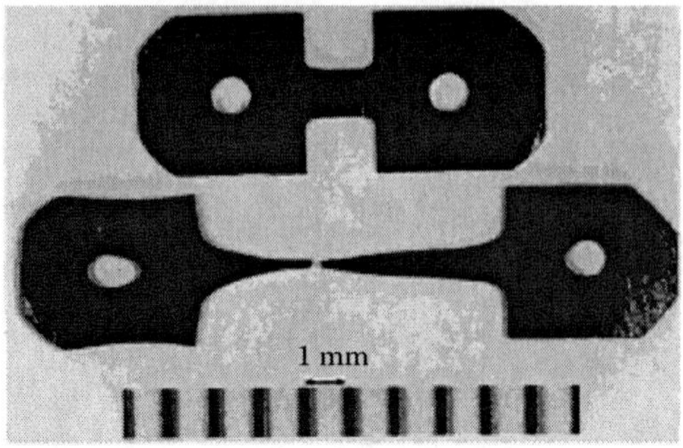

FIGURE 1.19
Nanocrystalline samples of Ni_3Al before and after tension test in 725°C and strain rate of $1 \times 10^{-3} s^{-1}$. (Reprinted from Mishra, R. S., Valiev, R. Z., McFadden, S. X., and Mukherjee, A. K., Tensile superplasticity in a nanocrystalline nickel aluminide, *Materials Science and Engineering A* 252 (2), 174–178, Copyright 1998, with permission from Elsevier.)

This temperature is about 450°C less than the superplastic temperature of microcrystalline Ni. Pure nanocrystalline copper shows a linear strain of 5000% without any hardening, whereas microcrystalline copper samples can have a linear strain of only 800%. A fascinating example on this issue is superplastic behavior of an intermetallic compound of Ni_3Al (Figure 1.19). These materials (with grain size of 50 nm) manifest a superplastic behavior in temperature more than 650°C, while the tensional sample of Ni_3Al has a linear strain of 560% in a temperature of 725°C. Figure 1.20 exhibits superplastic behavior in 7034 aluminum alloy under a tension test. It must be mentioned that thermal stability of nanostructures and lack of grain growth should be considered for use of their superplastic behavior. This thermal stability will be discussed in the next section.

Mechanical properties of nanostructures can be summed as follows:

1. Elastic modulus of nanocrystalline materials almost equals that of microcrystalline materials. However, in very fine grain size (less than 50 nm) elastic properties have not fully been studied.
2. Once grain size decreases to 15 to 25 nm, hardness and yield strength of nanostructured materials rise according to the Hall–Petch equation. In some cases, in grain sizes less than 15 to 25 nm, the curves gradient is zero, whereas in some others it is negative.
3. Creation of nanostructures in metals leads to a substantial rate of flexibility. Also, it results in high flexibility for brittle materials (such as ceramics and intermetallic compounds) in low temperatures.
4. Superplastic deformation of nanostructured materials takes place in lower temperatures and with higher rates.

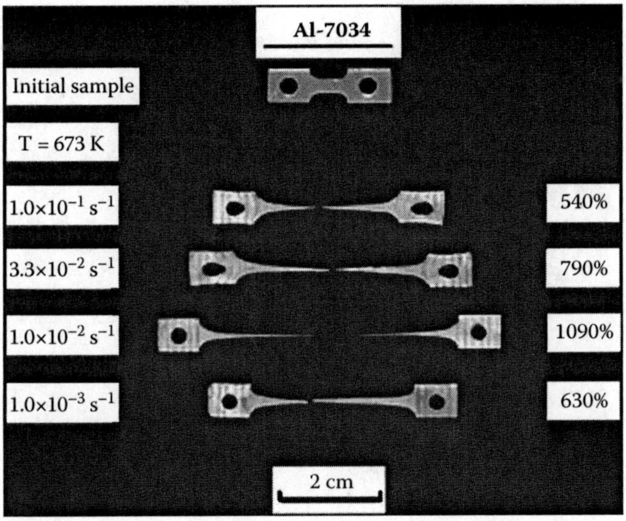

FIGURE 1.20
Appearance of the specimens after equal-channel angular pressing (ECAP) through six passes at 473°K and then pulling to failure at different initial strain rates at 673°K: the upper specimen is untested. (Reprinted from Xu, C., Furukawa, M., Horita, Z., and Langdon, T. G., Using ECAP to achieve grain refinement, precipitate fragmentation and high strain rate superplasticity in a spray-cast aluminum alloy, *Acta Materialia* 51 (20), 6139–6149, Copyright 2003, with permission from Elsevier.)

1.5 Thermal Stability of Nanostructures

Due to having extensive interfaces, nanostructured materials have more freedom than the common materials (microcrystals). As a result, nanostructured materials, disregarding their production method, are more stable in terms of thermodynamic aspects. Nanostructured materials lose their free energies by decreasing their interfaces, which result in growth of grains and phase particles.

The study of thermal stability and growth of grains and phases of nanostructures has two main aspects:

1. If the objective is production of a structure with the least grain size, we must stop grain and phase growth during the production processes (such as sintering), because enlarging the structure eliminates so many desirable properties of the nanostructures.
2. Through grain size and phase change through the growth process, it is feasible to obtain a wide range of mechanical and physical properties. In this way, through the control of growth process, one can create a fine structure in favorable dimension and size and given properties.

According to the Gibbs–Thomson equation, the thermodynamic motive force for grain growth in polycrystalline materials can be stated as follows:

$$\Delta\mu = \frac{2\gamma V}{d} \tag{1.10}$$

where $\Delta\mu$ is the chemical potential difference in both sides of the grain boundary caused by their face curve, γ is the grain boundary energy in the surface unit, V is the atomic volume, and d is the grain size.

Equation (1.10) shows that once the grain size is smaller, the thermodynamic motive force ($\Delta\mu$) for their growth would be higher. This means that the growth rate of the grains is inversely correlated with their size. On the other hand, grain growth is a diffusive process, so an increase in temperature leads to its increase. Then, one might say,

$$\frac{\delta d}{\delta t} = \frac{K}{d} \tag{1.11}$$

where K is a quantity that depends on energy and structure of grain boundary and varied with temperature according to the following equation:

$$K = K_0 \exp(-Q/RT) \tag{1.12}$$

where Q is the activation energy of the grain growth, T is the temperature (in Kelvin), R is the gas constant, and K_0 is a constant value.

If d is the grain size in the moment in which grains start to grow, integrating from Equation (1.11), we have

$$d^2 - d_0^2 = 2\text{Kt} \tag{1.13}$$

If we consider that the d_0 is small enough, then

$$d^2 = 2\text{Kt} \tag{1.14}$$

Equation (1.14) is a parabola that presents grain growth in ideal conditions. This shows that passing the time, grain growth rate (curve's gradient) shrinks. However, due to the presence of alloy elements, impurities, second-phase particles, voids, and so forth, on grain boundaries, Equation (1.14) does not match data obtained in the laboratory, so the grain growth equation is presented as

$$d^n = \text{Kt} \tag{1.15}$$

For standard polycrystalline materials, n is greater than two. Once the materials purity increases and temperature goes up, n approaches 2. The growth mechanism of the grains can be explained by two parameters of n and Q. In common polycrystalline materials activation energy of grain growth is rather equal to activation energy through grain boundary diffusion. This implies that grain growth is performed by their movement and transition through the grain boundary.

The presence of second-phase particles or impurities on the grain boundary causes a decrease of the grain growth rate. The presence of second-phase particles on the grain boundary leads to a decrease of its surface area. Consequently, grain boundary movement and its separation from second-phase particles requires a new grain boundary. To simplify, second-phase particles inhibit grain boundary movements by inserting a force against thermodynamic motive force. This is known as Zener force. When grain boundary energy in the surface unit and radius of second-phase particles are, respectively, γ and r, it can be inferred from Figure 1.21 that the magnitude of Zener force in a length unit of the grain boundary equals $\gamma\sin\theta$. If we multiply this by distance between intersection of the grain boundary with a second-phase particle ($2\pi\cos\theta$), the total value of Zener force would be

$$F = 2\pi r\gamma\sin\theta\cos\theta \tag{1.16}$$

The maximum amount of this force (F_m) is when $\theta = 45°$, so

$$F_m = \pi r\gamma \tag{1.17}$$

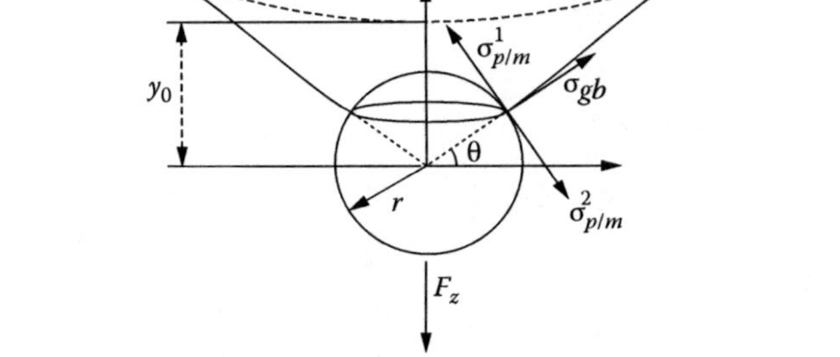

FIGURE 1.21
Effect of second-phase particles on movement of grain boundary. (Reprinted from Moelans, N., Blanpain, B., and Wollants, P., A phase field model for the simulation of grain growth in materials containing finely dispersed incoherent second-phase particles, *Acta Materialia* 53 (6), 1771–1781, Copyright 2005, with permission from Elsevier.)

If f is the volume fraction of the second phase, the average number of particles in volumetric unit of second phase (N) is

$$N = \frac{f}{\frac{4}{3}\pi r^3} \tag{1.18}$$

Because it is only possible for particles with a distance of $\pm r$ to touch the grain boundary, the number of particles in the second phase in the surface unit of a grain boundary would be

$$N' = 2rN = \frac{3f}{2\pi r^2} \tag{1.19}$$

Then, the total Zener force inserted on a surface unit of the grain boundary can be obtained from the following equation:

$$P = F_m \times N' = \frac{3f\gamma}{2r} \tag{1.20}$$

Because the grain size is small enough, the prohibiting force against grain growth (P) is very unsubstantial in comparison with the thermodynamic motive force for grain growth, so grain starts to be enlarged. But, once the grain size goes up thermodynamic motive force decreases, and whenever it is equal to the P grain growth would be stopped. Hence, the maximum grain diameter (d_m) achieved in the growth process in the presence of second-phase particles is as follows:

$$\frac{2\gamma}{d_m} = \frac{3f\gamma}{2r} \tag{1.21}$$

$$d_m = \frac{4r}{3f} \tag{1.22}$$

The most important result obtained from Equation (1.21) is that the finer particle size of second phase and their higher volumetric fraction leads to their greater influence on grain growth rate.

With respect to discussions above, it is expected that grain growth in nanocrystalline materials can be done even in low temperatures, though data and results of laboratory tests do not support this. In fact, many nanocrystalline materials including metals, ceramics, and intermetallic compounds produced in different methods are stable even in fairly high temperatures. Thermal stability of nanomaterials has been reported even at 40% to 50% of their melting point, which is equal to that for common polycrystalline materials.

The results show that activation energy for grain growth (Q) in nanocrystalline materials is in agreement with activation energy of grain boundary diffusion (Q_b), though different values have been reported for n (2 to 10). Some other outcomes imply that for nanocrystalline materials, activation energy of grain growth in high temperatures is in agreement with activation energy of lattice diffusion; however, in low temperatures it is close to the activation energy of grain boundary diffusion.

Thermal stability of nanocrystalline materials and lack of grain growth can originate from several factors such as the presence of coaxial grains with a flat surface, the presence of impurities and their separation in the grain boundary, and the occurrence of voids in their structure. For example, it was reported that grain size of nanocrystalline titanium oxide (TiO_2) with a porosity of 25%, after thermal operation in a temperature of 700°C for 20 hours, shifts from 14 to 30 nm. However, once the porosity drops to 10%, after performing the same thermal operation the grain growth exceeds 500 nm. This high thermal stability of the nanocrystals was assessed with several models. These models show that in grain sizes lower than a critical value (d_c), grain growth has a linear relationship with time, but in grain sizes larger than d_c, grain growth relationship with the time, like common polycrystalline materials, is a parabolic equation. Because $d < d_c$ grain growth is not controlled by thermodynamic motive force caused by grain boundaries curve, it might be possible that some other mechanisms, such as elimination of vacancies in grain boundaries, are responsible for grain growth. Because grain growth leads to a decrease of total area of grain boundary surface, then vacancies in grain boundaries must be located in other places or transferred to sample surfaces. Computational models indicate that during the growth process, these vacancies are mainly removed in areas close to grain boundaries and in the form of atomic vacancies. As a result, this leads to creation of an unbalanced concentration in atomic vacancies of the sample and an increase of free energy, which can compensate for the decrease of energy caused by grain growth. Based on this model, once the grain growth is controlled by redistribution of the vacancies, activation energy of the grain must be a larger number (close to activation energy of lattice diffusion). The obtained results from this model fairly match laboratory data. It is worth mentioning that the presence of second-phase particles, impurities, and voids created by production operations is a source of contradictions between the results of different researchers and laboratory data with mathematical models. In the presence of the above-mentioned agents, grain growth can mainly occur by atomic lattice diffusion.

Nanocomposites contribute to growth of reinforced phase particles, as well as grain growth. Similarly, grain growth is due to thermodynamic motive force of the second-phase particles and the extensive interface between second-phase particles and the matrix. Hence, the particles of nanophases gradually start to grow, and their number in a volumetric unit drops and the free energy of the system drops. This process leads to removal of favorable features of a nanocomposite-like decrease of its hardness and strength. Moreover, according to Equation (1.20), once the particles of the second phase expand in size, their effect on stopping grain growth will be less significant.

Like growth process of the grains, second-phase particles are controlled by atomic diffusion and then they will grow with an increase in the temperature. Atomic diffusion can occur across the grains, from grain boundary, or through the interface of second-phase particles and the matrix. Assuming that the diffusion only occurs across the grains, growth rate of second-phase particles (dr/dt) is offered as

$$dr/dt = K/r^2 \tag{1.23}$$

where r is the average size of second-phase particles in each moment, t is time, and K is a value related to the diffusion coefficient (D) and interface contact energy (γ).

If the particles' initial size is z_0, integrating from Equation (1.23) gives

$$r^3 - r_0^3 = 3\text{Kt} \tag{1.24}$$

From Equation (1.23) it can be found that finer grains have a higher growth rate. Because K increases with a increase of γ and D, then the growth rate of the second-phase particles will be low in the cases of small γ and D. Decrease of γ demands using a second phase that can create a shared interface with the interface. In addition, if particles of the second phase have elements with low solubility in the matrix, the diffusion coefficient (D) will be extremely small which, in turn, leads to thermal stability of the second phase.[149–154]

1.6 Nanotechnology and Future Perspectives

Nanotechnology mainly involves four key aspects including

1. Miniaturizing systems
2. Obtaining new materials with nanodimensions
3. Enhancing the efficiency and automation of the systems and equipment
4. Increasing informatics storage capability

Nanotechnology is used in several aspects ranging from electronic, visual communication, and biological systems to new intelligent materials. The reasons for the wide use of nanostructures and nanomaterials are listed below:

1. Unusual physical properties of the nanostructured materials, such as Au nanoparticles as an inorganic color used for glass dyeing
2. Expanded surfaces, like Au nanoparticles that are placed on the metallic oxides and used as low temperature catalyzers or nanoparticles for several types of the sensors
3. Extremely fine dimension, which is of great importance

Table 1.6 introduces some current and future applications of the produced materials by the nanotechnology using the new technologies in this field. In this text, some nanostructures and nanomaterials' applications as well as capabilities and potentials of the nanomaterials will be discussed in detail.[155–160]

TABLE 1.6

Some Current and Future Applications of the Produced Materials by the Nanotechnology

Technology	Current Impacts	Future Impacts
Distribution and coating	Heat-resistant coatings	Optimum heat resistance coatings
	Enhancing images' quality	Multiuse nanometric coatings
	Antiabrasive coatings	Fine-grain structures
	Information storing layers	Information storing layers with higher storage capacity
High contact surface	Molecular sifter	Specific molecules sensor
	Medicine carrier in the human body	Stimulated spotter particles
	Providing appropriate catalyzers	Energy storage (fuel cells)
	Absorptive and repellant materials	Gas sensors
Reinforced materials	Soft magnetic materials	Superplastic ceramics shaping
	Ductile and hard cutting tools	Ductile materials with high strength
	Nanocomposite cements	Composites filled with nanostructured fillers
Biological-medical aspects	Activated nanoparticles	Cellular marking
		Local heating
Nanotools	Reading heads made of giant magnetic resistance (GMR)	Memories and microprocessors with capacity of several thousands of Gigabites
		Biological sensors
		Nanotubes, highly bright monitors

1.7 Some Applications of Nanostructures

Applying the nanomaterials, it is possible to achieve the following in the nanomaterials and nanosystems:

1. Ability to control the systems' size
2. Ability to determine the accurate needed composition (not approximate one) without any structural defects such as voids
3. Possibility to control frequency dimension
4. During the assemblage of the constituting pieces, control of the mutual effect level between the pieces as well as materials designs

1.7.1 Nanoelectronic and Molecular Electronic

Many advances in the molecular electronic and nanoelectronic field have been made. In the molecular electronic, single molecules are designed for control of electron transfer. The control on the surface of common metals and semiconductors is conducted by surface arrangement and solid materials molecules via designing an electrical circuit. Once the molecules are biologically active the bioelectronics pieces would be obtained.

Many nanoelectronic pieces, such as adjustable connections, electronic switches, carbon nanotube transistors, single-molecular transistors, and so on, were fabricated until now. Au nanoparticles are widely used in nanoelectronics and molecular electronics, due to their surface characteristics and unified sizes. For instance, Au nanoparticles can serve as carriers through several types of applied organic molecules or biological components. Au nanoparticles can be used as connectors in many practical connections in the structure of the electronic nanoparticles for employing in the sensors and detectors.[161–165]

1.7.2 Nanorobots

Future applications of nanotechnology in medical sciences, mainly in nanodrugs, will be very broad. One interesting application of nanomedicine is creating nanotools for promotion of curing and recognition level. These types of nanoparticles are known as nanorobots. These nanorobots can be used as transportation tools for carrying therapeutic factors and early detectors of the diseases and maybe metabolic or genetic defect treatments.[166–168]

1.7.3 Biological Applications

Among the most important applications of the colloid biological nanocrystals, one can name molecules detection. Some biomolecules are able to detect one another with high precision and selectivity and can connect one to another. Antibodies as receptors have several applications in molecular detection applications. For instance, if a virus enters a live creature's body, antibodies detect it as an invader and attach to it, so the other sections of the immune system would be able to kill the virus.

Nanocrystals attached to a molecular receptor are directly connected to the position of ligand molecules. In this way, molecular labeling would be possible and completed by the receptor. For instance, in the very sensitive chromatic analysis method of DNA, color change of the Au nanoparticles from the rubious to the blue is used (because of its accumulation).

Other biological capabilities of the nanostructures and nanomaterials include employing semiconductor nanocrystalline colloids as a fluorescence … llular and chemical detection and use of nanostructured mate…

1.7.4

Each … sorption activity certainly includes a high … ease in nanomaterials size leads to an incr… terials are vastly used in catalysts. For inst… ameter of 10 nm, 15% of the atoms are pla… ocrystal with a diameter of 1 nm, 100% of … ce. Thus, a tiny nanocrystal with a vast su… activities.

… high catalytic potential and selective re… s. For example, the smooth surface of g… he temperature of 473°K in the reaction of … es (>10 nm) deposited on the oxide surfa… ity in the hydrogenation and oxidation r… his potential is up to an extent in which c… onoxide is even possible at a temperature …

… gaard, T., General properties of slow-plasmon … antennas and resonators, *Optics Express* 15 (17), …

… ral strategy for nano-encapsulation via inter… rolled radical miniemulsion polymerization, … *cations* 27 (1), 21–25, 2006.

3. Mokari, T., Sertchook, H., Aharoni, A., Ebenstein, Y., Avnir, D., and Banin, U., Nano@micro: General method for entrapment of nanocrystals in sol-gel-derived composite hydrophobic silica spheres, *Chemistry of Materials* 17 (2), 258–263, 2005.
4. Davis, S. S., Biomedical applications of nanotechnology—Implications for drug targeting and gene therapy, *Trends in Biotechnology* 15 (6), 217–224, 1997.
5. Pum, D., and Sleytr, U. B., The application of bacterial S-layers in molecular nanotechnology, *Trends in Biotechnology* 17 (1), 8–12, 1999.
6. Seeman, N. C., DNA engineering and its application to nanotechnology, *Trends in Biotechnology* 17 (11), 437–443, 1999.
7. Hughes, M. P., AC electrokinetics: Applications for nanotechnology, *Nanotechnology* 11 (2), 124–132, 2000.
8. Andersson, H., and Van Den Berg, A., Microtechnologies and nanotechnologies for single-cell analysis, *Current Opinion in Biotechnology* 15 (1), 44–49, 2004.
9. Khaled, A., Guo, S., Li, F., and Guo, P., Controllable self-assembly of nanoparticles for specific delivery of multiple therapeutic molecules to cancer cells using RNA nanotechnology, *Nano Letters* 5 (9), 1797–1808, 2005.
10. Choi, H., and Mody, C. C. M., The long history of molecular electronics: Microelectronics origins of nanotechnology, *Social Studies of Science* 39 (1), 11–50, 2009.
11. Carter, A., Learning from history: Understanding the carcinogenic risks of nanotechnology, *Journal of the National Cancer Institute* 100 (23), 1664–1665, 2008.
12. Marcovich, A., and Shinn, T., Socio/intellectual patterns in nanoscale research: Feynman nanotechnology prize laureates, 1993–2007, *Social Science Information* 49 (4), 615–638, 2010.
13. Ball, P., Feynman's fancy: Richard Feynman's famous talk on atom-by-atom assembly is often credited with kick-starting nanotechnology: Fifty years on, Philip Ball investigates how influential it really was, *Chemistry World* 6 (1), 58–62, 2009.
14. Devreese, J. T., Importance of nanosensors: Feynman's vision and the birth of nanotechnology, *MRS Bulletin* 32 (9), 718–724, 2007.
15. Devreese, J. T., Importance of nanosensors: Feynman's vision and the birth of nanotechnology, in *Materials Research Society Symposium Proceedings*, 2006, pp. 78–88.
16. Peterson, C. L., Nanotechnology: From Feynman to the grand challenge of molecular manufacturing, *IEEE Technology and Society Magazine* 23 (4), 9–15, 2004.
17. Drexler, K. E., Nanotechnology: From Feynman to funding, *Bulletin of Science, Technology and Society* 24 (1), 21–27, 2004.
18. Pokropivny, V. V., and Skorokhod, V. V., Classification of nanostructures by dimensionality and concept of surface forms engineering in nanomaterial science, *Materials Science and Engineering C* 27 (5-8 SPEC. ISS.), 990–993, 2007.
19. Hla, S. W., Meyer, G., and Rieder, K. H., Inducing single-molecule chemical reactions with a UHV-STM: A new dimension for nano-science and technology, *ChemPhysChem* 2 (6), 361–366, 2001.
20. Ritchie, R. O., Kruzic, J. J., Muhlstein, C. L., Nalla, R. K., and Stach, E. A., Characteristic dimensions and the micro-mechanisms of fracture and fatigue in "nano" and "bio" materials, *International Journal of Fracture* 128 (1), 1–15, 2004.
21. Cave, G. W. V., Ferrarelli, M. C., and Atwood, J. L., Nano-dimensions for the pyrogallol[4]arene cavity, *Chemical Communications* (22), 2787–2789, 2005.

22. Carpinteri, A., and Pugno, N., Strength and toughness of micro- and nanostructured materials: Unified influence of composition, grain size and structural dimension, *Reviews on Advanced Materials Science* 10 (4), 320–324, 2005.
23. Park, S., An, J., Potts, J. R., Velamakanni, A., Murali, S., and Ruoff, R. S., Hydrazine-reduction of graphite- and graphene oxide, *Carbon* 49 (9), 3019–3023, 2011.
24. Kim, J., Yun, J. H., and Han, C. S., Nanomaterial-embedded gas sensor fabrication, *Current Applied Physics* 9 (2 Suppl.), E38–E41, 2009.
25. Lu, Y., and Knize, R. J., Modified laser ablation process for nanostructured thermoelectric nanomaterial fabrication, *Applied Surface Science* 254 (4), 1211–1214, 2007.
26. Wu, J., Cheng, Y., Lin, J., Huang, Y., Huang, M., and Hao, S., Fabrication and photocatalytic properties of $HLaNb_2O$ 7(Pt, Fe_2O_3) pillared nanomaterial, *Journal of Physical Chemistry C* 111 (9), 3624–3628, 2007.
27. Lazzari, M., and Arturo López-Quintela, M., Block copolymers as a tool for nanomaterial fabrication, *Advanced Materials* 15 (19), 1583–1594, 2003.
28. Balandin, A. A., Thermal properties of graphene and nanostructured carbon materials, *Nature Materials* 10 (8), 569–581, 2011.
29. Khan, M. I., Aydemir, K., Siddiqui, M. R. H., Alwarthan, A. A., and Marshall, C. L., Oxidative dehydrogenation properties of novel nanostructured polyoxovanadate based materials, *Catalysis Letters* 141 (4), 538–543, 2011.
30. Szczech, J. R., Higgins, J. M., and Jin, S., Enhancement of the thermoelectric properties in nanoscale and nanostructured materials, *Journal of Materials Chemistry* 21 (12), 4037–4055, 2011.
31. Caruso, F., Nanoengineering of particle surfaces, *Advanced Materials* 13 (1), 11–22, 2001.
32. Lauhon, L. J., Gudlksen, M. S., Wang, D., and Lieber, C. M., Epitaxial core-shell and core-multishell nanowire heterostructures, *Nature* 420 (6911), 57–61, 2002.
33. Yin, Y., and Alivisatos, A. P., Colloidal nanocrystal synthesis and the organic-inorganic interface, *Nature* 437 (7059), 664–670, 2005.
34. Pileni, M. P., Nanosized particles made in colloidal assemblies, *Langmuir* 13 (13), 3266–3276, 1997.
35. Aksay, I. A., Trau, M., Manne, S., Honma, I., Yao, N., Zhou, L., Fenter, P., Eisenberger, P. M., and Gruner, S. M., Biomimetic pathways for assembling inorganic thin films, *Science* 273 (5277), 892–898, 1996.
36. Ruoff, R. S., and Lorents, D. C., Mechanical and thermal properties of carbon nanotubes, *Carbon* 33 (7), 925–930, 1995.
37. Li, M., and Xu, T., Topological and atomic scale characterization of grain boundary networks in polycrystalline and nanocrystalline materials, *Progress in Materials Science* 56 (6), 864–899, 2011.
38. Du, Z., and De Leeuw, N. H., A combined density functional theory and interatomic potential-based simulation study of the hydration of nano-particulate silicate surfaces, *Surface Science* 554 (2–3), 193–210, 2004.
39. Pellenq, R. J. M., and Van Damme, H., Why does concrete set?: The nature of cohesion forces in hardened cement-based materials, *MRS Bulletin* 29 (5), 319–323, 2004.
40. Brus, J., Solid-state NMR study of phase separation and order of water molecules and silanol groups in polysiloxane networks, *Journal of Sol-Gel Science and Technology* 25 (1), 17–28, 2002.

41. French, R. H., Parsegian, V. A., Podgornik, R., Rajter, R. F., Jagota, A., Luo, J., Asthagiri, D., Chaudhury, M. K., Chiang, Y. M., Granick, S., Kalinin, S., Kardar, M., Kjellander, R., Langreth, D. C., Lewis, J., Lustig, S., Wesolowski, D., Wettlaufer, J. S., Ching, W. Y., Finnis, M., Houlihan, F., Von Lilienfeld, O. A., Van Oss, C. J., and Zemb, T., Long range interactions in nanoscale science, *Reviews of Modern Physics* 82 (2), 1887–1944, 2010.
42. Berner, A., Fuks, D., Ellis, D. E., Mundim, K., and Dorfman, S., Formation of nano-crystalline structure at the interface in Cu-C composite, *Applied Surface Science* 144–145, 677–681, 1999.
43. Beznosjuk, S. A., Beznosjuk, M. S., and Mezentzev, D. A., Electron swarming in nanostructures, *Computational Materials Science* 14 (1–4), 209–214, 1999.
44. Gao, F., Wang, Y., Shi, D., Zhang, J., Wang, M., Jing, X., Humphry-Baker, R., Wang, P., Zakeeruddin, S. M., and Grätzel, M., Enhance the optical absorptivity of nanocrystalline TiO_2 film with high molar extinction coefficient ruthenium sensitizers for high performance dye-sensitized solar cells, *Journal of the American Chemical Society* 130 (32), 10720–10728, 2008.
45. Jennings, J. R., Ghicov, A., Peter, L. M., Schmuki, P., and Walker, A. B., Dye-sensitized solar cells based on oriented TiO_2 nanotube arrays: Transport, trapping, and transfer of electrons, *Journal of the American Chemical Society* 130 (40), 13364–13372, 2008.
46. Chen, H., Zhou, X., and Ding, C., Investigation of the thermomechanical properties of a plasma-sprayed nanostructured zirconia coating, *Journal of the European Ceramic Society* 23 (9), 1449–1455, 2003.
47. Fu, L. J., Liu, H., Zhang, H. P., Li, C., Zhang, T., Wu, Y. P., Holze, R., and Wu, H. Q., Synthesis and electrochemical performance of novel core/shell structured nanocomposites, *Electrochemistry Communications* 8 (1), 1–4, 2006.
48. Liz-Marzán, L. M., and Philipse, A. P., Synthesis and optical properties of gold-labeled silica particles, *Journal of Colloid and Interface Science* 176 (2), 459–466, 1995.
49. Nakade, S., Saito, Y., Kubo, W., Kanzaki, T., Kitamura, T., Wada, Y., and Yanagida, S., Enhancement of electron transport in nano-porous TiO_2 electrodes by dye adsorption, *Electrochemistry Communications* 5 (9), 804–808, 2003.
50. Vernon, D. R., Meng, F., Dec, S. F., Williamson, D. L., Turner, J. A., and Herring, A. M., Synthesis, characterization, and conductivity measurements of hybrid membranes containing a mono-lacunary heteropolyacid for PEM fuel cell applications, *Journal of Power Sources* 139 (1–2), 141–151, 2005.
51. Bohn, R., Klassen, T., and Bormann, R., Room temperature mechanical behavior of silicon-doped TiAl alloys with grain sizes in the nano- and submicron-range, *Acta Materialia* 49 (2), 299–311, 2001.
52. Wang, L., Bai, J., Huang, P., Wang, H., Zhang, L., and Zhao, Y., Self-assembly of gold nanoparticles for the voltammetric sensing of epinephrine, *Electrochemistry Communications* 8 (6), 1035–1040, 2006.
53. Fukai, Y., Kondo, Y., Mori, S., and Suzuki, E., Highly efficient dye-sensitized SnO_2 solar cells having sufficient electron diffusion length, *Electrochemistry Communications* 9 (7), 1439–1443, 2007.
54. Cao, Q., Zhang, H. P., Wang, G. J., Xia, Q., Wu, Y. P., and Wu, H. Q., A novel carbon-coated $LiCoO_2$ as cathode material for lithium ion battery, *Electrochemistry Communications* 9 (5), 1228–1232, 2007.

55. Emeline, A. V., Ryabchuk, V. K., and Serpone, N., Spectral dependencies of the quantum yield of photochemical processes on the surface of nano-/microparticulates of wide-band-gap metal oxides. 1. Theoretical approach, *Journal of Physical Chemistry B* 103 (8), 1316–1324, 1999.
56. Adebahr, J., Byrne, N., Forsyth, M., MacFarlane, D. R., and Jacobsson, P., Enhancement of ion dynamics in PMMA-based gels with addition of TiO_2 nanoparticles, *Electrochimica Acta* 48 (14–16 Spec.), 2099–2103, 2003.
57. Gupta, D., and Asai, K. W., Grain boundary self-diffusion in evaporated Au films at low temperatures, *Thin Solid Films* 22 (1), 121–130, 1974.
58. Schumacher, S., Birringer, R., Strauß, R., and Gleiter, H., Diffusion of silver in nanocrystalline copper between 303 and 373 K, *Acta Metallurgica* 37 (9), 2485–2488, 1989.
59. Lee, S. J., and Kriven, W. M., Crystallization and densification of nano-size amorphous cordierite powder prepared by a PVA solution-polymerization route, *Journal of the American Ceramic Society* 81 (10), 2605–2612, 1998.
60. Wong, W. L. E., and Gupta, M., Development of Mg/Cu nanocomposites using microwave assisted rapid sintering, *Composites Science and Technology* 67 (7–8), 1541–1552, 2007.
61. Huang, J., Xiao, Y., Mya, K. Y., Liu, X., He, C., Dai, J., and Siow, Y. P., Thermomechanical properties of polyimide-epoxy nanocomposites from cubic silsesquioxane epoxides, *Journal of Materials Chemistry* 14 (19), 2858–2863, 2004.
62. Huang, J., He, C., Liu, X., Xu, J., Tay, C. S. S., and Chow, S. Y., Organic-inorganic nanocomposites from cubic silsesquioxane epoxides: Direct characterization of interphase, and thermomechanical properties, *Polymer* 46 (18), 7018–7027, 2005.
63. Lai, Y. H., Kuo, M. C., Huang, J. C., and Chen, M., On the PEEK composites reinforced by surface-modified nano-silica, *Materials Science and Engineering A* 458 (1–2), 158–169, 2007.
64. Oshima, Y., Eri, T., Shibata, M., Sunakawa, H., and Usui, A., Fabrication of free-standing GaN wafers by hydride vapor-phase epitaxy with void-assisted separation, *Physica Status Solidi (A) Applied Research* 194 (2 Spec.), 554–558, 2002.
65. Goyal, R. K., Tiwari, A. N., Mulik, U. P., and Negi, Y. S., Novel high performance Al_2O_3/poly(ether ether ketone) nanocomposites for electronics applications, *Composites Science and Technology* 67 (9), 1802–1812, 2007.
66. Zhi, C., Bando, Y., Terao, T., Tang, C., Kuwahara, H., and Golberg, D., Towards thermoconductive, electrically insulating polymeric composites with boron nitride nanotubes as fillers, *Advanced Functional Materials* 19 (12), 1857–1862, 2009.
67. Hassan, S. F., and Gupta, M., Effect of type of primary processing on the microstructure, CTE and mechanical properties of magnesium/alumina nanocomposites, *Composite Structures* 72 (1), 19–26, 2006.
68. Liu, Y. L., Lin, Y. L., Chen, C. P., and Jeng, R. J., Preparation of epoxy resin/silica hybrid composites for epoxy molding compounds, *Journal of Applied Polymer Science* 90 (14), 4047–4053, 2003.
69. Levitas, V. I., Asay, B. W., Son, S. F., and Pantoya, M., Mechanochemical mechanism for fast reaction of metastable intermolecular composites based on dispersion of liquid metal, *Journal of Applied Physics* 101 (8), art no. 083524, 2007.
70. Lu, N., Song, X., and Zhang, J., Microstructure and fundamental properties of nanostructured gadolinium (Gd), *Materials Letters* 63 (12), 1089–1092, 2009.

71. Wang, Y. Z., Qiao, G. W., Liu, X. D., Ding, B. Z., and Hu, Z. Q., Electrical resistivity of nanocrystalline Fe-Cu-Si-B alloys obtained by crystallization of the amorphous alloy, *Materials Letters* 17 (3–4), 152–154, 1993.
72. Sakai, G., Baik, N. S., Miura, N., and Yamazoe, N., Gas sensing properties of tin oxide thin films fabricated from hydrothermally treated nanoparticles: Dependence of CO and H_2 response on film thickness, *Sensors and Actuators, B: Chemical* 77 (1–2), 116–121, 2001.
73. Knite, M., Teteris, V., Kiploka, A., and Kaupuzs, J., Polyisoprene-carbon black nanocomposites as tensile strain and pressure sensor materials, *Sensors and Actuators, A: Physical* 110 (1–3), 142–149, 2004.
74. Teoh, L. G., Hon, Y. M., Shieh, J., Lai, W. H., and Hon, M. H., Sensitivity properties of a novel NO_2 gas sensor based on mesoporous WO_3 thin film, *Sensors and Actuators, B: Chemical* 96 (1–2), 219–225, 2003.
75. Zhang, W., Suhr, J., and Koratkar, N., Carbon nanotube/polycarbonate composites as multifunctional strain sensors, *Journal of Nanoscience and Nanotechnology* 6 (4), 960–964, 2006.
76. Fu, Y., Wang, Y., Wang, X., Liu, J., Lai, Z., Chen, G., and Willander, M., Experimental and theoretical characterization of electrical contact in anisotropically conductive adhesive, *IEEE Transactions on Advanced Packaging* 23 (1), 15–21, 2000.
77. Wang, L., and Li, D. Y., Mechanical, electrochemical and tribological properties of nanocrystalline surface of brass produced by sandblasting and annealing, *Surface and Coatings Technology* 167 (2–3), 188–196, 2003.
78. Chen, X. J., Khor, K. A., Chan, S. H., and Yu, L. G., Preparation yttria-stabilized zirconia electrolyte by spark-plasma sintering, *Materials Science and Engineering A* 341 (1–2), 43–48, 2003.
79. Kida, T., Nishiyama, A., Yuasa, M., Shimanoe, K., and Yamazoe, N., Highly sensitive NO_2 sensors using lamellar-structured WO_3 particles prepared by an acidification method, *Sensors and Actuators, B: Chemical* 135 (2), 568–574, 2009.
80. Fedtke, P., Wienecke, M., Bunescu, M. C., Pietrzak, M., Deistung, K., and Borchardt, E., Hydrogen sensor based on optical and electrical switching, *Sensors and Actuators, B: Chemical* 100 (1–2), 151–157, 2004.
81. Branagan, D. J., and Tang, Y., Developing extreme hardness (>15 GPa) in iron based nanocomposites, *Composites—Part A: Applied Science and Manufacturing* 33 (6), 855–859, 2002.
82. Liu, W. S., Zhang, B. P., Zhao, L. D., and Li, J. F., Improvement of thermoelectric performance of $CoSb_{3-x}Te_x$ skutterudite compounds by additional substitution of IVB-group elements for Sb, *Chemistry of Materials* 20 (24), 7526–7531, 2008.
83. van Dalen, M. E., Karnesky, R. A., Cabotaje, J. R., Dunand, D. C., and Seidman, D. N., Erbium and ytterbium solubilities and diffusivities in aluminum as determined by nanoscale characterization of precipitates, *Acta Materialia* 57 (14), 4081–4089, 2009.
84. Shimizu, H., Li, Y., Kaito, A., and Sano, H., High-shear effects on the nanodispersed structure of the PVDF/PA11 blends, *Journal of Nanoscience and Nanotechnology* 6 (12), 3923–3928, 2006.
85. Santos-Beltrán, A., Estrada-Guel, I., Miki-Yoshida, M., Barajas-Villaruel, J. I., and Martínez-Sánchez, R., Microstructural and mechanical characterization of aluminum-graphite composites, *Journal of Metastable and Nanocrystalline Materials* 20–21, 133–138, 2004.

86. Beeri, O., Dunand, D. C., and Seidman, D. N., Roles of impurities on precipitation kinetics of dilute Al-Sc alloys, *Materials Science and Engineering A* 527 (15), 3501–3509, 2010.
87. Rios, O., and Ebrahimi, F., Spinodal decomposition of the gamma-phase upon quenching in the Ti-Al-Nb ternary alloy system, *Intermetallics* 19 (1), 93–98, 2011.
88. Meng, Q. P., Rong, Y. H., and Hsu, T. Y., Distribution of solute atoms in nanocrystalline materials, *Materials Science and Engineering A* 471 (1–2), 22–27, 2007.
89. Roy, S., Dubenko, I., Edorh, D. D., and Ali, N., Size induced variations in structural and magnetic properties of double exchange $La_{0.8}Sr_{0.2}MnO_{3-\sigma}$ nano-ferromagnet, *Journal of Applied Physics* 96 (2), 1202–1208, 2004.
90. Cui, Z. L., Dong, L. F., and Hao, C. C., Microstructure and magnetic property of nano-Fe particles prepared by hydrogen arc plasma, *Materials Science and Engineering A* 286 (1), 205–207, 2000.
91. Cheng, Y., Zheng, Y., Wang, Y., Bao, F., and Qin, Y., Synthesis and magnetic properties of nickel ferrite nano-octahedra, *Journal of Solid State Chemistry* 178 (7), 2394–2397, 2005.
92. Yoshikawa, H., Hayashida, K., Kozuka, Y., Horiguchi, A., Awaga, K., Bandow, S., and Iijima, S., Preparation and magnetic properties of hollow nano-spheres of cobalt and cobalt oxide: Drastic cooling-field effects on remnant magnetization of antiferromagnet, *Applied Physics Letters* 85 (22), 5287–5289, 2004.
93. Panda, R. N., Shih, J. C., and Chin, T. S., Magnetic properties of nano-crystalline Gd- or Pr-substituted $CoFe_2O_4$ synthesized by the citrate precursor technique, *Journal of Magnetism and Magnetic Materials* 257 (1), 79–86, 2003.
94. Thakur, S., Katyal, S. C., and Singh, M., Structural and magnetic properties of nano nickel-zinc ferrite synthesized by reverse micelle technique, *Journal of Magnetism and Magnetic Materials* 321 (1), 1–7, 2009.
95. Ohnuma, S., and Masumoto, T., High frequency magnetic properties and GMR effect of nano-granular magnetic thin films, *Scripta Materialia* 44 (8–9), 1309–1313, 2001.
96. Mishra, D., Anand, S., Panda, R. K., and Das, R. P., Studies on characterization, microstructures and magnetic properties of nano-size barium hexa-ferrite prepared through a hydrothermal precipitation-calcination route, *Materials Chemistry and Physics* 86 (1), 132–136, 2004.
97. Makino, A., Inoue, A., and Masumoto, T., Nanocrystalline soft magnetic Fe-M-B (M = Zr, Hf, Nb), Fe-M-O (M = Zr, Hf, rare earth) alloys and their applications, *Nanostructured Materials* 12 (5), 825–828, 1999.
98. Crobu, M., Scorciapino, A., Elsener, B., and Rossi, A., The corrosion resistance of electroless deposited nano-crystalline Ni-P alloys, *Electrochimica Acta* 53 (8), 3364–3370, 2008.
99. Kimura, M., Kihira, H., Ohta, N., Hashimoto, M., and Senuma, T., Control of Fe(O,OH)6 nano-network structures of rust for high atmospheric-corrosion resistance, *Corrosion Science* 47 (10), 2499–2509, 2005.
100. Feng, Q., Li, T., Teng, H., Zhang, X., Zhang, Y., Liu, C., and Jin, J., Investigation on the corrosion and oxidation resistance of Ni-Al_2O_3 nano-composite coatings prepared by sediment co-deposition, *Surface and Coatings Technology* 202 (17), 4137–4144, 2008.
101. Shi, H., Liu, F., Han, E., and Wei, Y., Effects of nano pigments on the corrosion resistance of alkyd coating, *Journal of Materials Science and Technology* 23 (4), 551–558, 2007.

102. Ye, X., Chen, M., Yang, M., Wei, J., and Liu, D., In vitro corrosion resistance and cytocompatibility of nano-hydroxyapatite reinforced Mg-Zn-Zr composites, *Journal of Materials Science: Materials in Medicine* 21 (4), 1321–1328, 2010.
103. Shen, G. X., Du, R. G., Chen, Y. C., Lin, C. J., and Scantlebury, D., Study on hydrophobic nano-titanium dioxide coatings for improvement in corrosion resistance of type 316L stainless steel, *Corrosion* 61 (10), 943–950, 2005.
104. Aliofkhazraei, M., Sabour Rouhaghdam, A., and Heydarzadeh, A., Strong relation between corrosion resistance and nanostructure of compound layer of treated 316 austenitic stainless steel, *Materials Characterization* 60 (2), 83–89, 2009.
105. Zaluska, A., Zaluski, L., and Ström-Olsen, J. O., Structure, catalysis and atomic reactions on the nano-scale: A systematic approach to metal hydrides for hydrogen storage, *Applied Physics A: Materials Science and Processing* 72 (2), 157–165, 2001.
106. Fujii, H., and Orimo, S. I., Hydrogen storage properties in nano-structured magnesium- and carbon-related materials, *Physica B: Condensed Matter* 328 (1–2), 77–80, 2003.
107. Kojima, Y., Kawai, Y., and Haga, T., Magnesium-based nano-composite materials for hydrogen storage, *Journal of Alloys and Compounds* 424 (1–2), 294–298, 2006.
108. Zhu, M., Gao, Y., Che, X. Z., Yang, Y. Q., and Chung, C. Y., Hydriding kinetics of nano-phase composite hydrogen storage alloys prepared by mechanical alloying of Mg and $MmNi_{5-x}$ $(CoAlMn)_x$, *Journal of Alloys and Compounds* 330–332, 708–713, 2002.
109. Yang, J., Ciureanu, M., and Roberge, R., Hydrogen storage properties of nanocomposites of Mg and Zr-Ni-Cr alloys, *Materials Letters* 43 (5), 234–239, 2000.
110. Xu, X., and Song, C., Improving hydrogen storage/release properties of magnesium with nano-sized metal catalysts as measured by tapered element oscillating microbalance, *Applied Catalysis A: General* 300 (2), 130–138, 2006.
111. Liu, Z., and Lei, Z., Cyclic hydrogen storage properties of Mg milled with nickel nano-powders and MnO_2, *Journal of Alloys and Compounds* 443 (1–2), 121–124, 2007.
112. Zhenglong, L., Zuyan, L., and Yanbin, C., Cyclic hydrogen storage properties of Mg milled with nickel nano-powders and NiO, *Journal of Alloys and Compounds* 470 (1–2), 470–472, 2009.
113. Chang, J. K., Tsai, H. Y., and Tsai, W. T., Effects of post-treatments on microstructure and hydrogen storage performance of the carbon nano-tubes prepared via a metal dusting process, *Journal of Power Sources* 182 (1), 317–322, 2008.
114. Keun Kwon, I., Kidoaki, S., and Matsuda, T., Electrospun nano- to microfiber fabrics made of biodegradable copolyesters: Structural characteristics, mechanical properties and cell adhesion potential, *Biomaterials* 26 (18), 3929–3939, 2005.
115. Hong, Z., Zhang, P., He, C., Qiu, X., Liu, A., Chen, L., Chen, X., and Jing, X., Nanocomposite of poly(L-lactide) and surface grafted hydroxyapatite: Mechanical properties and biocompatibility, *Biomaterials* 26 (32), 6296–6304, 2005.
116. Park, J. H., and Jana, S. C., The relationship between nano- and micro-structures and mechanical properties in PMMA-epoxy-nanoclay composites, *Polymer* 44 (7), 2091–2100, 2003.
117. Gao, L., Wang, H. Z., Hong, J. S., Miyamoto, H., Miyamoto, K., Nishikawa, Y., and Torre, S. D. D. L., Mechanical properties and microstructure of nano-SiC-Al_2O_3 composites densified by spark plasma sintering, *Journal of the European Ceramic Society* 19 (5), 609–613, 1999.

118. Fong, H., Sarikaya, M., White, S. N., and Snead, M. L., Nano-mechanical properties profiles across dentin-enamel junction of human incisor teeth, *Materials Science and Engineering C* 7 (2), 119–128, 2000.
119. Mukai, T., Kawazoe, M., and Higashi, K., Dynamic mechanical properties of a near-nano aluminum alloy processed by equal-channel-angular-extrusion, *Nanostructured Materials* 10 (5), 755–765, 1998.
120. He, G., Eckert, J., Löser, W., and Hagiwara, M., Composition dependence of the microstructure and the mechanical properties of nano/ultrafine-structured Ti-Cu-Ni-Sn-Nb alloys, *Acta Materialia* 52 (10), 3035–3046, 2004.
121. Cho, J., Joshi, M. S., and Sun, C. T., Effect of inclusion size on mechanical properties of polymeric composites with micro and nano particles, *Composites Science and Technology* 66 (13), 1941–1952, 2006.
122. Kumar, R., Prakash, K. H., Cheang, P., and Khor, K. A., Microstructure and mechanical properties of spark plasma sintered zirconia-hydroxyapatite nano-composite powders, *Acta Materialia* 53 (8), 2327–2335, 2005.
123. Šajgalík, P., Hnatko, M., Lofaj, F., Hvizdoš, P., Dusza, J., Warbichler, P., Hofer, F., Riedel, R., Lecomte, E., and Hoffmann, M. J., SiC/Si_3N_4 nano/micro-composite—Processing, RT and HT mechanical properties, *Journal of the European Ceramic Society* 20 (4), 453–462, 2000.
124. Robertson, A., Erb, U., and Palumbo, G., Practical applications for electrodeposited nanocrystalline materials, *Nanostructured Materials* 12 (5), 1035–1040, 1999.
125. Swain, M. V., and Menčík, J., Mechanical property characterization of thin films using spherical tipped indenters, *Thin Solid Films* 253 (1–2), 204–211, 1994.
126. Constantinides, G., Ravi Chandran, K. S., Ulm, F. J., and Van Vliet, K. J., Grid indentation analysis of composite microstructure and mechanics: Principles and validation, *Materials Science and Engineering A* 430 (1–2), 189–202, 2006.
127. Saha, R., and Nix, W. D., Solt films on hard substrates—Nanoindentation of tungsten films on sapphire substrates, *Materials Science and Engineering A* 319–321, 898–901, 2001.
128. Chen, T., Dvorak, G. J., and Yu, C. C., Size-dependent elastic properties of unidirectional nano-composites with interface stresses, *Acta Mechanica* 188 (1–2), 39–54, 2007.
129. Yang, Q., Lengauer, W., Koch, T., Scheerer, M., and Smid, I., Hardness and elastic properties of $Ti(C_xN_{1-x})$, $Zr(C_xN_{1-x})$ and $Hf(C_xN_{1-x})$, *Journal of Alloys and Compounds* 309 (1–2), L5–L9, 2000.
130. Atanacio, A. J., Latella, B. A., Barbé, C. J., and Swain, M. V., Mechanical properties and adhesion characteristics of hybrid sol-gel thin films, *Surface and Coatings Technology* 192 (2–3), 354–364, 2005.
131. Siegel, R. W., Cluster-assembled nanophase materials, *Annual Review of Materials Science* 21 (1), 559–578, 1991.
132. Espinosa, H. D., Peng, B., Moldovan, N., Friedmann, T. A., Xiao, X., Mancini, D. C., Auciello, O., Carlisle, J., Zorman, C. A., and Merhegany, M., Elasticity, strength, and toughness of single crystal silicon carbide, ultrananocrystalline diamond, and hydrogen-free tetrahedral amorphous carbon, *Applied Physics Letters* 89 (7), art no. 073111, 2006.
133. Remediakis, I. N., Kopidakis, G., and Kelires, P. C., Softening of ultra-nanocrystalline diamond at low grain sizes, *Acta Materialia* 56 (18), 5340–5344, 2008.

134. Nowak, R., Pessa, M., Suganuma, M., Leszczynski, M., Grzegory, I., Porowski, S., and Yoshida, F., Elastic and plastic properties of GaN determined by nano-indentation of bulk crystal, *Applied Physics Letters* 75 (14), 2070–2072, 1999.
135. He, J. L., Li, W. Z., Li, H. D., and Liu, C. H., Plastic properties of nano-scale ceramic-metal multilayers, *Surface and Coatings Technology* 103–104, 276–280, 1998.
136. Lamagnere, P., Girodin, D., Meynaud, P., Vergne, F., and Vincent, A., Study of elasto-plastic properties of microheterogeneities by means of nano-indentation measurements: Application to bearing steels, *Materials Science and Engineering A* 215 (1–2), 134–142, 1996.
137. Nieman, G. W., Weertman, J. R., and Siegel, R. W., Tensile strength and creep properties of nanocrystalline palladium, *Scripta Metallurgica et Materiala* 24 (1), 145–150, 1990.
138. El-Sherik, A. M., Erb, U., Palumbo, G., and Aust, K. T., Deviations from Hall-Petch behaviour in as-prepared nanocrystalline nickel, *Scripta Metallurgica et Materiala* 27 (9), 1185–1188, 1992.
139. Chokshi, A. H., Rosen, A., Karch, J., and Gleiter, H., On the validity of the Hall-Petch relationship in nanocrystalline materials, *Scripta Metallurgica* 23 (10), 1679–1683, 1989.
140. Qi, Z. B., Sun, P., Zhu, F. P., Wang, Z. C., Peng, D. L., and Wu, C. H., The inverse Hall-Petch effect in nanocrystalline ZrN coatings, *Surface and Coatings Technology* 205 (12), 3692–3697, 2011.
141. Malow, T. R., and Koch, C. C., Mechanical properties in tension of mechanically attrited nanocrystalline iron by the use of the miniaturized disk bend test, *Acta Materialia* 46 (18), 6459–6473, 1998.
142. Van Swygenhoven, H., Footprints of plastic deformation in nanocrystalline metals, *Materials Science and Engineering A* 483–484 (1–2 C), 33–39, 2008.
143. Jia, K., and Fischer, T. E., Sliding wear of conventional and nanostructured cemented carbides, *Wear* 203–204, 310–318, 1997.
144. Xue, J. X., Zhao, B., and Gao, G. F., Study on the plastic removal mechanismof nano-ZrO_2 ceramics by ultrasonic grinding, in *Key Engineering Materials* 455, 686–689, 2011.
145. Tjong, S. C., and Chen, H., Nanocrystalline materials and coatings, *Materials Science and Engineering R: Reports* 45 (1–2), 1–88, 2004.
146. Mishra, R. S., Valiev, R. Z., McFadden, S. X., and Mukherjee, A. K., Tensile superplasticity in a nanocrystalline nickel aluminide, *Materials Science and Engineering A* 252 (2), 174–178, 1998.
147. Xu, C., Furukawa, M., Horita, Z., and Langdon, T. G., Using ECAP to achieve grain refinement, precipitate fragmentation and high strain rate superplasticity in a spray-cast aluminum alloy, *Acta Materialia* 51 (20), 6139–6149, 2003.
148. Moelans, N., Blanpain, B., and Wollants, P., A phase field model for the simulation of grain growth in materials containing finely dispersed incoherent second-phase particles, *Acta Materialia* 53 (6), 1771–1781, 2005.
149. Nikitin, I., Altenberger, I., Maier, H. J., and Scholtes, B., Mechanical and thermal stability of mechanically induced near-surface nanostructures, *Materials Science and Engineering A* 403 (1–2), 318–327, 2005.

150. Umek, P., Korošec, R. C., Jančar, B., Dominko, R., and Arčon, D., The influence of the reaction temperature on the morphology of sodium titanate 1 D nanostructures and their thermal stability, *Journal of Nanoscience and Nanotechnology* 7 (10), 3502–3508, 2007.
151. Salavati-Niasari, M., Loghman-Estarki, M. R., and Davar, F., Synthesis, thermal stability and photoluminescence of new cadmium sulfide/organic composite hollow sphere nanostructures, *Inorganica Chimica Acta* 362 (10), 3677–3683, 2009.
152. Majumder, P., and Takoudis, C., Thermal stability of Ti/Mo and Ti/MoN nanostructures for barrier applications in Cu interconnects, *Nanotechnology* 19 (20), art no. 205202, 2008.
153. Li, Y., Ding, D., Ning, C., Bai, S., Huang, L., Li, M., and Mao, D., Thermal stability and in vitro bioactivity of Ti-Al-V-O nanostructures fabricated on Ti_6Al_4V alloy, *Nanotechnology* 20 (6), art no. 065708, 2009.
154. Guo, Y., Lee, N. H., Oh, H. J., Park, K. S., Jung, S. C., and Kim, S. J., Formation of 1-D nanostructures of titanate thin films and thermal stability study, *Journal of Nanoscience and Nanotechnology* 8 (10), 5316–5320, 2008.
155. Li, D., and Xia, Y., Electrospinning of nanofibers: Reinventing the wheel?, *Advanced Materials* 16 (14), 1151–1170, 2004.
156. Katz, E., and Willner, I., Biomolecule-functionalized carbon nanotubes: Applications in nanobioelectronics, *ChemPhysChem* 5 (8), 1084–1104, 2004.
157. Oberdorster, G., Stone, V., and Donaldson, K., Toxicology of nanoparticles: A historical perspective, *Nanotoxicology* 1 (1), 2–25, 2007.
158. Sardar, R., Funston, A. M., Mulvaney, P., and Murray, R. W., Gold nanoparticles: Past, present, and future, *Langmuir* 25 (24), 13840–13851, 2009.
159. Dresselhaus, M. S., Jorio, A., Hofmann, M., Dresselhaus, G., and Saito, R., Perspectives on carbon nanotubes and graphene Raman spectroscopy, *Nano Letters* 10 (3), 751–758, 2010.
160. Hutchings, G. J., and Haruta, M., A golden age of catalysis: A perspective, *Applied Catalysis A: General* 291 (1–2), 2–5, 2005.
161. Kudernac, T., Katsonis, N., Browne, W. R., and Feringa, B. L., Nano-electronic switches: Light-induced switching of the conductance of molecular systems, *Journal of Materials Chemistry* 19 (39), 7168–7177, 2009.
162. Maki, W. C., Mishra, N. N., Rastogi, S. K., Cameron, E., Filanoski, B., Winterrowd, P., and Maki, G. K., Universal bio-molecular signal transduction-based nanoelectronic bio-detection system, *Sensors and Actuators, B: Chemical* 133 (2), 547–554, 2008.
163. Kim, D. H., Song, C. K., Kang, Y., Lee, C., Lee, H., So, H. M., and Kim, J., Fabrication of an electronic test platform containing nano via holes for molecular devices, *Journal of the Korean Physical Society* 45 (4), 983–987, 2004.
164. Kim, D. H., Song, C. K., Kang, Y., Lee, C., Lee, H., So, H. M., and Kim, J., Characterization of phenylene-based molecular switches self-assembled in nano via holes of an electronic test platform, *Journal of the Korean Physical Society* 45 (2), 470–474, 2004.
165. Ziegler, M. M., and Stan, M. R., CMOS/nano co-design for crossbar-based molecular electronic systems, *IEEE Transactions on Nanotechnology* 2 (4), 217–230, 2003.

166. Lou, M., and Jonckheere, E., Magnetically levitated nano-robots: An application to visualization of nerve cells injuries, *Studies in Health Technology and Informatics* 125, 310–312, 2007.
167. Li, G., Xi, N., and Wang, D. H., Functionalized nano-robot end effector for in situ sensing and manipulation of biological specimen, in *Proceedings—IEEE International Conference on Robotics and Automation*, 2005, pp. 448–453.
168. Kim, S. H., Choi, J. S., and Kim, B. K., Development of BEST nano-robot soccer team, *Proceedings—IEEE International Conference on Robotics and Automation* 4, 2680–2685, 1999.
169. Du, M., Song, W., Cui, Y., Yang, Y., and Li, J., Fabrication and biological application of nano-hydroxyapatite (nHA)/alginate (ALG) hydrogel as scaffolds, *Journal of Materials Chemistry* 21 (7), 2228–2236, 2011.
170. Boturyn, D., Defrancq, E., Dolphin, G. T., Garcia, J., Labbe, P., Renaudet, O., and Dumy, P., RAFT Nano-constructs: Surfing to biological applications, *Journal of Peptide Science* 14 (2), 224–240, 2008.
171. Fu, Y. Q., Gu, Y. W., Shearwood, C., Luo, J. K., Flewitt, A. J., and Milne, W. I., Spark plasma sintering of TiNi nano-powders for biological application, *Nanotechnology* 17 (21), 5293–5298, 2006.
172. Brasuel, M. G., Miller, T. J., Kopelman, R., and Philbert, M. A., Liquid polymer nano-PEBBLEs for Cl- analysis and biological applications, *Analyst* 128 (10), 1262–1267, 2003.
173. Wang, J., Liu, Y., Zhang, X., Mi, Z., and Wang, L., Facile preparation of hydrocarbon fuel-soluble nano-catalyst and its novel application in catalytic combustion of JP-10, *Catalysis Communications* 10 (11), 1518–1522, 2009.
174. Shironita, S., Mori, K., and Yamashita, H., Synthesis of nano-sized Pt metal particle on Ti-containing mesoporous silica and efficient catalytic application for hydrogenation of nitrobenzene, in *Studies in Surface Science and Catalysis* 174, 1291–1294, 2008.
175. Mori, K., Shironita, S., Shimizu, T., Sakata, T., Mori, H., Ohmichi, T., and Yamashita, H., Design of nano-sized Pt metals synthesized on Ti-containing mesoporous silicas and efficient catalytic application for NO reduction, *Materials Transactions* 49 (3), 398–401, 2008.
176. Jongsomjit, B., Kittiruangrayub, S., and Praserthdam, P., Study of cobalt dispersion onto the mixed nano-SiO_2-ZrO_2 supports and its application as a catalytic phase, *Materials Chemistry and Physics* 105 (1), 14–19, 2007.
177. Li, G., Li, W., Zhang, M., and Tao, K., Characterization and catalytic application of homogeneous nano-composite oxides ZrO_2-Al_2O_3, *Catalysis Today* 93–95, 595–601, 2004.

2

Classification of Two-Dimensional Nanostructures

2.1 Introduction

Conversion of a microstructure state to a nanostructure state may lead to some changes in a material's physical properties. An increase in surface-to-volume ratio and particles' transfer to a domain with dominant quantum effects are two main factors in characteristics' change.

An increase in surface-to-volume ratio, which is gradually intensified by particles' fining, leads to the dominance of surface atoms' behavior compared with internal atoms. This factor affects both the particle's properties and its interaction with the other materials. Vast surface is a critical factor in catalyst and structure (such as electrodes) efficiency and improves technologies such as fuel cells and batteries. Also, the consequences of vast surfaces of the nanoparticles in the interactions of the nanocomposites lead to specific features such as enhanced strength and chemical or thermal resistance. Transition from classic to quantum mechanics is not gradual. As the particles are fine enough, they start to reveal quantum mechanics behaviors. Today nanoparticles are made from various sources. Most nanoparticles are ceramic, metallic, or polymer. However, ceramics are the most prevalent and are mainly made of metallic oxides such as Ti, Zn, Al, and Fe.[1–5]

2.2 Various Methods for Production of Nanostructures

Nanoparticles are the basic structural blocks in nanotechnology. They are the starting point in producing nanostructured materials and tools. The production of nanoparticles is an important factor in the development of research in nanoengineering and nanoscience.

Various nanoparticles are produced by several methods. Some precursors in liquid, solid, and gas states are used in the production and arrangement of nanoparticles and nanomaterials. Based on their chemical reactivity or physical compression, these materials align the nanostructures as structural blocks side by side. Principally, common techniques in the production of the nanomaterials are grouped into two main categories: "bottom-up" and "top-down" methods. Bottom-up methods result in substance development from the bottom (atom by atom, molecule by molecule, or cluster by cluster). Colloid dispersion is a good example of this nanoparticle producing method. The top-down method initiates with a mass bulky material and progresses toward the ideal state by its designing or abrasion. This method is similar to the use of patterns (e.g., method of electron beam lithography).

Both mentioned methods have a significant role in industry, particularly in nanotechnology. Both methods involve advantages and disadvantages. The main problem in the top-down method is creation of extremely tiny structures with efficient purity, the same as those produced in the bottom-up method. The bottom-up method offers a better chance for obtaining nanostructures with less defects, more homogeneous composition, and better short-term and long-term orders. This is due to the fact that in the bottom-up method, free energy of Gibbs decreases and as a result nanostructures and the materials are thermodynamically in balance. In contrary, in the top-down method, in most cases the material is subjected to internal stress, which leads to an increase of surface defects.[6–10]

2.2.1 Producing Nanoparticles by the Sol-Gel Process

In this method the agents (metallic ion solution) are rapidly added to a reaction container that includes a hot solution such as alkyl phosphate, alkyl phosphide, pyridine, alkyl amide, or furan. This quick introducing of the agents to the reaction container increases the precursors' concentration to an extent higher than the nucleation threshold concentration; also the solution would be saturated due to high temperatures. As a result, a sudden nucleation would occur followed by a concentration of agents that would reach a level less than that for critical nucleation. If the growth time of nanoclusters during the nucleation is less than that of successive growth processes, nanoclusters can be formed in a more unified manner.[11–15]

2.2.2 Production of Nanoparticles by the Chemical Deposition Method

Nucleation kinetic and particles' growth in the homogeneous solutions can be adjusted by controlled release of the anions and cations. Precise control of the deposition rate leads to development of same-size nanoparticles. Thus, control of factors, such as pH, constituents' concentration, and ions, which are determiners of the deposition process, is necessary. Organic molecules are used for release control of the reactants and ions during the participation. Reactant concentration, pH, and temperature have a considerable influence on particles' size. Through engineering of these factors, particles could be produced with narrow particle size distribution. Nanoparticles such as $Zr(OH)_4$, $BaTiO_2$, CdS, HgTe, CdTe, and AgAu are produced by this method.

Some researchers reported that urea can be used for control of the nucleation process in nanoparticles' production. They explained deposition of barium titanate nanoparticles by adding ethanolic oxalate acid solution to a mixture of titanium and barium aqueous solutions at room temperature. The deposition method for producing nanoparticles is very practical. This method can be used as a substitute for complex nanostructures such as quantum systems and other core/shell structures.[16–21] A detailed discussion about chemical methods for the fabrication of two-dimensional nanostructures is presented Chapter 5.

2.2.3 Production of Nanoparticles by the Hydrothermal Method

The hydrothermal method is a common technique for producing zeolite molecules. This method is used to solve all inorganic materials in the water at high pressure and temperature conditions and leads to production of solved materials crystals. Water plays a key role in this method because its structure in vapor state changes by room temperature, and reactants' properties such as solubility and reactivity change at high temperatures. These changes lead to production of high-quality nanoparticles and nanotubes, which cannot be produced at low temperatures. Through adjustment and change of factors such as temperature, pressure, reaction time, and system of precursor-products, it would be possible to obtain simultaneous nucleation and, in turn, narrow distribution of the particles size with high speed. Various types of nanoparticles such as ZrO_2, $LaCrO_3$, TiO_2, PbS, Ni_2P, SnS_2, CrN, Sb_2S_3, $Y_2Si_2O_7$, $SrTiO_3$, $BaTiO_3$, Bi_2O_3 nanorods and SiC nanowires are successfully produced by this method.[22–29]

2.2.4 Production of Nanoparticles by the Pyrolysis Method

Pyrolysis is a chemical process in which chemical precursors are degraded to a solid product accessory product in the vapor phase, and the needed product is produced by the end of the process. In general, pyrolysis synthesis produces a powder product with micron range grain size distribution. To obtain nanoscale unified materials it is required to change some reaction and experiment conditions. For instance, through decrease of reaction velocity or decomposition of the precursor into the solution, it is possible to improve the product. The pyrolysis method is used in the production of various types of nanoparticles, such as metals; metallic oxides; semiconductors; composite materials such as SnO_2, Al_2O_3, ZrO_2, Ni, $YBa_2Cu_3O_{7-x}$, ZnS,

AgGaN, TiO_2, and Au; and carbon nanotubes. Aqueous materials pyrolysis is a direct effective method for production of various types of regular carbon nanotubes.[30–36]

2.2.5 Production of Nanoparticles by Chemical Vapor Deposition (CVD)

In the CVD method, the evaporated precursor is entered into a reactor and absorbed on a surface through high temperature. These molecules are pyrolyzed or react with gases and other vapors to develop the crystals. The CVD process includes three steps: (1) transfer of reactants' mass to the grown surface through boundary layer diffusion, (2) chemical reaction on the grown surface, and (3) removal of the accessory products from the grown surface.

There are many examples of using this method in the literature. The key factor highlighting the importance of the CVD method is the capability of producing nanoparticles with dual or multiple compositions from several initial materials.[37–45]

2.2.6 Solid-State Processes

Milling or abrasion can be used for producing the nanoparticles. Mills materials (balls and body of the mill), milling time, and atmosphere of the mill affect the obtained nanoparticles. This method can be used for production of nanoparticles that cannot be produced by the two above-mentioned methods. Polluting by the mills' materials is among the problems of this method.[46–51]

2.2.7 Advanced Methods

Increased demands from different industries necessitate producing nanoparticles with definite shape and size in a large number and at low cost. During the past years, research was conducted on the method for use of supercritical solvents as an environment for production of metallic nanoparticles. The process of deposition in supercritical fluid can produce a product with narrow grain distribution. A gas placed at the upper part of a critical point turned into a supercritical state in a specific temperature T_c and P_c. Supercritical fluids have characteristics between gas and liquid states. CO_2—because of its accessible supercritical conditions (T_c = 31°C, P_c = 73 bar) and because it is nonpoisonous, noncorrosive, incombustible, and inflammable—is typically used in this method. Here, surfactants are mixed with a metal's salt solution at supercritical fluid.

Another method is electrical deposition of the solved metallic ions in an aqueous solution that is finally reduced to metallic nanoparticles. The other novel methods for production of nanoparticles are based on the use of microwaves and ultrasonic methods.[25,52–56]

2.3 Physical and Chemical Analysis of Nanoparticles

Nanoparticles can be analyzed and detected by different methods that offer significant information to understand their physicochemical differences. This method can be classified as follows:

1. X-ray diffraction (XRD)
2. Scanning electron microscopy (SEM)
3. Transmission electron microscopy (TEM)
4. Optical spectrometry
   a. Visible light spectrometry, ultraviolet
   b. Fluorescence spectrometry
   c. Fourier transform infrared spectrometry
5. X-ray photoelectron spectrometry
6. Atomic absorption spectrometry

2.3.1 X-Ray Diffraction (XRD)

XRD is a very important method and has been used for a long time to detect the crystalline structure of the solids, network constant and geometry, materials detection, defining monocrystals' order, defects, and so on. An x-ray diffraction pattern can be obtained by measuring the angles at which x-rays are diffracted by a sample crystal's phases. The Bragg equation describes the relationship between the distance of two plates (d) and diffraction angle (2θ) as follows:

$$n\lambda = 2d\sin\theta$$

where λ is the wavelength of the x-ray, and n is an integer expressing the diffraction order.

From the pure diffraction pattern of the crystalline level, the crystal structure's uniformity and size of a single cell of nanocrystalline material can be defined. X-ray diffraction is a nondestructive method and requires no complicated sample preparation steps. This reason has led to vast employment of XRD for materials detection. Diffraction analysis is mainly used to define nanocrystals sizes. Nanoparticle size can be estimated by the Debye–Scherrer equation:

$$D = k\lambda/\beta\cos\theta$$

where D is the nanocrystal's dimension, k is a constant, λ is the x-ray's wavelength, and β is the full width at half maximum (FWHM) diffraction at the angle of 2θ.[57–63]

2.3.2 Scanning Electron Microscopy (SEM)

SEM is a frequently used method for detection and analysis of nanomaterials and nanostructures. The resolution of SEM reaches up to several nanometers. This tool can produce magnifications of 10 to 1,000,000 times. This method not only offers information about the topography (same as an optical microscope) but also produces information about elemental composition of the surface. A SEM is able to produce an electron beam for scanning the surface of a solid sample. Interaction of this beam and sample produces several signals that introduce detailed information about surface structure and sample shape.[64–68]

2.3.3 Transmission Electron Microscopy (TEM)

TEM is the most common and useful tool for nanoparticles analysis. This microscope provides some information about nanoparticles' size, shape, structure, and phase. TEM is typically used for producing high-resolution photos from thin layers of a solid sample for compositional and nanostructural analysis. This method includes the following stages:

1. Radiation of high-energy electron beam to a very thin sample: This radiation is diffracted by networks of a crystalline or semicrystalline material and propagated in different directions.
2. Imaging and analysis of angular distribution of the electrons transmitted from the sample—unlike SEM in which diffracted electrons are detected—are used for structural analysis and determining the various phases of nanomaterials.
3. Analysis of the x-ray energy propagated from the sample.

One disadvantage of TEM is that electron diffraction is developed from a three-dimensional (3D) photo taken by TEM but is gathered in a two-dimensional (2D) detector. Thus, the obtained data from electronic directions are collected on a 2D photo screen. The selected area diffraction pattern has unique advantages for determining the crystalline structure of specific nanomaterials such as nanocrystals and nanorods, and can sample different parts of the crystalline structure.

In addition to being capable of structural detection and chemical analyses, TEM has other applications in the nanoscience field. For instance, estimation of nanocrystal's melt point, in which an electron beam is used for nanocrystal heating and melting point is defined by the disappeared electron diffraction. Another application of TEM is measuring mechanical and electrical characteristics of nanowires and nanotubes.[66,69–76]

2.3.4 Optical Spectrometry

Optical spectrometry is widely used for nanomaterial detection and analysis. This method is categorized into two groups: absorption spectrometry-fluorescence propagation, and infrared vibration spectrometry. The first method is used to determine the electron structure of atoms, ions, molecules, or crystals through electron excitation from a basic level to excited levels (absorption) and then resting from the excited state to the ground level (propagation). The vibration method includes photon interaction with various parts of the sample, which results in transition to or from the sample by vibration excitation. Vibration frequencies provide information from the chemical bounds present in the sample.[77,78]

2.3.5 Ultraviolet (UV)-Visible Spectroscopy

This method studies electron transfer between orbitals or atomic bounds, ions, or molecules in the gas, liquid, or solid state. Metallic nanoparticles can be detected by their unique colors. Some researchers theoretically explained the source of these colors through solving the Maxwell equation for absorption and diffraction of electromagnetic radiation by tiny metallic particles. This absorption by metallic nanoparticles is due to systematic vibrations of their capacity layer electrons. These vibrations are created by interactions of the electrons by electromagnetic fields. These "surface plasmon" resonances are developed in nanoparticles and cannot be seen in mass state metals and bulky metallic particles. Therefore, the UV-visible method can be applied to the study of a particle's specific optical properties. Using this method, a substance's spectral characteristics can be analyzed, the absorption spectrum is plotted for all suspended particles in the colloid solution, and then the plotted spectrum is investigated. Symmetric plasmon absorption bound implies that particles are not considerably concentrated in the solution.[79–81]

2.3.6 Fluorescence

In this method the light is radiated to a sample by some wavelengths and causes electron transmission from the ground level to an excited level, which is moved to a more stable level by performing a nonradiation test to the excited electron. After a particular period of material's life, the electron returns to the ground level from the excited level; this is accompanied by propagation of a substance-specific wavelength of the light. This propagated energy can be used to obtain qualitative and quantitative information of chemical composition, structure, impurities, process kinetic, and transfer energy.[82–88]

2.3.7 Fourier Transform Infrared Spectrometry (FTIR)

FTIR spectrometry detects resonance of the chemical bounds present in a molecule under different frequencies, which depends on the elements and bound type. After absorption of electromagnetic radiation, resonance frequencies are enhanced from a bound and lead to transmission between ground level and several excited levels. These absorbed frequencies show vibration excitation of the chemical bounds, and as a result bound-type and agent group participate in the vibration. The energy of these frequencies is related to the infrared zone of the electromagnetic spectrum (400 to 4000 cm^{-1}). *Fourier transform* implies the recent advances in the data collecting methods and their data conversion from an interference pattern to an infrared absorption spectrum. Estimation of the FTIR is used for the study of protein molecules' presence in the solution, because FTIR provides some information about the presence of -C-O and -N-H groups in the zone of 1400 to 1700 cm^{-1}.[89–95]

2.4 Different Forms of Growth

Basically, type of growth depends on the interaction between deposited and substrate atoms so that it may be divided into three major classes:[96–105]

1. **Layer Growth (LG) or (2D, Frank–van der Merwe):** This model that is an interaction among deposited atoms and substrate atoms is stronger than the same-type atoms, so that after completing one single film, the next film starts to grow (Figure 2.1a).

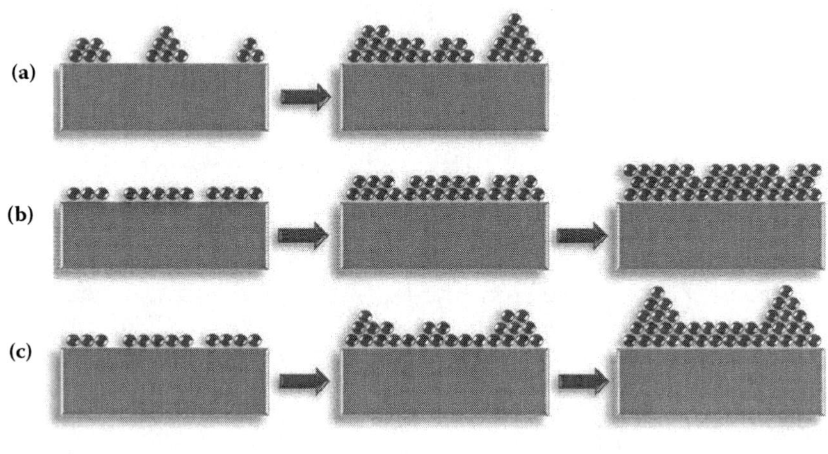

FIGURE 2.1
Basic growth modes including (a) Volmer–Weber (island), (b) Frank–van der Merwe (layer-by-layer), and (c) Stranski–Krastanov growth. (Reprinted from Martin, L. W., Chu, Y. H., and Ramesh, R., Advances in the growth and characterization of magnetic, ferroelectric, and multiferroic oxide thin films, *Materials Science and Engineering R: Reports* 68 (4–6), 89–133, Copyright 2010, with permission from Elsevier.)

2. **Stranski–Krastanov (SK) Growth:** In this model one or more complete films are formed and then the growth type is changed and islands start to grow in previous single films (Figure 2.1b).
3. **Island Growth (IG):** If the interaction energy between deposited atoms is higher than their interaction energy with substrate atoms, the growth will continue in the form of island or Velmer–Weber (Figure 2.1c).

2.5 Relation between Growth and Energy Level

For interfilm energy of substrate and deposition, related to one drop of deposition, in the substrate surface the following equation is obtained:

$$\gamma_B = \gamma^* + \gamma_A.\cos\varphi \tag{2.1}$$

where γ_B is substrate surface tension, γ^* is deposition and substrate surface tension, and γ_A is deposition surface tension. Figure 2.2 represents the relation among γ_B, γ°, and γ_A, so the relation determines the growth type.

Therefore,

$$\text{IG:} \quad \gamma_B < \gamma_A + \gamma^\circ, \quad \varphi > 0 \tag{2.2}$$

$$\text{LG:} \quad \gamma_B \geq \gamma_A + \gamma^\circ, \quad \varphi = 0 \tag{2.3}$$

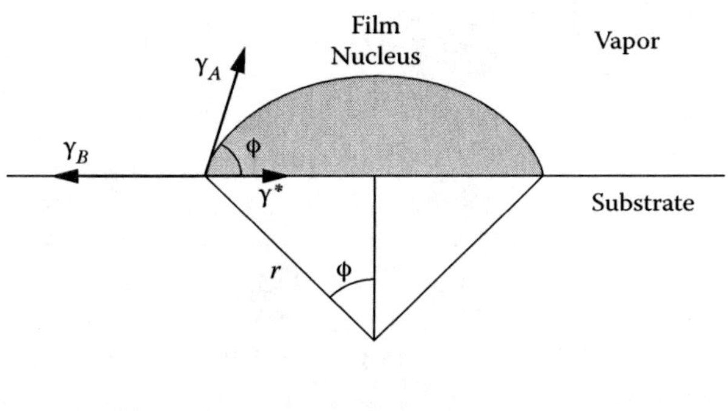

FIGURE 2.2
Relation among γ_B, γ^*, and γ_A. (Reprinted from Martin, L. W., Chu, Y. H., and Ramesh, R., Advances in the growth and characterization of magnetic, ferroelectric, and multiferroic oxide thin films, *Materials Science and Engineering R: Reports* 68 (4–6), 89–133, Copyright 2010, with permission from Elsevier.)

SK growth would be described in this model very easily. If elastic distortions are considered in the film energy, then the misfit between substrate and deposition will decrease in the first stage of the film growth. After completion of the first film, if adhesion force is weaker than elastic force, total energy will decrease, so a transition from 2D to 3D will occur. Meanwhile, any misfit is defined as $\frac{a-a_s}{a_s}$, where a and a_s are deposition networks and substrate parameters, respectively.[107–112]

2.6 Overaturation Effect on Growth

It must be mentioned that the deposition growth type in the substrate depends on oversaturation, and also its changeability is related to oversaturation rates. The oversaturation is defined with the following equation:

$$\xi = \frac{p}{p_e} = \frac{R}{R_e} \tag{2.4}$$

where P is evaporation ray vapor pressure, P_e is equilibrium vapor pressure of the deposition in base temperature, and R and R_e are subsidence and return rates, respectively. The vapor pressure of a ray of molecules is a motion that takes place during a unit of time in a square centimeter of the substrate surface. For deposition of films, oversaturation depends on the entrance rate and the substrate temperature. The oversaturation for the film condensation phase is equal to 10^{20}. Considering this fact, Equations (2.2) and (2.3) would be written as

$$\text{for IG:} \quad \gamma_B < \gamma_A + \gamma^{\circ} - \Delta\mu / \text{Const} \tag{2.5}$$

$$\text{for LG:} \quad \gamma_B \geq \gamma_A + \gamma^{\circ} - \Delta\mu / \text{Const} \tag{2.6}$$

where $\Delta\mu$ = KB.Ts.Lnξ is the changing process of free energy during transition from the vapor phase to the solid phase. These two equations show that increased oversaturation is better for LG. Figure 2.3 shows the growth changes from 3D to 2D in critical oversaturation, thus Equation (2.5) changes.

The transition from island growth of 3D to film growth of 2D due to oversaturation is depicted in Figure 2.4 by computer simulation. In Figure 2.4 film growth morphology is depicted in a convergent coverage 20% from film surface for two various oversaturation states.

If $\Delta\mu > \Delta\mu_c$, the free enthalpy for formation of 2D nuclei is lower than the free enthalpy for the formation of 3D nuclei.

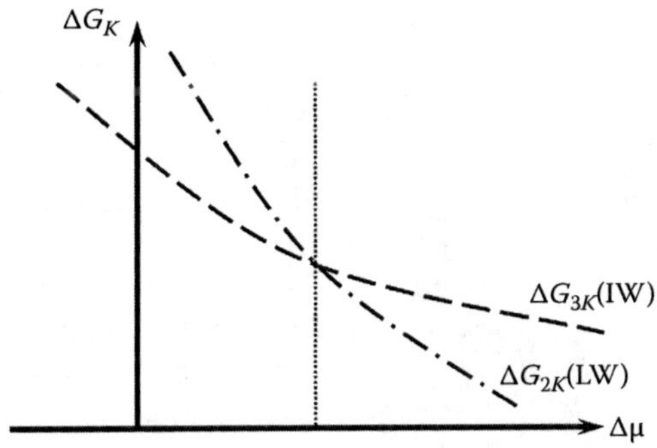

FIGURE 2.3
Relation between formation of nucleus enthalpy ΔG_k and chemical potential $\Delta\mu$ (2K means growth in two dimensions, and 3K means growth in three dimensions).

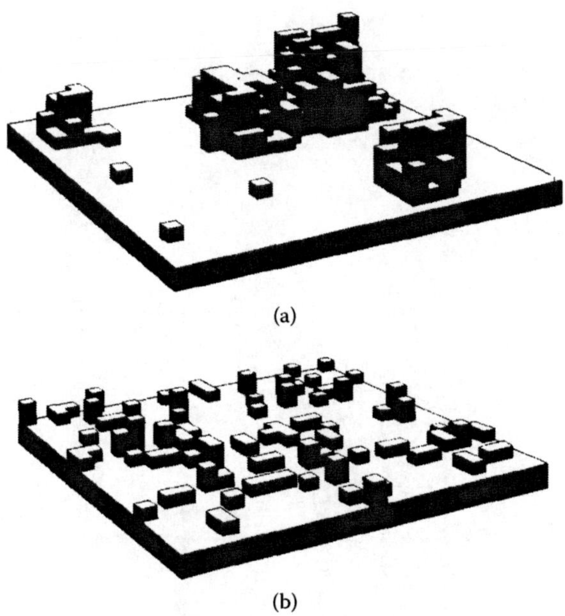

FIGURE 2.4
Computer simulation of a film growth with 20% coating: (a) island growth, (b) film growth. (Reprinted from Kashchiev, D., van der Eerden, J. P., and van Leeuwen, C., Transition from island to layer growth of thin films: A Monte Carlo simulation, *Journal of Crystal Growth* 40 (1), 47–58, Copyright 1977, with permission from Elsevier.)

2.7 Quantitative Description of Initial Stages of Film Growth

2.7.1 Volmer–Weber Theory

The theory of incongruous nucleation of atomic clusters in 2D and 3D on a perfect surface is similar to bulk nucleation.

Enthalpy (ΔG) to form a nucleus can be measured based on a function of its size (j) (the number of atoms). Free enthalpy is the sum of enthalpy due to vapor condensation and required enthalpy to form a film–substrate interface.

ΔG_K has a maximum ($G_{K_{max}}$) because of an exponential difference of surface and volume portion (Figure 2.5).

The atomic clusters under this maximum value may grow only randomly. Clusters with $G_{K_{max}}$ enthalpy are called critical clusters. If j is the atoms of a cluster and i is the atoms of a critical cluster, critical cluster enthalpy is $i = j$, and enthalpy is $\Delta G_k(i)$.

Clusters, which are bigger than critical value, grow according to Equations (2.5) and (2.6), regarding relations among nucleus and substrate surface tension.[114–119]

2.7.2 Three-Dimensional (3D) Island Growth

If j is the number of atoms in an atomic cluster, i is the number of atoms in a critical cluster, and $\Delta\mu$ is changes of free enthalpy during transition from the vapor phase to the solid phase, we have the following equation:

$$\Delta G_K(j) = -j\ \Delta\mu + j^{3/2}\ X \tag{2.7}$$

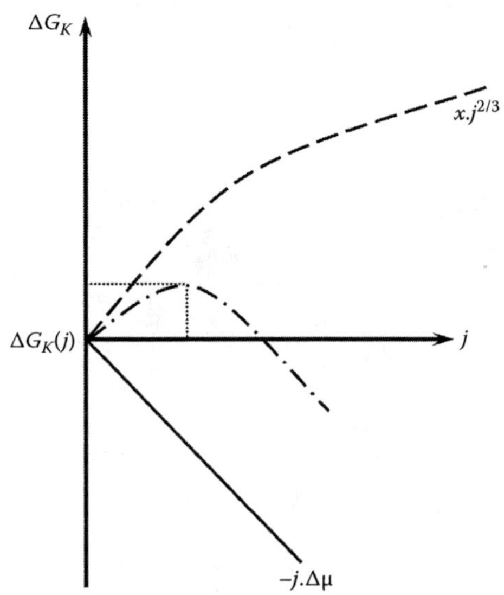

FIGURE 2.5
Formation of $\Delta G_k(j)$ enthalpy of a three-dimensional nucleus.

The first term is the result of free enthalpy due to vapor phase condensation of atoms, and the second term indicates lost free enthalpy because of formation of the atomic cluster surface and the surface between substrate and cluster.

X is defined as

$$X = \sum C_K \gamma_{AK} + C_K \gamma_{AB}(\gamma^* - \gamma_B) \tag{2.8}$$

where C_k and C_{AB} are geometric factors.

If $i = j$, then

$$\left.\frac{\partial \Delta G_K}{\partial j}\right|_{j=i} = 0 \tag{2.9}$$

and critical cluster is calculated as

$$\Delta G_K(i) = \frac{4}{27}\frac{X^3}{(\Delta\mu)^2}, \; i = \left(\frac{2X}{3\Delta\mu}\right)^3 \tag{2.10}$$

2.7.3 Two-Dimensional (2D) Film Growth

In this kind of growth, enthalpy changes are

$$\Delta G_K(j) = -j\Delta\mu + j(\gamma_A + \gamma^* - \gamma_B)\Omega^{3/2} + j^{1/2}Y \tag{2.11}$$

where Ω is the atomic volume, $Y = \Sigma C_e\ \gamma_e$ is an edge term in which γ_e is the tension energy in edges, and C_e is a geometric factor.

The first term shows changed free enthalpy due to vapor atom condensation. The second term shows changed free enthalpy for formation of surface and intermediate film, which is negative for film nucleation (2D) and also it is more effective than the first term.

$$\gamma_A + \gamma^{\circ} - \gamma_B \leq 0 \tag{2.12}$$

The third term represents decreased free enthalpy due to the formation of edges. If $i = j$ in a critical condition, then $\frac{\partial \Delta G_K}{\partial j} = 0$; thus, enthalpy and critical cluster size are calculated as

$$\Delta G_K(i) = \frac{1}{4}\frac{Y^2}{\Delta\mu'}, \; i = \left(\frac{Y}{2\Delta\mu'}\right)^2 \tag{2.13}$$

where $\Delta\mu' = \Delta\mu - \Delta\mu_e$.

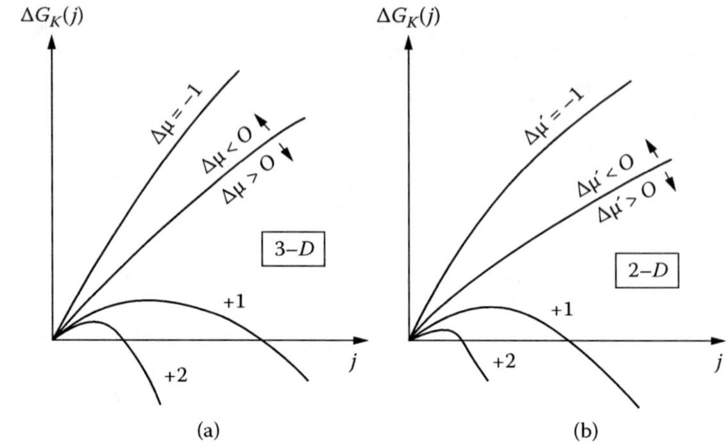

FIGURE 2.6
(a) $\Delta G_k(j)$ enthalpy for $\Delta\mu$ in three-dimensional growth and (b) $\Delta G_k(j)$ enthalpy for $\Delta\mu'$ in two-dimensional growth. (Reprinted from Reichelt, K., Nucleation and growth of thin films, *Vacuum* 38 (12), 1083–1099, Copyright 1988, with permission from Elsevier.)

According to Figure 2.6, 3D growth takes place when $\Delta\mu > 0$. However, 2D growth occurs when $\Delta\mu' = \Delta\mu - \Delta\mu_e > 0$.

2.8 Kinetic Theory of Growth

The size of critical nuclei is typically about atomic dimensions and includes some atoms. Therefore, using droplet models and thermodynamic quantities of macroscopic states for such small atomic sets is somehow questionable. Hence, growth phenomenon is analyzed from a different perspective.

With regard to definitions of parameters such as the number of singularities on the surface (n_1) and incident rate of particles (R), rate equations are written as

$$\frac{dn_1}{dt} = R - \frac{n_1}{\tau_A} - \sum_{j=1} \lambda_j n_j + \sum_{j=2} K_j n_j \tag{2.14}$$

$$\frac{dn_2}{dt} = R - \frac{\lambda_1 n_1}{2} - (\lambda_2 + K_2) n_2 + K_3 n_3 \tag{2.15}$$

$$\frac{dn_j}{dt} = -\lambda_{j-1} - n_{j-1} - (\lambda_j + K_j) n_j + K_{j+1} n_{j+1} \tag{2.16}$$

where K is the cluster disintegration possibility ($n_j \rightarrow n_1 + n_{j+1}$) and λj is the cluster growth possibility ($n_j + n_1 \rightarrow n_{j+1}$).

If we presume that the disintegration possibility for each cluster is zero ($k_j = 0$), constant = ω_j, and $\lambda_j = n_1\omega_1$, where ω_j is the jump frequency of the jth atom, and ω is the jump frequency of the atoms in the trapping area of an island.

Thus, we have

$$\frac{dn_1}{dt} = R - \frac{n_1}{\tau_A} - n_1.\omega \sum_{i=1} n_i, \ (\omega_j = \omega) \tag{2.17}$$

$$\frac{dn_2}{dt} = \frac{\omega}{2} n_1^2 - n_1.n_2\omega \tag{2.18}$$

$$\frac{dn_3}{dt} = n_1.n_2\omega - n_1.n_3\omega \tag{2.19}$$

$$\frac{dn_j}{dt} = n_1\omega n_{j-1} - n_1\omega n_j \tag{2.20}$$

Figure 2.7 indicates differential equations for a state in which n_1 is equal to a constant value. It is observed that after conducting an initial rotation, number of nuclei (n_1) will reach a saturation state, but smaller grains reach a possible maximum value in a shorter time. In the next stage, considering the related term of coalescence, the stable integrated cluster intensity will decline after reaching a maximum state. Meanwhile, it must be mentioned that in kinetic theory, critical nuclei matter are discussed as well.

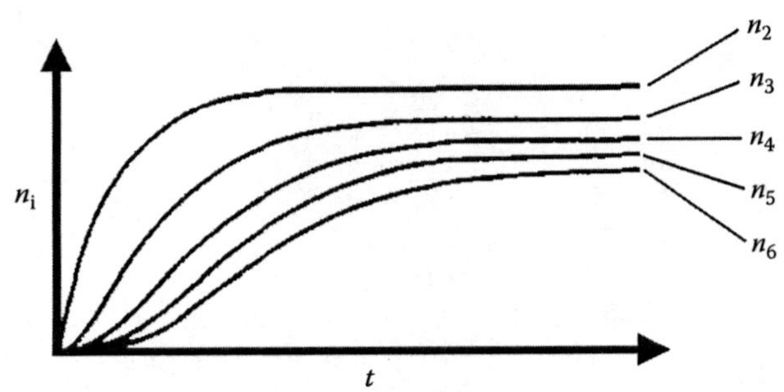

FIGURE 2.7
Solution of differential equations (2.20) for different values of n.

If islands are divided into two classes of instable clusters ($j \leq i$) and stable clusters ($j > i$), the precedent equations for stable clusters will change into the following equations:

$$\frac{dn_1}{dt} = R - \frac{n_1}{\tau_A} - \frac{d(n_x.W_x)}{dt} \tag{2.21}$$

$$\frac{dn_j}{dt} = 0 \quad (2 \leq j \leq i) \tag{2.22}$$

$$\frac{dn_x}{dt} = U_i - U_c - U_m \tag{2.23}$$

$$\frac{d(n_x.W_x)}{dt} = (i+1)U_i + n_x\sigma_x D.n_1 + R.Z \tag{2.24}$$

where W_x is the average of atoms in a cluster, n_x is the density of stable clusters, v_i is the nucleation rate and $J = \sigma_i.D.n_1.n_j$ is the trapping rate of surface atoms by critical nuclei, $U_c = 2n_x.[dz/dt]$ is the decreased rate due to enlarging and closing nuclei, Z is the coating, r_x is the average radius of stable clusters, $R.Z$ is the rate of direct collision of vapor atoms in clusters, and U_m is a decrease in rate due to coalescence of motile clusters.

Regarding the relations between Z and n_xW_x for 2D islands with atomic volume = Ω and φ = 0, we have

$$\frac{dz}{dt} = \Omega^{3/2}\frac{d(n_x.W_x)}{dt} \tag{2.25a}$$

$n_1(t)$, $n_x(t)$, and $Z(t)$ would be measured using Equations (2.21), (2.23), and (2.25a) in the case of having E_a and E_d and σ_x values.

For 3D island (φ = 90), the following equation is used:

$$\frac{dz}{dt} = \Omega\frac{d(n_x.W_x)}{dt}.\left(\frac{\pi R_x}{Z}\right)^{1/2} \tag{2.25b}$$

In Figure 2.8, the gold growth curve is depicted on a NaCl substrate. In the first phase of growth, density of singularities will increase in a short time and reach a maximum rate and then decrease slowly.

In the case that the substrate temperature is high (T_s = 350°C), an equilibrium state occurs between absorption and reevaporation for a short moment (n_1 = constant) and n_1 increasing is stopped at $\tau_A = n_1/R$. The equilibrium state between absorption and reevaporation will continue for a longer time through enhancing substrate temperature (T_s), so if the substrate

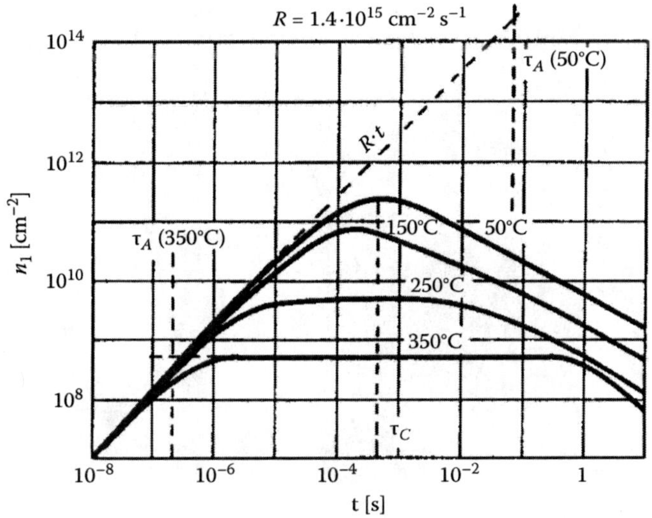

FIGURE 2.8
The $n_1(t)$ curve for gold growth on NaCl substrate. (Reprinted from Reichelt, K., Nucleation and growth of thin films, *Vacuum* 38 (12), 1083–1099, Copyright 1988, with permission from Elsevier.)

temperature is low (T_s = 50°C), such equilibrium would not be obtained ($\tau_A = 5.4 \times 10^{-2}$ s).

Because of high intensity of clusters (n_x) in Equation (3.21) and regarding Equation (2.24), it can be concluded that

$$\frac{n_1}{\tau_A} < n_1 \sigma_x D.n_x \quad (2.26)$$

Thus, n_1 intensities are limited because they are trapped by stable clusters.

Meanwhile, if the substrate temperature is high, the adhesion coefficient (β) is zero for the area in absorption and the reevaporation equilibrium (Figure 2.9).

2.9 Orientation of Thin Films

Thin films may grow as amorphous, polycrystalline, and monocrystalline forms depending on the substrate nature, evaporation rate, and film materials. Phenomena that increase the motion of adsorbed atoms in the surface will result in elongation of film grains. It would occur due to the increased substrate temperature, and vice versa, it would decrease the grain size. Likewise, misfits and impurities on the surface may decrease atom motility.

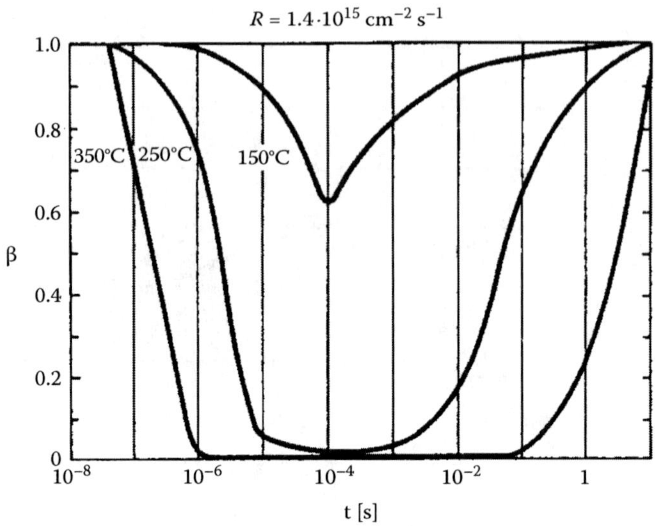

FIGURE 2.9
Three-dimensional growth curve of gold on NaCl with different adhesion coefficients β(*t*). (Reprinted from Reichelt, K., Nucleation and growth of thin films, *Vacuum* 38 (12), 1083–1099, Copyright 1988, with permission from Elsevier.)

During growth of films, unstable phases and special structures are formed, which are rare in volume. Such phases disappear in line with the increase in film thickness.

Under certain conditions, thick films may grow with an unstable phase (e.g., formation of diamond-like carbons [DLCs] through carbon deposition concurrent with ionic bombardment). If a film grows on a monocrystalline substrate, an internal relation is seen between orientation of substrate materials and the thin film; it is called epitaxy. Under desired conditions, a whole film would grow as monocrystalline. Epitaxy is a complicated phenomenon, and there is no theory to predict epitaxy growth by using film and substrate physical features.

2.10 Film Growth with a Certain Orientation

The Frank–Van der Merwe theory describes oriented growth of some films. In this theory, the surface potential is described as a sinus form that changes in line with period parameters of the film substrate network (a_s). Here, deposition is introduced by the spring model with neutral period (a).

Springs have to move horizontally in the period field of a substrate. Equilibrium state is achievable through minimizing the energy of the whole system.

Results show that a trivial misfit ($\frac{a_s-a}{a}$) will lead to a primary growth of a tensioned thin film in the substrate (which is in the form of amorphous growth). Tension energy rises through increasing film thickness. Tension energy is inclined to decrease after passing a threshold thickness, which takes place through production of continuous misfits. For a thick film, non-compliance of two films appears as misfits and film tension. Then growth occurs naturally in terms of film parameters.

2.11 Film–Substrate Interfaces

Shapes/forms of film–substrate interfaces depend on the substrate morphology, chemical reactions, distribution rate, and nucleation stages. Film–substrate interfaces are classified as discussed below.

2.11.1 Abrupt Interface (Single Layer on Single Layer)

Abrupt interface is identified by the sudden change from the film to the substrate material with an atomic distance (1 to 5 eV) (Figure 2.10a).

In this interface, tensions and defects make up a very thin area in which tension gradient is enhanced. Film adhesion to substrate is low in this case. Such interfaces are developed when (1) there is no deep portion, (2) the chemical reaction is very weak, and (3) the substrate surface is completely smooth and dense.[121,122]

2.11.2 Interface with Chemical Bond

The interface with chemical bonds may either be monofilm or multifilm, and its dimensions are equal to several atomic distances and are developed through chemical reactions and distribution among film and substrate atoms (Figure 2.10b).

The formation of this interface film is the result of film atomic reactions with substrate atoms, and penetration of residual gases to these chemical reactions is possible. Metal films of active oxygen on oxide substrates are examples of this interface. Adhesion to this interface is relatively good. However, less thickness for such an interface is preferred.

2.11.3 Distributed Film–Substrate Interfaces

The characteristic of this interface is distribution of film and substrate materials in two reciprocal areas as its density declines from interface area gradually (Figure 2.10c). Different atomic dynamicity for film and substrate atoms

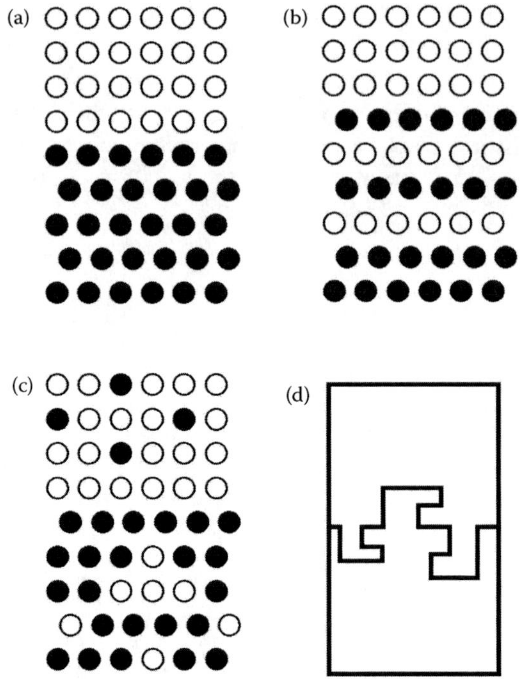

FIGURE 2.10
Film–substrate interfaces: (a) abrupt interface, (b) interface film with chemical bond, (c) distribution film interface, and (d) mechanical film interface.

develops the Kir–Kendall effect. The effect will weaken the interface; however, generally it has good adhesion.

The required energy to develop such an interface is about 1 to 5 eV, which would be supplied through increasing substrate temperatures by the condensation energy of descending atoms. For example, the energy resulting from condensation is enough for penetration during gold evaporation on copper substrate in the room temperature. Sedimentation of very different materials on each other is the most important property of distribution thin film interface, so it is possible to decrease mechanical tension due to a large difference between film and substrate thermal expansion coefficient.

2.11.4 Semidistribution Film–Material Interface

A semidistribution film–material interface may be developed through subsidence in high energies or in ion sputtering through transmittance. Some features and advantages of this interface are similar to that of the distribution film interface; however, there is a difference. A bilateral or mutual distribution takes place in the distribution interface while films and substrates in the semidistribution version are in such a fashion that distribution is done unilaterally from film to a substrate.

Upon formation of such an interface that is developed by ion bombardment concurrent with sputtering, diffused atoms from a substrate will be vaporized along with film atoms and then are deposited in the substrate.

2.11.5 Mechanical Film–Substrate Interface

Interlocking of deposited material with a rough substrate surface is the characteristic of such interfaces. Adhesion intensity depends on mechanical features of films and substrates as well as interface film geometry (Figure 2.10d). Actually a single interface film is not observed and instead combinations of different film interfaces are found simultaneously.

References

1. Brown, S. C., Palazuelos, M., Sharma, P., Powers, K. W., Roberts, S. M., Grobmyer, S. R., and Moudgil, B. M., Nanoparticle characterization for cancer nanotechnology and other biological applications, *Methods in Molecular Biology* 624, 39–65, 2010.
2. Leroueil, P. R., Hong, S., Mecke, A., Baker Jr., J. R., Orr, B. G., and Holl, M. M. B., Nanoparticle interaction with biological membranes: Does nanotechnology present a janus face? *Accounts of Chemical Research* 40 (5), 335–342, 2007.
3. Roco, M. C., Reviews of national research programs in nanoparticle and nanotechnology research, *Journal of Aerosol Science* 29 (5–6), 749–760, 1998.
4. Thomas, D. G., Pappu, R. V., and Baker, N. A., Nanoparticle ontology for cancer nanotechnology research, *Journal of Biomedical Informatics* 44 (1), 59–74, 2011.
5. Zahn, M., Magnetic fluid and nanoparticle applications to nanotechnology, *Journal of Nanoparticle Research* 3 (1), 73–78, 2001.
6. Shimomura, M., and Sawadaishi, T., Bottom-up strategy of materials fabrication: A new trend in nanotechnology of soft materials, *Current Opinion in Colloid and Interface Science* 6 (1), 11–16, 2001.
7. Shu, D., Moll, W. D., Deng, Z., Mao, C., and Guo, P., Bottom-up assembly of RNA arrays and superstructures as potential parts in nanotechnology, *Nano Letters* 4 (9), 1717–1723, 2004.
8. Teo, B. K., and Sun, X. H., From top-down to bottom-up to hybrid nanotechnologies: Road to nanodevices, *Journal of Cluster Science* 17 (4), 529–540, 2006.
9. Powell, M. C., Griffin, M. P. A., and Tai, S., Bottom-up risk regulation? How nanotechnology risk knowledge gaps challenge federal and state environmental agencies, *Environmental Management* 42 (3), 426–443, 2008.
10. Balzani, V., Nanoscience and nanotechnology: The bottom-up construction of molecular devices and machines, *Pure and Applied Chemistry* 80 (8), 1631–1650, 2008.

11. Ennas, G., Musinu, A., Piccaluga, G., Zedda, D., Gatteschi, D., Sangregorio, C., Stanger, J. L., Concas, G., and Spano, G., Characterization of iron oxide nanoparticles in an Fe_2O_3-SiO_2 composite prepared by a sol-gel method, *Chemistry of Materials* 10 (2), 495–502, 1998.
12. Chen, D. H., and He, X. R., Synthesis of nickel ferrite nanoparticles by sol-gel method, *Materials Research Bulletin* 36 (7–8), 1369–1377, 2001.
13. Gu, F., Wang, S. F., Lü, M. K., Zhou, G. J., Xu, D., and Yuan, D. R., Photoluminescence properties of SnO_2 nanoparticles synthesized by sol-gel method, *Journal of Physical Chemistry B* 108 (24), 8119–8123, 2004.
14. Niederberger, M., Nonaqueous sol-gel routes to metal oxide nanoparticles, *Accounts of Chemical Research* 40 (9), 793–800, 2007.
15. Epifani, M., Giannini, C., Tapfer, L., and Vasanelli, L., Sol-gel synthesis and characterization of Ag and Au nanoparticles in SiO_2, TiO_2, and ZrO_2 thin films, *Journal of the American Ceramic Society* 83 (10), 2385–2393, 2000.
16. Goebbert, C., Nonninger, R., Aegerter, M. A., and Schmidt, H., Wet chemical deposition of ATO and ITO coatings using crystalline nanoparticles redispersable in solutions, *Thin Solid Films* 351 (1–2), 79–84, 1999.
17. Yao, W., Yang, J., Wang, J., and Nuli, Y., Chemical deposition of platinum nanoparticles on iridium oxide for oxygen electrode of unitized regenerative fuel cell, *Electrochemistry Communications* 9 (5), 1029–1034, 2007.
18. Goebbert, C., Bisht, H., Al-Dahoudi, N., Nonninger, R., Aegerter, M. A., and Schmidt, H., Wet chemical deposition of crystalline, redispersable ATO and ITO nanoparticles, *Journal of Sol-Gel Science and Technology* 19 (1–3), 201–204, 2000.
19. Wang, G., Shi, C., Zhao, N., and Du, X., Synthesis and characterization of Ag nanoparticles assembled in ordered array pores of porous anodic alumina by chemical deposition, *Materials Letters* 61 (18), 3795–3797, 2007.
20. Xu, P., Han, X., Wang, C., Zhang, B., Wang, X., and Wang, H. L., Facile synthesis of polyaniline-polypyrrole nanofibers for application in chemical deposition of metal nanoparticles, *Macromolecular Rapid Communications* 29 (16), 1392–1397, 2008.
21. Kordás, K., Tóth, G., Levoska, J., Huuhtanen, M., Keiski, R., Härkönen, M., George, T. F., and Vähäkangas, J., Room temperature chemical deposition of palladium nanoparticles in anodic aluminium oxide templates, *Nanotechnology* 17 (5), 1459–1463, 2006.
22. Song, H., Rioux, R. M., Hoefelmeyer, J. D., Komor, R., Niesz, K., Grass, M., Yang, P., and Somorjai, G. A., Hydrothermal growth of mesoporous SBA-15 silica in the presence of PVP-stabilized Pt nanoparticles: Synthesis, characterization, and catalytic properties, *Journal of the American Chemical Society* 128 (9), 3027–3037, 2006.
23. Lu, Q., Gao, F., and Zhao, D., One-step synthesis and assembly of copper sulfide nanoparticles to nanowires, nanotubes, and nanovesicles by a simple organic amine-assisted hydrothermal process, *Nano Letters* 2 (7), 725–728, 2002.
24. Jeon, S., and Braun, P. V., Hydrothermal synthesis of Er-doped luminescent TiO_2 nanoparticles, *Chemistry of Materials* 15 (6), 1256–1263, 2003.
25. Hakuta, Y., Haganuma, T., Sue, K., Adschiri, T., and Arai, K., Continuous production of phosphor YAG:Tb nanoparticles by hydrothermal synthesis in supercritical water, *Materials Research Bulletin* 38 (7), 1257–1265, 2003.
26. Daou, T. J., Pourroy, G., Bégin-Colin, S., Grenèche, J. M., Ulhaq-Bouillet, C., Legaré, P., Bernhardt, P., Leuvrey, C., and Rogez, G., Hydrothermal synthesis

of monodisperse magnetite nanoparticles, *Chemistry of Materials* 18 (18), 4399–4404, 2006.
27. Baruwati, B., Kumar, D. K., and Manorama, S. V., Hydrothermal synthesis of highly crystalline ZnO nanoparticles: A competitive sensor for LPG and EtOH, *Sensors and Actuators, B: Chemical* 119 (2), 676–682, 2006.
28. Jing, Z., and Wu, S., Synthesis and characterization of monodisperse hematite nanoparticles modified by surfactants via hydrothermal approach, *Materials Letters* 58 (27–28), 3637–3640, 2004.
29. Chiu, H. C., and Yeh, C. S., Hydrothermal synthesis of SnO_2 nanoparticles and their gas-sensing of alcohol, *Journal of Physical Chemistry C* 111 (20), 7256–7259, 2007.
30. Tsai, S. C., Song, Y. L., Tsai, C. S., Yang, C. C., Chiu, W. Y., and Lin, H. M., Ultrasonic spray pyrolysis for nanoparticles synthesis, *Journal of Materials Science* 39 (11), 3647–3657, 2004.
31. Wang, J. N., Zhang, L., Yu, F., and Sheng, Z. M., Synthesis of carbon encapsulated magnetic nanoparticles with giant coercivity by a spray pyrolysis approach, *Journal of Physical Chemistry B* 111 (8), 2119–2124, 2007.
32. Sort, J., Suriñach, S., Baró, M. D., Muraviev, D., Dzhardimalieva, G. I., Golubeva, N. D., Pomogailo, S. I., Pomogailo, A. D., Macedo, W. A. A., Weller, D., Skumryev, V., and Nogués, J., Direct synthesis of isolated L10 FePt nanoparticles in a robust TiO_2 matrix via a combined sol-gel/pyrolysis route, *Advanced Materials* 18 (4), 466–470, 2006.
33. Panatarani, C., Lenggoro, I. W., and Okuyama, K., Synthesis of single crystalline ZnO nanoparticles by salt-assisted spray pyrolysis, *Journal of Nanoparticle Research* 5 (1–2), 47–53, 2003.
34. Itoh, Y., Wuled Lenggoro, I., Okuyama, K., Mädler, L., and Pratsinis, S. E., Size tunable synthesis of highly crystalline $BaTiO_3$ nanoparticles using salt-assisted spray pyrolysis, *Journal of Nanoparticle Research* 5 (3–4), 191–198, 2003.
35. Tani, T., Mädler, L., and Pratsinis, S. E., Synthesis of zinc oxide/silica composite nanoparticles by flame spray pyrolysis, *Journal of Materials Science* 37 (21), 4627–4632, 2002.
36. Strobel, R., and Pratsinis, S. E., Direct synthesis of maghemite, magnetite and wustite nanoparticles by flame spray pyrolysis, *Advanced Powder Technology* 20 (2), 190–194, 2009.
37. Okumura, M., Tsubota, S., Iwamoto, M., and Haruta, M., Chemical vapor deposition of gold nanoparticles on MCM-41 and their catalytic activities for the low-temperature oxidation of CO and of H_2, *Chemistry Letters* (4), 315–316, 1998.
38. Etzkorn, J., Therese, H. A., Rocker, F., Zink, N., Kolb, U., and Tremel, W., Metal-organic chemical vapor deposition synthesis of hollow inorganic-fullerene-type MoS_2 and $MoSe_2$ nanoparticles, *Advanced Materials* 17 (19), 2372–2375, 2005.
39. Kim, S. W., Fujita, S., and Fujita, S., ZnO nanowires with high aspect ratios grown by metalorganic chemical vapor deposition using gold nanoparticles, *Applied Physics Letters* 86 (15), 1–3, 2005.
40. Yu, J., Wang, E. G., and Bai, X. D., Electron field emission from carbon nanoparticles prepared by microwave-plasma chemical-vapor deposition, *Applied Physics Letters* 78 (15), 2226–2228, 2001.
41. Xia, W., Su, D., Birkner, A., Ruppel, L., Wang, Y., Wöll, C., Qian, J., Liang, C., Marginean, G., Brandl, W., and Muhler, M., Chemical vapor deposition and

synthesis on carbon nanofibers: Sintering of ferrocene-derived supported iron nanoparticles and the catalytic growth of secondary carbon nanofibers, *Chemistry of Materials* 17 (23), 5737–5742, 2005.

42. Zhang, Z., Wei, B. Q., and Ajayan, P. M., Self-assembled patterns of iron oxide nanoparticles by hydrothermal chemical-vapor deposition, *Applied Physics Letters* 79 (25), 4207–4209, 2001.
43. Huh, Y., Lee, J. Y., Cheon, J., Hong, Y. K., Koo, J. Y., Lee, T. J., and Lee, C. J., Controlled growth of carbon nanotubes over cobalt nanoparticles by thermal chemical vapor deposition, *Journal of Materials Chemistry* 13 (9), 2297–2300, 2003.
44. Leach, W. T., Zhu, J., and Ekerdt, J. G., Cracking assisted nucleation in chemical vapor deposition of silicon nanoparticles on silicon dioxide, *Journal of Crystal Growth* 240 (3–4), 415–422, 2002.
45. Liang, C., Xia, W., Soltani-Ahmadi, H., Schlüter, O., Fischer, R. A., and Muhler, M., The two-step chemical vapor deposition of Pd(allyl)Cp as an atom-efficient route to synthesize highly dispersed palladium nanoparticles on carbon nanofibers, *Chemical Communications* (2), 282–284, 2005.
46. Li, Y. B., Wei, B. Q., Liang, J., Yu, Q., and Wu, D. H., Transformation of carbon nanotubes to nanoparticles by ball milling process, *Carbon* 37 (3), 493–497, 1999.
47. Part, Nicoara, G., Fratiloiu, D., Nogues, M., Dormann, J. L., and Vasiliu, F., Ni-Zn ferrite nanoparticles prepared by ball milling, in *Materials Science Forum* 145–150, 1997.
48. Wang, Y., Li, Y., Rong, C., and Liu, J. P., Sm-Co hard magnetic nanoparticles prepared by surfactant-assisted ball milling, *Nanotechnology* 18 (46), art no. 465701, 2007.
49. Damonte, L. C., Mendoza Zélis, L. A., Marí Soucase, B., and Hernández Fenollosa, M. A., Nanoparticles of ZnO obtained by mechanical milling, *Powder Technology* 148 (1), 15–19, 2004.
50. Zhang, D. W., Chen, C. H., Zhang, J., and Ren, F., Novel electrochemical milling method to fabricate copper nanoparticles and nanofibers, *Chemistry of Materials* 17 (21), 5242–5245, 2005.
51. Lam, C., Zhang, Y. F., Tang, Y. H., Lee, C. S., Bello, I., and Lee, S. T., Large-scale synthesis of ultrafine Si nanoparticles by ball milling, *Journal of Crystal Growth* 220 (4), 466–470, 2000.
52. Ji, M., Chen, X., Wai, C. M., and Fulton, J. L., Synthesizing and dispersing silver nanoparticles in a water-in-supercritical carbon dioxide microemulsion, *Journal of the American Chemical Society* 121 (11), 2631–2632, 1999.
53. Ohde, H., Hunt, F., and Wai, C. M., Synthesis of silver and copper nanoparticles in a water-in-supercritical-carbon dioxide microemulsion, *Chemistry of Materials* 13 (11), 4130–4135, 2001.
54. Ye, X. R., Lin, Y., Wang, C., Engelhard, M. H., Wang, Y., and Wai, C. M., Supercritical fluid synthesis and characterization of catalytic metal nanoparticles on carbon nanotubes, *Journal of Materials Chemistry* 14 (5), 908–913, 2004.
55. Ohde, H., Wai, C. M., Kim, H., Kim, J., and Ohde, M., Hydrogenation of olefins in supercritical CO_2 catalyzed by palladium nanoparticles in a water-in-CO_2 microemulsion, *Journal of the American Chemical Society* 124 (17), 4540–4541, 2002.
56. Yeung, L. K., Lee Jr., C. T., Johnston, K. P., and Crooks, R. M., Catalysis in supercritical CO_2 using dendrimer-encapsulated palladium nanoparticles, *Chemical Communications* (21), 2290–2291, 2001.

57. Waddon, A. J., and Coughlin, E. B., Crystal structure of polyhedral oligomeric silsequioxane (POSS) nano-materials: A study by x-ray diffraction and electron microscopy, *Chemistry of Materials* 15 (24), 4555–4561, 2003.
58. Uvarov, V., and Popov, I., Metrological characterization of X-ray diffraction methods for determination of crystallite size in nano-scale materials, *Materials Characterization* 58 (10), 883–891, 2007.
59. Rolo, A. G., Vasilevskiy, M. I., Conde, O., and Gomes, M. J. M., Structural properties of Ge nano-crystals embedded in SiO_2 films from X-ray diffraction and Raman spectroscopy, *Thin Solid Films* 336 (1–2), 58–62, 1998.
60. Ahmadzadi, H., Marandi, F., and Morsali, A., Structural and X-ray powder diffraction studies of nano-structured lead(II) coordination polymer with η^2 Pb…C interactions, *Journal of Organometallic Chemistry* 694 (22), 3565–3569, 2009.
61. Jones, J. L., Hung, J. T., and Meng, Y. S., Intermittent X-ray diffraction study of kinetics of delithiation in nano-scale $LiFePO_4$, *Journal of Power Sources* 189 (1), 702–705, 2009.
62. Labat, S., Chamard, V., and Thomas, O., Local strain in a 3D nano-crystal revealed by 2D coherent X-ray diffraction imaging, *Thin Solid Films* 515 (14 Spec. Iss.), 5557–5562, 2007.
63. Itoh, K., Sasaki, H., Takeshita, H. T., Mori, K., and Fukunaga, T., Structure of nano-crystalline $FeTiD_x$ by neutron and X-ray diffraction, *Journal of Alloys and Compounds* 404-406 (Spec. Iss.), 95–98, 2005.
64. Antognozzi, M., Sentimenti, A., and Valdrè, U., Fabrication of nano-tips by carbon contamination in a scanning electron microscope for use in scanning probe microscopy and field emission, *Microscopy Microanalysis Microstructures* 8 (6), 355–368, 1997.
65. Nagase, M., and Kurihara, K., Imaging of Si nano-patterns embedded in SiO_2 using scanning electron microscopy, *Microelectronic Engineering* 53 (1), 257–260, 2000.
66. Tanaka, N., Yamasaki, J., Mitani, S., and Takanashi, K., High-angle annular dark-field scanning transmission electron microscopy and electron energy-loss spectroscopy of nano-granular Co-Al-O alloys, *Scripta Materialia* 48 (7), 909–914, 2003.
67. Kim, H., Negishi, T., Kudo, M., Takei, H., and Yasuda, K., Quantitative backscattered electron imaging of field emission scanning electron microscopy for discrimination of nano-scale elements with nm-order spatial resolution, *Journal of Electron Microscopy* 59 (5), 379–385, 2010.
68. Delobelle, B., Courvoisier, F., and Delobelle, P., Morphology study of femtosecond laser nano-structured borosilicate glass using atomic force microscopy and scanning electron microscopy, *Optics and Lasers in Engineering* 48 (5), 616–625, 2010.
69. Ikeda, Y., Katoh, A., Shimanuki, J., and Kohjiya, S., Nano-structural observation of in situ silica in natural rubber matrix by three dimensional transmission electron microscopy, *Macromolecular Rapid Communications* 25 (12), 1186–1190, 2004.
70. Hoppe, H., Drees, M., Schwinger, W., Schaffler, F., and Sariciftcia, N. S., Nano-crystalline fullerene phases in polymer/fullerene bulk-heterojunction solar cells: A transmission electron microscopy study, *Synthetic Metals* 152 (1–3), 117–120, 2005.
71. Luo, Q., and Hovsepian, P. E., Transmission electron microscopy and energy dispersive X-ray spectroscopy on the worn surface of nano-structured TiAlN/VN multilayer coating, *Thin Solid Films* 497 (1–2), 203–209, 2006.

72. Kohjiya, S., Katoh, A., Shimanuki, J., Hasegawa, T., and Ikeda, Y., Nanostructural observation of carbon black dispersion in natural rubber matrix by three-dimensional transmission electron microscopy, *Journal of Materials Science* 40 (9–10), 2553–2555, 2005.
73. Lee, B. T., Han, J. K., and Saito, F., Microstructure of sol-gel synthesized Al_2O_3-ZrO $2(Y_2O_3)$ nano-composites studied by transmission electron microscopy, *Materials Letters* 59 (2–3), 355–360, 2005.
74. Oshima, Y., Nangou, T., Hirayama, H., and Takayanagi, K., Face centered cubic indium nano-particles studied by UHV-transmission electron microscopy, *Surface Science* 476 (1–2), 107–114, 2001.
75. Ji, H., Li, M., Kim, J. M., Kim, D. W., and Wang, C., Nano features of Al/Au ultrasonic bond interface observed by high resolution transmission electron microscopy, *Materials Characterization* 59 (10), 1419–1424, 2008.
76. Tsurui, T., Kawamura, J., and Suzuki, K., Nano-scale structural inhomogeneities of CuI-Cu_2MoO_4 superionic conducting glass observed by high resolution transmission electron microscopy, *Journal of Non-Crystalline Solids* 353 (3), 302–307, 2007.
77. Sahayam, A. C., Venkateswarlu, G., and Chaurasia, S. C., Nano platinum-catalyzed dry ashing of flour samples for the determination of trace metals by inductively coupled plasma optical emission spectrometry, *Atomic Spectroscopy* 30 (4), 139–142, 2009.
78. Hu, X., Zhan, L., and Xia, Y., Compact optical filter for dual-wavelength fluorescence-spectrometry based on enhanced transmission through metallic nano-slit array, *Applied Physics B: Lasers and Optics* 94 (4), 629–633, 2009.
79. Singh, S. K., Singh, A. K., Kumar, D., Prakash, O., and Rai, S. B., Efficient UV-visible up-conversion emission in Er^{3+}/Yb^{3+} co-doped La_2O_3 nano-crystalline phosphor, *Applied Physics B: Lasers and Optics* 98 (1), 173–179, 2010.
80. Sharif Sh, M., Khatibi, E., Sarpoolaki, H., and Fard, F. G., An investigation of dispersion and stability of carbon black nano particles in water via UV-Visible spectroscopy, *International Journal of Modern Physics B* 22 (18–19), 3172–3178, 2008.
81. Remita, S., Fontaine, P., Lacaze, E., Borensztein, Y., Sellame, H., Farha, R., Rochas, C., and Goldmann, M., X-ray radiolysis induced formation of silver nano-particles: A SAXS and UV-visible absorption spectroscopy study, *Nuclear Instruments and Methods in Physics Research, Section B: Beam Interactions with Materials and Atoms* 263 (2), 436–440, 2007.
82. Egelhaaf, H. J., Gierschner, J., and Oelkrug, D., Characterization of oriented oligo(phenylenevinylene) films and nano-aggregates by UV/Vis-absorption and fluorescence spectroscopy, *Synthetic Metals* 83 (3), 221–226, 1996.
83. Gouanvé, F., Schuster, T., Allard, E., Méallet-Renault, R., and Larpent, C., Fluorescence quenching upon binding of copper ions in dye-doped and ligand-capped polymer nanoparticles: A simple way to probe the dye accessibility in nano-sized templates, *Advanced Functional Materials* 17 (15), 2746–2756, 2007.
84. Thompson, W. H., Simulations of time-dependent fluorescence in nano-confined solvents, *Journal of Chemical Physics* 120 (17), 8125–8133, 2004.
85. Wang, L. Y., Kan, X. W., Zhang, M. C., Zhu, C. Q., and Wang, L., Fluorescence for the determination of protein with functionalized nano-ZnS, *Analyst* 127 (11), 1531–1534, 2002.

86. Sharma, P. K., Jilavi, M. H., Varadan, V. K., and Schmidt, H., Influence of initial pH on the particle size and fluorescence properties of the nano scale Eu(III) doped yttria, *Journal of Physics and Chemistry of Solids* 63 (1), 171–177, 2001.
87. Wenger, J., Lenne, P. F., Popov, E., Rigneault, H., Dintinger, J., and Ebbesen, T. W., Single molecule fluorescence in rectangular nano-apertures, *Optics Express* 13 (18), 7035–7044, 2005.
88. Bruemmel, Y., Chan, C. P. Y., Renneberg, R., Thuenemann, A., and Seydack, M., On the influence of different surfaces in nano- and submicrometer particle based fluorescence immunoassays, *Langmuir* 20 (21), 9371–9379, 2004.
89. Panda, R. N., Hsieh, M. F., Chung, R. J., and Chin, T. S., FTIR, XRD, SEM and solid state NMR investigations of carbonate-containing hydroxyapatite nano-particles synthesized by hydroxide-gel technique, *Journal of Physics and Chemistry of Solids* 64 (2), 193–199, 2003.
90. Jensen, H., Soloviev, A., Li, Z., and Søgaard, E. G., XPS and FTIR investigation of the surface properties of different prepared titania nano-powders, *Applied Surface Science* 246 (1–3), 239–249, 2005.
91. Pradeep, A., and Chandrasekaran, G., FTIR study of Ni, Cu and Zn substituted nano-particles of $MgFe\ 2O_4$, *Materials Letters* 60 (3), 371–374, 2006.
92. Dioumaev, A. K., and Braiman, M. S., Nano- and microsecond time-resolved FTIR spectroscopy of the halorhodopsin photocycle, *Photochemistry and Photobiology* 66 (6), 755–763, 1997.
93. Battisha, I. K., El Beyally, A., El Mongy, S. A., and Nahrawi, A. M., Development of the FTIR properties of nano-structure silica gel doped with different rare earth elements, prepared by sol-gel route, *Journal of Sol-Gel Science and Technology* 41 (2), 129–137, 2007.
94. Roonasi, P., and Holmgren, A., A Fourier transform infrared (FTIR) and thermogravimetric analysis (TGA) study of oleate adsorbed on magnetite nano-particle surface, *Applied Surface Science* 255 (11), 5891–5895, 2009.
95. Bhattacharyya, K., Varma, S., Tripathi, A. K., Bharadwaj, S. R., and Tyagi, A. K., Mechanistic insight by in situ FTIR for the gas phase photo-oxidation of ethylene by V-doped titania and nano titania, *Journal of Physical Chemistry B* 113 (17), 5917–5928, 2009.
96. Forrest, S. R., Burrows, P. E., Haskal, E. I., and So, F. F., Ultrahigh-vacuum quasiepitaxial growth of model van der Waals thin films. II. Experiment, *Physical Review B* 49 (16), 11309–11321, 1994.
97. Knuyt, G., Quaeyhaegens, C., D'Haen, J., and Stals, L. M., A quantitative model for the evolution from random orientation to a unique texture in PVD thin film growth, *Thin Solid Films* 258 (1–2), 159–169, 1995.
98. Forrest, S. R., and Zhang, Y., Ultrahigh-vacuum quasiepitaxial growth of model van der Waals thin films. I. Theory, *Physical Review B* 49 (16), 11297–11308, 1994.
99. Sun, C. J., Kung, P., Saxler, A., Ohsato, H., Haritos, K., and Razeghi, M., A crystallographic model of (00.1) aluminum nitride epitaxial thin film growth on (00.1) sapphire substrate, *Journal of Applied Physics* 75 (8), 3964–3967, 1994.
100. Lou, Y., and Christofides, P. D., Estimation and control of surface roughness in thin film growth using kinetic Monte-Carlo models, *Chemical Engineering Science* 58 (14), 3115–3129, 2003.
101. Stearns, D. G., Stochastic model for thin film growth and erosion, *Applied Physics Letters* 62 (15), 1745–1747, 1993.

102. Dubois, L. H., Model studies of low temperature titanium nitride thin film growth, *Polyhedron* 13 (8), 1329–1336, 1994.
103. Karpenko, O. P., Bilello, J. C., and Yalisove, S. M., Growth anisotropy and self-shadowing: A model for the development of in-plane texture during polycrystalline thin-film growth, *Journal of Applied Physics* 82 (3), 1397–1403, 1997.
104. Lo, A., and Skodje, R. T., Kinetic and Monte Carlo models of thin film coarsening: Cross over from diffusion-coalescence to Ostwald growth modes, *Journal of Chemical Physics* 112 (4), 1966–1974, 2000.
105. Habuka, H., Nagoya, T., Mayusumi, M., Katayama, M., Shimada, M., and Okuyama, K., Model on transport phenomena and epitaxial growth of silicon thin film in $SiHCl_3$-H_2 system under atmospheric pressure, *Journal of Crystal Growth* 169 (1), 61–72, 1996.
106. Martin, L. W., Chu, Y. H., and Ramesh, R., Advances in the growth and characterization of magnetic, ferroelectric, and multiferroic oxide thin films, *Materials Science and Engineering R: Reports* 68 (4–6), 89–133, 2010.
107. Chen, Y., Bagnall, D. M., Koh, H. J., Park, K. T., Hiraga, K., Zhu, Z., and Yao, T., Plasma assisted molecular beam epitaxy of ZnO on c-plane sapphire: Growth and characterization, *Journal of Applied Physics* 84 (7), 3912–3918, 1998.
108. Choy, K. L., Chemical vapour deposition of coatings, *Progress in Materials Science* 48 (2), 57–170, 2003.
109. Chen, Y., Bagnall, D. M., Zhu, Z., Sekiuchi, T., Park, K. T., Hiraga, K., Yao, T., Koyama, S., Shen, M. Y., and Goto, T., Growth of ZnO single crystal thin films on c-plane (0 0 0 1) sapphire by plasma enhanced molecular beam epitaxy, *Journal of Crystal Growth* 181 (1–2), 165–169, 1997.
110. Butko, V. Y., Chi, X., Lang, D. V., and Ramirez, A. P., Field-effect transistor on pentacene single crystal, *Applied Physics Letters* 83 (23), 4773–4775, 2003.
111. Muller, D. A., Tzou, Y., Raj, R., and Silcox, J., Mapping sp2 and sp3 states of carbon at sub-nanometre spatial resolution, *Nature* 366 (6457), 725–727, 1993.
112. Wu, J., Walukiewicz, W., Li, S. X., Armitage, R., Ho, J. C., Weber, E. R., Haller, E. E., Lu, H., Schaff, W. J., Barcz, A., and Jakiela, R., Effects of electron concentration on the optical absorption edge of InN, *Applied Physics Letters* 84 (15), 2805–2807, 2004.
113. Kashchiev, D., van der Eerden, J. P., and van Leeuwen, C., Transition from island to layer growth of thin films: A Monte Carlo simulation, *Journal of Crystal Growth* 40 (1), 47–58, 1977.
114. Freund, L. B., and Chason, E., Model for stress generated upon contact of neighboring islands on the surface of a substrate, *Journal of Applied Physics* 89 (9), 4866–4873, 2001.
115. Renaud, G., Lazzari, R., and Leroy, F., Probing surface and interface morphology with grazing incidence small angle X-ray scattering, *Surface Science Reports* 64 (8), 255–380, 2009.
116. Hu, W. S., Liu, Z. G., and Feng, D., Low electric field induced (001) oriented growth of $LiNbO_3$ films by pulsed laser ablation, *Solid State Communications* 97 (6), 481–485, 1996.
117. Müller, P., and Kern, R., Equilibrium shape of epitaxially strained crystals (Volmer-Weber case), *Journal of Crystal Growth* 193 (1–2), 257–270, 1998.
118. Gautier, F., and Stoeffler, D., Electronic structure, magnetism and growth of ultrathin films of transition metals, *Surface Science* 249 (1–3), 265–280, 1991.

119. Komsiyska, L., and Staikov, G., Electrocrystallization of Au nanoparticles on glassy carbon from $HClO_4$ solution containing [$AuCl_4$], *Electrochimica Acta* 54 (2), 168–172, 2008.
120. Reichelt, K., Nucleation and growth of thin films, *Vacuum* 38 (12), 1083–1099, 1988.
121. Lorenz, M., Hochmuth, H., Grüner, C., Hilmer, H., Lajn, A., Spemann, D., Brandt, M., Zippel, J., Schmidt-Grund, R., Von Wenckstern, H., and Grundmann, M., Oxide thin film heterostructures on large area, with flexible doping, low dislocation density, and abrupt interfaces: Grown by pulsed laser deposition, *Laser Chemistry* 2010, art no. 140976, 2010.
122. Lien, W. C., Cheng, K. B., Senesky, D. G., Carraro, C., Pisano, A. P., and Maboudian, R., Growth of 3C-SiC thin film on AlN/Si(100) with atomically abrupt interface via tailored precursor feeding procedure, *Electrochemical and Solid-State Letters* 13 (7), D53–D56, 2010.

3

Characterization and Fabrication Methods of Two-Dimensional Nanostructures

3.1 Introduction

The empirical techniques applied in the study of two-dimensional nanostructures mainly include auger electron spectroscopy (AES), x-ray photoelectron spectroscopy (XPS), ellipsometry, synchrotron radiation, or high-resolution photoemission spectroscopy (HRPS), and a wide range of other techniques with similar functions. In some of these techniques, distribution of energy transitioned from the particles surface is measured. This measurement may be performed on some parameters such as energy distribution, angle, temperature, and so forth, and the particles can be those that are backscattered or particles produced by exciting processes induced by x-ray, ions, or electrons. A special interest in AES, XPS, and synchrotron is due to the fact that they are of great importance in the study of very thin films, even those with grain sizes lower than 1 nm. So, it was decided to study these techniques; however, because they have common features it was decided to mainly underline their differences and further applications. In general, using these techniques one can define the presented elements in the surface of a solid.

The ranges of detecting with AES and XPS techniques are somehow similar. Although measuring with these techniques are highly accurate, making it possible to calculate the thickness of superthin layers, in thick films it is necessary to use other techniques because the signals decrease exponentially with penetration depth. Both techniques include analysis of transmitted electrons from the material's surface. Electron excitation in AES and XPS techniques is carried out through descended electrons and x-rays (or photons), respectively.

In addition to having the peaks in the spectrums, one can find out about the presented elements in the materials composition or even determine the atomic ratio of the components. In the following sections some of their functions are discussed. The elements are famous in the industrial world, so some of them, including silicon and silicon oxide, are studied, as well as the mentioned techniques.[1–6]

3.2 Silicon (Si)

As far as we know, silicon (Si) is a semiconductor capable of having four bounds with atoms of other elements or Si elements. Si's surface is very sensitive and must be extremely immaculate. Once it is exposed to free air, it can easily react with oxygen or carbon or any other gas and create bounds. For this reason, despite the availability of various methods for silicon cleaning, such as putting it in a beaker containing some ethanol or placing it in a supersonic bath for 2 hours, it is probable that the initial oxide may be developed on it and must be cleaned from Si's surface with any available method.

The cleaning process can be performed even in the oven; the Si specimen is placed in an oven with pressure of 1 atm and then when the specimen is heated it is cleaned in the presence of Ar. Another method for silicon cleaning is to put it in a vacuum or ultravacuum container. Here, heating the specimen by the passing current makes it clean. Each of these pumps works in a particular range of pressure, and they can altogether lower the inner pressure of the container to a very low extent. In other words, they can create an ultravacuum space.

The rotary pump can diminish the pressure of a container from atmosphere pressure to 10^{-3} torr; a turbo pump can diminish it up to 7.6×10^{4}, and finally, ion and titanium pumps can create the pressure of 10^{-11} to 10^{-12} altogether. Thus, there would be an ultravacuum container that cleans the sample from impurities, artifacts, water, and so forth.

In addition, the samples such as Si, which are very reactive, can be placed in the container with elements such as thallium, and then pass the electrical current from both ends of the sample, and finally obtain a clean sample after heating the samples and releasing elements such as oxygen and carbon. However, it is needed to heat the ovens up to 100 to 150°C before putting each sample in it—this is known as baking.

Now, let's come back to the silicon; an element freely found in nature as well as in combination with an abundance of oxides and impurities and silicates. As well as its availability, another advantage of silicon is that one can find it in compounds such as sand, quartz, agglomerate crystals, and pearl. The silicates are also present in granite, asbestoses, mine dusts, and mica, and can be easily extracted.

Silicon constitutes about 25.7 weight percent of earth's composition and in terms of availability is the second element in the earth. It is worth mentioning that Si is even found in the sun and stars and is the main element of a group of celestial rocks known as aerolite. It is very important in industry, namely in microelectronics and nano-electronics. As previously mentioned, introduction of impurities such as phosphorous (P) or boron (B) results in the development of semiconductors such as *n* and *p*, respectively, where their binding in p-n form makes a diode, and three of them makes a bipolar transistor that is strongly dependent on their middle layer. One main feature of the Si is its capability of making stable silicon oxide (SiO_2).[7–9]

3.3 Dimer-Adatom-Stacking Fault (DAS) Model

Some researchers succeeded in explaining the geometric structure of Si 7 × 7 in their studies, using the scatter pattern and transmission electron diffraction (TEM). This model is known as DAS due to the presence of dimers, ad atoms, and rest atoms and refers to the fact that in the rebuilt structure of Si [111] 7 × 7, the mentioned pattern is available. This means that noting the structure of atoms in the unit cell in the form of two triangles or a parallelogram reveals that there are some ad atoms, dimers, and rest atoms in their structures. The noticeable point between two sides of the triangles is the atoms placed around the B-labeled atoms that are absent around A-labeled atoms; this leads to staking fault of this model.

In addition, the presence of a side hole must not be neglected. In general, there are nine dimer atoms along the boundaries. Also, there are six ad atoms between the ad atoms and the layer between them. Around the angular hole, the atoms are arranged as a 12-fold loop where the dimer atoms are bounded by eight loops.

One can generally claim that the DAS model constitutes 12 ad atoms, 42 rest atoms, and 48 atoms between the layers including stacking atoms. Also, the rebuilt 7 × 7 structure needs four extra atoms of Si, in contrast to an ideal structure of 1 × 1. Hence, there would be 102 atoms, in total, in a 7 × 7 structure.

As previously mentioned, due to heating the atomic structure of the Si surface changes and exhibits some patterns as they are rebuilt. However, during the cooling to, for example, a pattern of Si 7 × 7, the number of incomplete bounds would decrease, where in the case of a pattern structure of Si 7 × 7, this number diminishes from 49 incomplete bounds to 19. The remaining incomplete bounds are hybridized toward the outer surface and bound-counter bound stripes, which are known as $\pi - \pi^*$ bounds, are developed with an energy gap of about 0.5 to 0.7 eV. The maximum of π strip would correspond with a maximum of mass capacity strip, which is exactly placed beneath it.

The total density of the electrons for π and π^* strips is 6.78×10^{14}, and the effective mass is about 0.4 of free electron mass. But the effective mass is offered for light electrons, and it is supposed that is equal to the effective mass of light holes. This assumption can be applied for Si [100] rebuilt at 200°K. This structure, in spite of having a similar surface and electronic structure to that of Si 2 × 1, has a very narrow optical gap in this state.

Here we mention the techniques used for measuring the thickness of ultrathin films (even those thicknesses less than 1 nm). One of the most important techniques is AES. In the next section the spectrometry and spectrum receiver system of AES will be discussed.[10–14]

3.4 Auger Electron Spectrometry (AES)

Among the advantages of this method, one can name that using this technique it is possible to determine the chemical composition of the surface, usually up to a depth of 1 nm. First, the descended electron creates a hole through ionizing the nucleus level (*K* or *L*). Then, the electrons (both descended and those placed near to the nucleus) leave the atom. Electron transmission leads to energy drop; the energy loss is replaced by electron transfer from the 2l to 1s layer. So, the hole in nucleus level (such as *K*) is refilled by another electron that is placed in an upper layer with higher energy content.

During the electron promotion from an upper layer to a lower one, the energy difference causes transmission of one photon. This photon can grab and release another electron from this level or even an upper level. The auger-released electrons can leave the sample surface and even be detected. A system designed in such a way as to study the chemical structure or absence or presence of an element in a composition (using the auger electrons) also depends on its producing models.

If the initially created hole is from level *K* and the electron fills this hole from level L_1, and the electron from level L_2 is transmitted, then the energy of the auger electron would be shown as E_{KL1L2}. But if the electrons are capacity electrons, the energy is shown as E_{KVV}. Besides, if the atomic number of the element is *Z* and ΔE is the energy change created by rearrangement of the other electrons on auger electrons, then it would be possible to define all elements in a composition using the AES technique.

As far as we know, the signal of each element has the features and characteristics of the related element, shown in the elements spectrum, where the horizontal axis is its energy. For instance, imagine the Si that appears in the energy of 92 eV. Although several fine ranges of peaks are present in addition to the sharp and tall peak of the Si, which are all related to transition of higher energy levels, it is necessary to consider the K_β and K_γ transitions as well as K_α. Nevertheless, in the study of structural components of the materials, generally the taller peak (which is for transition of K_α) is considered.

As well as the mentioned points, it must be added that the ionization of the initial materials may occur by a number of radiation sources that often transmit synchrotron and x-rays and ionic rays, as well as electron rays. Then, the necessary requirement is equality of the threshold energy of the photons or electrons with ionization energy of the initial hole.

In addition to the energy issue, the possibility of ionization with a photoelectric section would also be provided. However, it is not required for the transmitter source to be monochrome, because the energy of the auger electron is independent from the transmitter source. Each electron with the energy higher than the threshold energy can lead to an energy drop of the auger electron. Then, the backscattered electrons may be involved throughout the ionization process and produce auger electrons in the sample. As seen in the electron transmission process, two electrons can scape; however, they are not in the relativity range, but as they contain electrical charge can interact with each other.[15–18]

3.5 Low Energy Electron Diffraction (LEED) Technique

This technique is mainly used in surface crystallography and detects that, for example, the surface is just made of Si atoms or has other atoms as well as Si. Using this technique it is possible to find out whether the developed film on the surface of a sample is crystalline or amorphous. Hence, the LEED technique can be used in detecting the atomic structure of the surface. The low-energy electron rays are vertically descended on the sample surface, and because the electrons' energy is very low (50 to 300 eV), they are not able to penetrate the sample. In other words, they can penetrate just several atomic layers in the solid state. The deeper penetration leads to multidirectional sputtering on the surface, which influences the LEED pattern. As a result, recognition and understating the details of the LEED intensity pattern in this state (which is a multiple scattering process) must be assumed in the highest atomic layers, which is out of kinetic theory range. Yet this can be regarded as a surface sensitive technique in which the electrons can be elastically scattered to develop the sputtering phenomenon according to the Bragg's Law on a phosphorous plate. These experiments are also performed in ultravacuum containers.

In practice, an LEED pattern must be perceived as very clear, visible points, and the intensity of the matrix should be low. Once the points are observed as widened and dim, it is possible some defect or other elements are present in the sample. In an empirical sense, the electron transmitted from hot filaments of the electronic gun reaches the given area after passing from several lenses and openings.[19–23]

3.6 X-Ray Phototransmission Spectrometry

An XPS rig consists of the spherical analyzer, ray gun, and vacuum container. The samples are put in the vacuum container and heated by a thallium fixer.

After cleaning the sample, it is required to use XPS diagrams to check its purity. Just Si peaks must be present and no other peaks implying the presence of other elements. However, one should not accept the presence of elements such as H_2 and He because they have very fine cross sections—other than H_2, which has one electron. Before showing the other spectrums of the XPS technique, it is necessary to describe its process, which is based on the occurred transitions.

The electrons are transmitted from the cathode and accelerated toward the anode, until they hit the anode. Through this impact and regarding the potential difference of 12.5 kV between the cathode and anode, it is possible to produce soft x-rays. Now this ray comes toward the sample and, similar to the photoelectric process, ray energy can be conveyed to the electron. Once the energy is enough, the electron would be released according to the three-stage model. There can be several virtual transitions involved in the process, so there would be smaller peaks in the diagram, as well as the Si2s and Si2p peaks. Also, the peaks related to plasmon and interstrip and interatomic transitions are present in the spectrum. One of the important parts of the machine is the soft x-ray source that includes two anodes.[24–28]

The methods for fabricating thin films are classified into five major groups:

1. Physical vapor deposition (PVD) methods
2. Chemical vapor deposition (CVD) methods
3. Ion-assisted beam deposition
4. Molecular beam epitaxy (MBE)
5. Pulsed laser deposition
6. Chemical bath deposition

Each of these methods may be implemented differently. They will be discussed in detail.

3.7 Physical Vapor Deposition (PVD) Methods

The most important physical methods for developing thin films are as follows:[29–35]

1. Thermal evaporation
2. Cathodic sputtering

3.7.1 Thermal Evaporation

The usage of this method in different cases is now on the increase.

It can be used in the construction of decorative coatings, and it has important applications in engineering, chemical-nuclear, microelectronics, and related industries. It is usually used in vacuum chambers and employed more than other methods to develop thin films, because it is a simple method and also is able to construct films with a very high purity degree and desired crystal architecture in proper conditions. A film formation process consists of three phases: evaporation or sublimation, deposition on substrates, and refining bonds among particles.

Vapor sources and substrates are put in vacuum chambers and then the chambers are discharged in a pressure that is less than 10^{-5} Torr. Then the stuff (to be deposited) will be evaporated or sublimated through methods discussed below. The first phase, heating, continues by the time we reach a condition at which the vapor pressure of the deposited stuff is more than the pressure of the chamber, so that the deposition of vapor on the substrate will be possible in practice.

PVD is a highly sophisticated process discussed here briefly. In order to fulfill the evaporation process, the kinetic energy of molecules that move out of the stuff (vertical component of velocity) should be more than the required latent heat (L_v) to cope with molecular adhesion bonds. Because kinetic energy of such molecules increases in line with temperatures, increased temperature raises the evaporation of the stuff.

The evaporated atoms or molecules move and deposit on the substrate. The substrate is placed on the evaporation source surface. It should be noted that the deposited stuff on the substrate is not uniform in all points. Deposition is thicker where the substrate is exactly located over the source, and it will be thinner in farther points. Thus, it is better to rotate the substrate around different directions so that the vapor flux is the same in all parts of the substrate; this problem is also solved by introducing pressured vapor into the chamber. This method is called pressure plating.

Constructing thin films via this method requires tools and devices that can initially create an adequate vacuum in the vacuum chamber. Usually, a rotator pump in association with diffusion, cryopumps, or other pumps are used to create a vacuum. Other parts include an evaporation source, which will be discussed later, and measuring tools that perform required measurements. Schematic and real images of this rig are depicted in Figure 3.1.

Generally, a thermal method is usable for all elements except refractory metals. For metals with melting points over 1500°C, very fine powders of metals are used instead of foils or small sheets in order to conduct evaporation.

The evaporation speeds of resistant sources considerably depend on various parameters such as heat concentration condition, wetting rate of element, and warm points. Therefore, for developing a certain thickness of a film, a certain weight of the evaporated stuff is put in the source and it will be heated up to full evaporation, or a speed or thickness monitor is used and it continues until reaching evaporation speed with a certain thickness.

3.7.2. Evaporation Sources

Thermal evaporation of stuffs may be fulfilled in various forms. The most important forms are discussed here.

3.7.2.1 Resistive Heating

The simplest method for evaporation of stuffs is resistive heating in which refractory metals including tungsten (W), tantalum (Ta), and molybdenum (Mo) or ceramic coatings with high melting points (Table 3.1).

Tungsten (W) may be used as a fiber or crucible. If the stuff reacts with the bush, using carbon crucibles or other nonmetal crucibles such as quartz, graphite, alumina, and zirconium will be necessary for melting and evaporation purposes.

3.7.2.2 Flash Evaporation

This method is used to manufacture thin films, whose structures are similar to that of the main composition, from multiple compounds. The stuff is powdered and then it is poured on the source surface slowly, which is very hot (burning surface), and it will be evaporated quickly (Figure 3.2). Likewise the method is used to develop ceramic films, Bi_2Te_3 and semiconductive compounds, and so forth. The disadvantage of the method is that the abrupt

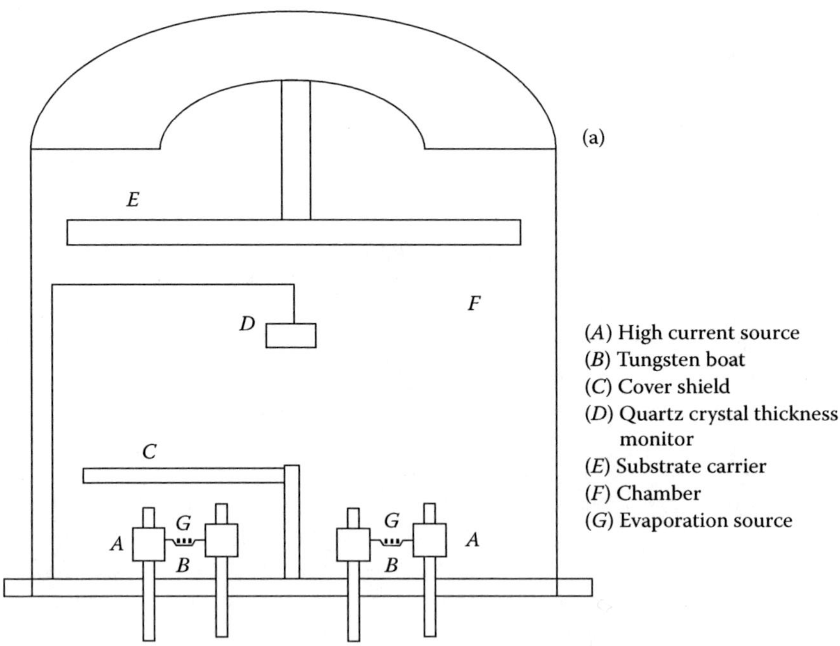

FIGURE 3.1
(a) Schematic and (b) real images of the main structure of a thermal evaporation rig. (Reprinted from Sung, M. F., Kuan, Y. D., Chen, B. X., and Lee, S. M., Design and fabrication of lightweight current collectors for direct methanol fuel cells using the micro-electro mechanical system technique, *Journal of Power Sources* 196 (14), 5897–5902, Copyright 2011, with permission from Elsevier.)

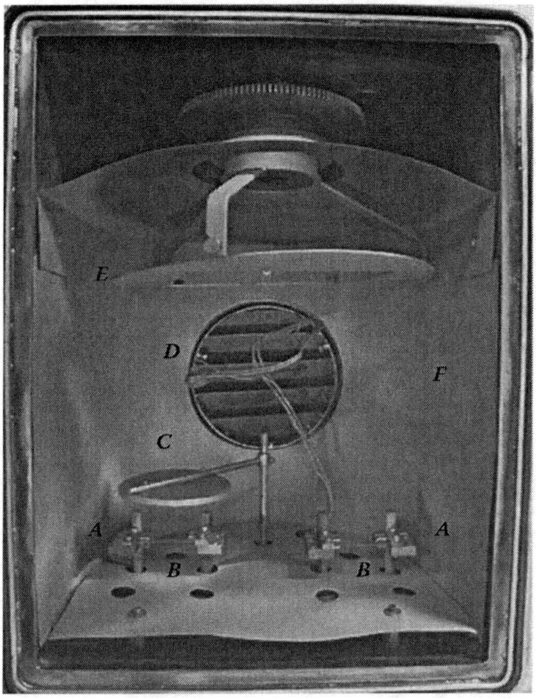

(b)

(*A*) High current source
(*B*) Tungsten boat
(*C*) Cover shield
(*D*) Quartz crystal thickness monitor
(*E*) Substrate carrier
(*F*) Chamber

FIGURE 3.1
(Continued)

release of the gas will lead to sputtering of evaporation, so this method is not simply controllable.

3.7.2.3 Arc Evaporation

Evaporation is conducted by putting certain stuff between electrodes of an electric arc and generation of extraordinary heat. It is used to evaporate Nb, Ta, C, and so forth. The deposition rate of such stuffs is about $50\mathring{A}$/sec. For creating an electric arc, electrodes are connected to a capacitor. An electric arc starts when electrodes approach each other, which lasts about a few seconds (Figure 3.3a).

TABLE 3.1
Refractory Metals

Refractory Metal	Melting Point (°C)	Torr Pressure in Melting Temperature
W	3410	10^{-2}
Ta	2996	10^{-2}
Mo	2620	10^{-2}

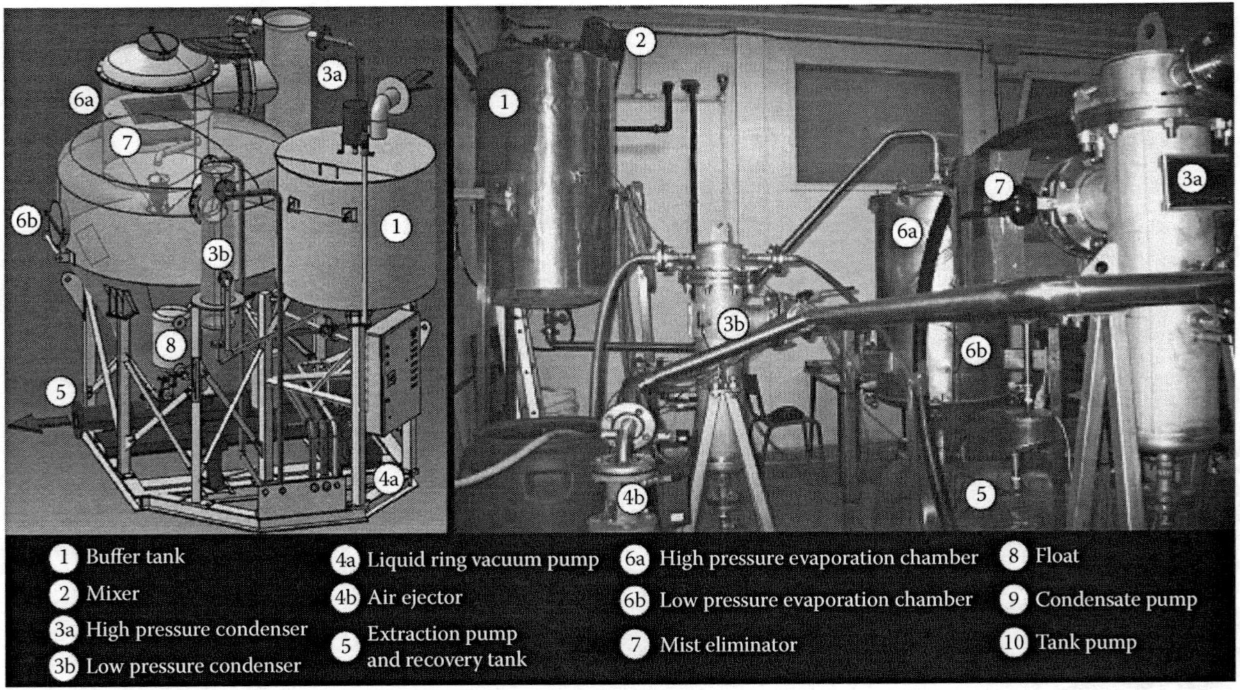

FIGURE 3.2
Two-stage flash evaporator: computer-aided design model of the industrial system (left) and its corresponding experimental pilot (right). (Reprinted from Ho Kon Tiat, V., Sebastian, P., and Quirante, T., Multiobjective optimization of the design of two-stage flash evaporators: Part 1. Process modeling, *International Journal of Thermal Sciences* 49 (12), 2453–2458, Copyright 2010, with permission from Elsevier.)

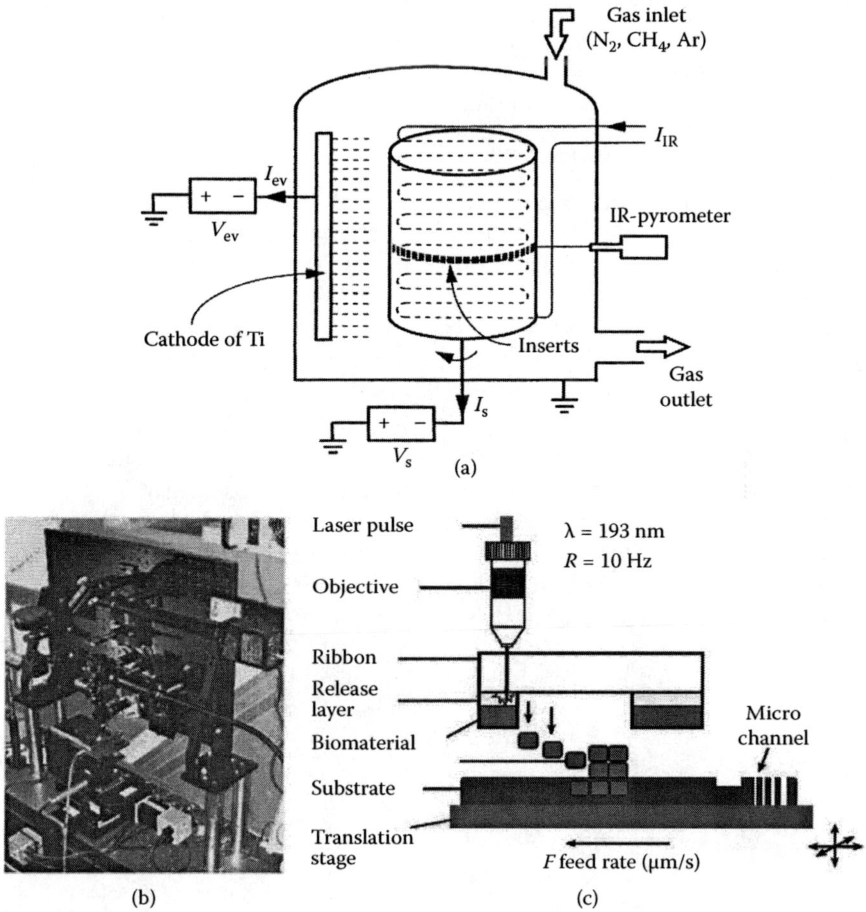

FIGURE 3.3
(a) Arc evaporation deposition system. (b) The matrix-assisted pulsed-laser evaporation direct-write (MAPLE DW) system. (c) The matrix-assisted pulsed-laser evaporation direct-write (MAPLE DW) process. (Reprinted from Harris, M. L., Doraiswamy, A., Narayan, R. J., Patz, T. M., and Chrisey, D. B., Recent progress in CAD/CAM laser direct-writing of biomaterials, *Materials Science and Engineering C* 28 (3), 359–365, Copyright 2008, with permission from Elsevier; and from Karlsson, L., Hultman, L., Johansson, M. P., Sundgren, J. E., and Ljungcrantz, H., Growth, microstructure, and mechanical properties of arc evaporated TiC_xN_{1-x} ($0 \leq x \leq 1$) films, *Surface and Coatings Technology* 126 (1), 1–14, Copyright 2000, with permission from Elsevier.)

3.7.2.4 Laser Evaporation

In this method, the intensity of the laser radiation on the stuff heats it up and vaporizes it. The laser radiation source is located out of the vacuum machine. With regard to the not-too-deep penetration of the laser (about 100$\mathring{A}$), the stuff is evaporated superficially. Schwarz and Tourtellotte succeeded in evaporating Sb_2S_3, $SrTiO_3$, and $BaTiO_3$ by using neodymium lasers (80 to 150 J). Several thousand angstrom deposition with a $10^6 \mathring{A}$/sec rate was achieved in this method (Figures 3.3b and 3.3c).

3.7.2.5 Application of Electron Bombardment

Because some solids including Si interact severely with thermal sources (crucible), electron bombarding is used to evaporate them and to avoid film doping. The stuff located in a crucible containing cold water is bombarded with accelerated electrons produced by tungsten fibers. Refractory metals such as Mo, Ta, and W may be evaporated by this method (Figure 3.4). The simplest type of this system includes W fibers to prepare electrons that are accelerated through a positive potential toward the stuff that would be evaporated. Electrons lose their energy quickly upon colliding with the stuff and then evaporate the stuff.

Another type of it is using electronic optics to concentrate and direct the beam toward the stuff. This was done by a commercial company. In some configurations, the beam is concentrated by an electrostatic preserver of fiber and also is used for deposition of Si films with a 3 μm/min rate. Also, it would be employed to direct a beam across magnetic choppers or a permanent magnet with 100 to 300 G. Tungsten fibers of electronic sources are usually heated with a weak transformer (6 to 12 V, $100 \mathring{A}$) through a central earth connection. The maximum density is limited because of spatial load and is measured as

$$I = 2.3 \times 10^{-6} \frac{V^{\frac{3}{2}}}{d^2} amp/cm^2 \quad (3.1)$$

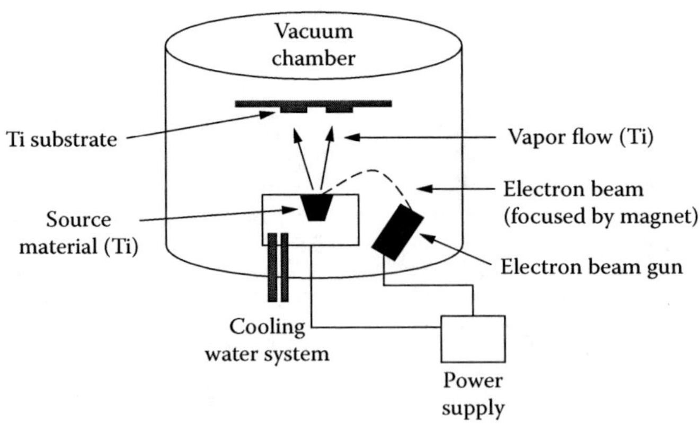

FIGURE 3.4
The electron beam evaporation process (high electron beam bombardment used to deposit a material). (Reprinted from Puckett, S. D., Lee, P. P., Ciombor, D. M., Aaron, R. K., and Webster, T. J., Nanotextured titanium surfaces for enhancing skin growth on transcutaneous osseointegrated devices, *Acta Biomaterialia* 6 (6), 2352–2362, Copyright 2010, with permission from Elsevier.)

where V is the acceleration voltage, and d is the distance between the cathode and the anode.

3.7.2.6 Evaporation through Radio Frequency (RF) Heating

In this method, radiation uses a proper radio frequency to transfer the required heating, which is called RF heating or induction heating, to the stuff either directly or indirectly and evaporates it. Because heating is brought about through induction, film may not be doped by the preserver. Some researchers used this method to evaporate aluminum, and they evaporated aluminum across the BN + TiB_3 of the crucible. They put a slim crucible near the surface, without any preserver against the RF field, and then aluminum was heated by a 200-KHz field.

3.7.3 Sputtering Method

Several sputtering coating techniques have emerged to develop thin films from different materials since the late nineteenth century. Here we deal with some coating techniques based on sputtering. The sputtering process enjoys some more unique advantages than other similar methods. Some of them are

1. Thickness uniformity
2. Coating hard stuff
3. Coating insulators
4. Targets with large surfaces
5. Lack of spitting like what is observed in evaporation
6. Lack of leakage like what is observed in electric arc coating

3.7.3.1 Reactive and Nonreactive Processes

There are two nonreactive and reactive aspects for film formation through sputtering. For nonreactive, the sputtering takes place using the plasma of inert gases such as Ar, which does not have any direct influence on formation of compounds on either targets or substrates. Ar is the most common gas compared with Kr and Xe because of its sufficient mass for high sputtering products and low costs. Although ions of inert gases are not combined with targets and substrates as the main components, any combination even in the trivial rates of inert gases may be very detrimental for specifications of films. In hard coatings, for example, Ar is able to expand lattice sites and enhance internal tension of films.

Furthermore, bombardment of both films and substrates by ions of inert gases may affect gas combination possibility, growth aspects, stoichiometry, and specifications of coating films. Stoichiometry refers to the ratio of film ingredients. Reactive sputtering films may be prepared through several methods, such as diode sputtering, DC, RF, triode, magnetron, and refined RF magnetron sputtering.

There are two main dimensions for sputtering. One is metal cathode, and the other is compound cathode. A pure metal target is used for the former, and compound formation is confined to a substrate and walls of the chamber. Actually, it depends on precise control of the process in order to prevent it from target doping and coating nonstoichiometric films. The other dimension, which is sputtering from a compound target, seems easier. The rate of sputtering is very slow for compounds because of a decrease in the product of sputtering and an increase in secondary electrons in most compound targets. Overall, with regard to the type of sputtering technique and materials and condition of coating, it is possible that the film and target stuff have different chemical compositions.

3.7.3.2 Diode Sputtering

Diode sputtering is the oldest version of such a process. Diode plasma is formed when in the presence of gas with sufficient concentration (1 to 50 × 10^{-3} mbar) a relatively great potential (300 to 500 V) is applied between the anode and cathode. A small fraction of these gas atoms are ionized, and the ions accelerate along a potential gradient of the cathode cap. They then collide with the target and lead to the sputtering of its surface. The DC diode sputtering may be used for compound and noncompound targets. However, the target should be an electric conductor.

For the diode sputtering, 75% to 95% of the capacity of the feeding source is consumed for heating the target, which is wasted by a water cooler. Thus, heat conduction of the target is of great significance. Coating thin films on samples of electron microscopes is a common example of DC diode sputtering. While its simplicity is the biggest advantage of the DC diode sputtering, it is rarely used because of its low coating rate, high temperature of the substrate, and energy inefficiency.[41–45]

3.7.3.3 RF Sputtering

Using a fluctuating feed source to develop sputtering plasma is preferred to the DC method because it is able to sputter targets by using low AC frequencies (50 Hz). Its industrial applications are very limited because both electrodes are being corroded during working. Low-frequency AC electric discharge is more similar to DC electrical discharge than RF electrical discharge in many aspects.

When frequency is more than 50 KHz, required electrons for stability of electrical discharge decrease slowly. Moreover, when frequency is over 50 KHz, it is not necessary that both electrodes should be conductive, because in this case the electrode is able to couple with impedance. In order to apply sputtering only to the coupled electrode or insulator, coupled electrodes should be considerably smaller than direct electrodes. It is usually carried out through an earth connection with a RF generator, chamber walls, and substrate preservers. A compatibilizer network of impedance is required between the RF generator and load to provide proper induction for reinforcement. In RF systems, removing unnecessary edge lumps is essential for minimizing capacitor and inductive wastes.

Other ions are not able to follow up potential fluctuations over a MHz domain because of their relative heavy mass, so ion aggregation will decrease during a certain phase of the cycle in which an electrode acts as a cathode. Generally, frequencies over 10 MHz may be used effectively for sputtering purposes. The most common frequencies, 13.56 and 27 MHz, are allowed frequencies for medical and industrial applications.

There are two main advantages for RF sputtering: ability to sputter insulators and good applicability in low pressures. Unfortunately, the sputtering rate of RF methods is limited because of trivial heat conduction of insulator targets. Thus, insulator films preferably are prepared from metal sources through reactive methods. In RF electrical discharge almost all materials may be sputtered in reactive or nonreactive ways, but produced films may not have the initial composition. Coating metals, metallic alloys, oxides, nitrides, and carbides are some of the applications of the RF sputtering.[46–49]

3.7.3.4 Triode Sputtering

There is a third electrode besides anodes and cathodes in this kind of sputtering that has been added for increasing ionization. The third electrode would be a bias simple conductor or a thermionic electron source. A schematic view of a triode sputtering system with a Langmuir probe can be seen in Figure 3.5. Thanks to this extra electrode, the electric discharge does not depend on the formation of secondary electrons in cathodes for continuing electrical discharge. Thus, electrical discharge takes place in pressures $<10^{-5}$ Torr and discharge voltages ~40 V.

The discharge current would be altered independent from voltage through changing the source excitation, so developing high ionic density in both targets and substrates using low discharge potentials will be possible. Triode electrical discharge like RF and DC has been used successfully for coating various semiconductive films, caps, optical resistances, and other coatings. The main advantages of triode sputtering are lower discharge pressure, lower discharge voltage, higher coating rate, and independent control of plasma density.

Unfortunately this system is intricate in practice. It may increase doping through thermionic radiation, and its conversion into an industrial-scale system is difficult. On the other hand, the presence of an electronic eye

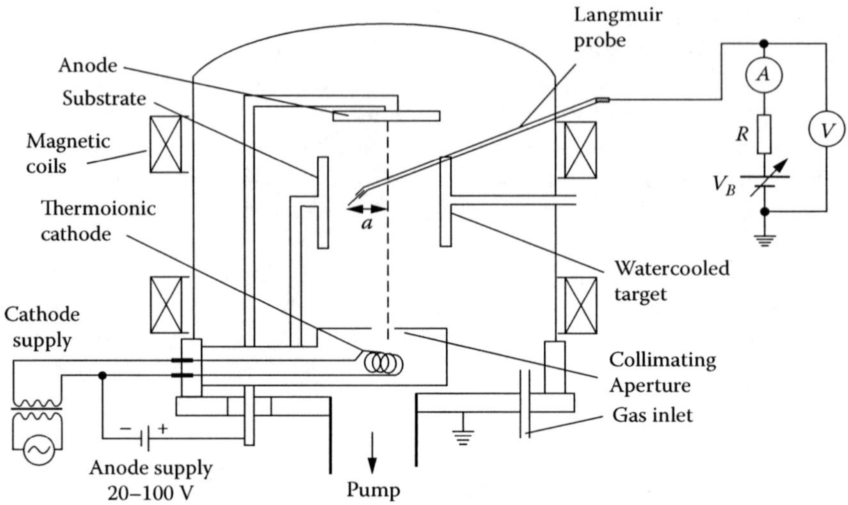

FIGURE 3.5
A triode sputtering system with a Langmuir probe. (Reprinted from Axelevitch, A., Gorenstein, B., Darawshe, H., and Golan, G., Investigation of thin solid ZnO films prepared by sputtering, *Thin Solid Films* 518 (16), 4520–4524, Copyright 2010, with permission from Elsevier.)

in some applications would be problematic, particularly in heat-sensitive reactive processes.

3.7.3.5 Magnetron Sputtering

Magnetron sputtering is different from other sputtering approaches because here plasma is limited only to areas near the target surface through applying a potent magnetic field. Such a magnetic field diverts the route of radiated secondary electrons from the target, and electrons move helically in a closed loop parallel with the cathode surface. Figure 3.6 illustrates different types of magnetron configurations.

This kind of arrangement puts a limited plasma near the cathode surface and increases ionization possibility and also increases plasma density (>1) near the target.

Advantages of limited plasma are as follows:

- Increasing the coating rate
- Decreasing sputtering on the substrate and chamber walls
- Decreasing the substrate temperature during the coating
- Decreasing the required pressure for the working gas

This method has been completely successful in developing high-quality and pure films with high coating rates.

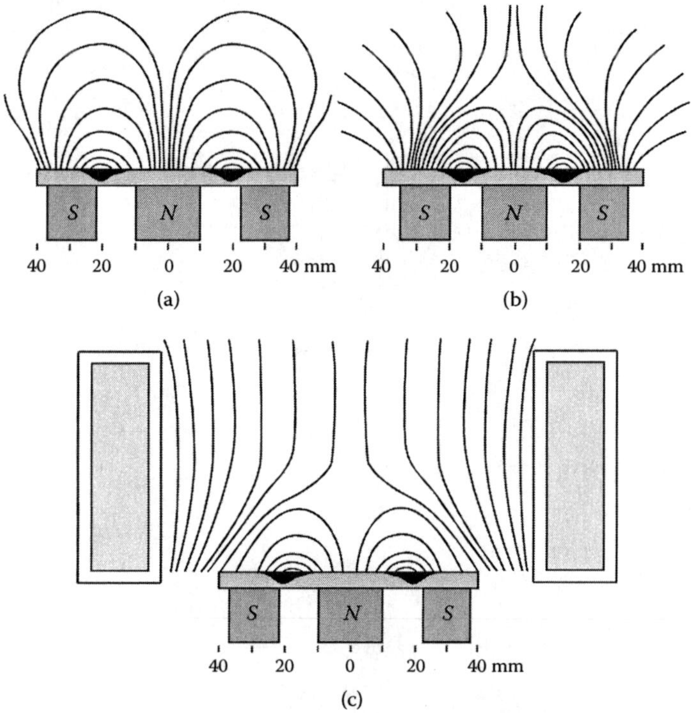

FIGURE 3.6
Types of magnetron configurations: (a) balanced magnetron, (b) unbalanced magnetron, and (c) magnetron with an additional electromagnetic coil. (Reprinted from Svadkovski, I. V., Golosov, D. A., and Zavatskiy, S. M., Characterisation parameters for unbalanced magnetron sputtering systems, *Vacuum* 68 (4), 283–290, Copyright 2002, with permission from Elsevier.)

Cylindrical and planar magnetron sources are the most popular magnetron sources (as shown in Figures 3.7 and 3.8). In a cylindrical magnetron, the target is not consumed very quickly, which is considered as an advantage. Electrons are imprisoned in the plasma coverage and surround the cylindrical cathode in a radius almost equal to the radius of the anode.

For cylindrical magnetron methods, solid and hollow cathodes would be used. But the most common sputtering source is the circular plane magnetron with magnetic limitation. As pointed out previously, targets usually are corroded in such magnetrons as race tracks and produce a great deal of wastage as well as circular distribution of density of the sputtered atoms from the target.

A magnetron sputtering system is able to act as a triode by adding an electron source. An example is the enhanced magnetron with the hollow cathode in which electrical discharge of the hollow cathode is used directly opposite the magnetron source for increasing ionization. Moreover, magnetron cathodes act with RF voltages as well. In this case, because the electric

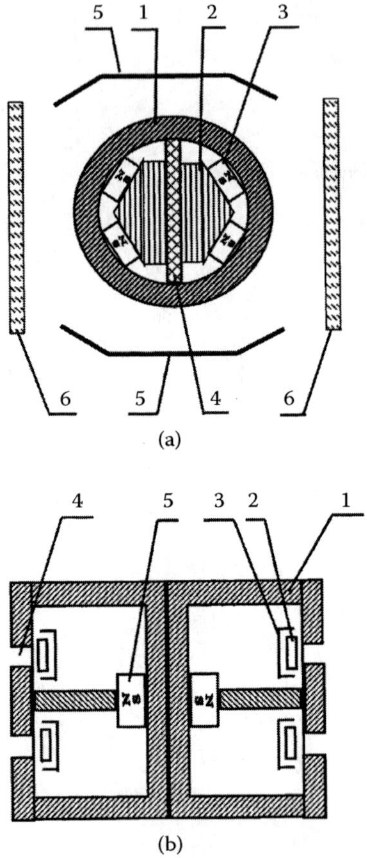

FIGURE 3.7
(a) Cross section of a cylindrical two-sided magnetron: 1, cathode tube; 2, magnetic circuit; 3, permanent magnets; 4, expending gear; 5, anode; 6, substrate. (b) Cross section of a magnetron-type ion source: 1, magnetic circuit cathode; 2, anode; 3, screen; 4, magnetic gap; and 5, permanent magnets. (Reprinted from Bugaev, S. P., and Sochugov, N. S., Production of large-area coatings on glasses and plastics, *Surface and Coatings Technology* 131 (1–3), 474–480, Copyright 2000, with permission from Elsevier.)

field vector alters in terms of both amplitude and direction, the applied force on electrons of plasma changes during the cycle so the plasma is no longer limited to the vicinity of its target. As a result, actual behavior of the magnetron is observable in a part of the cycle. Such a sputtering system is easily converted into an industrial-scale system.

3.7.3.6 Unbalanced Magnetron Sputtering

Numerous studies have shown that ionic bombardment of samples during the coating process will impose alterations on nucleation synthetic,

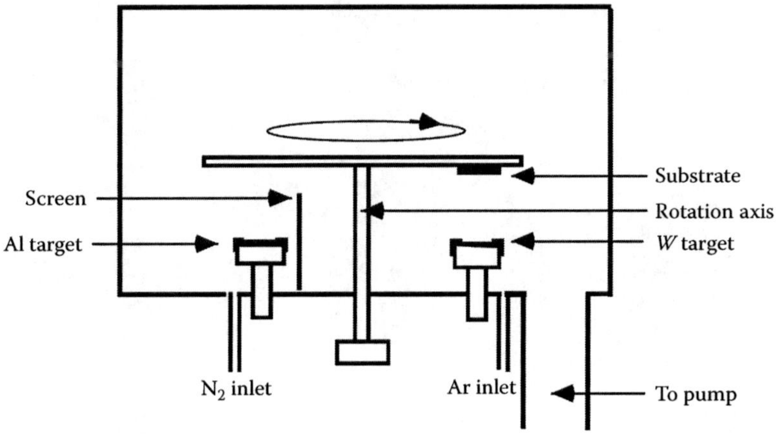

FIGURE 3.8
A planar DC magnetron sputter coater. (Reprinted from Zhang, Q. C., and Shen, Y. G., High performance W-AlN cermet solar coatings designed by modelling calculations and deposited by DC magnetron sputtering, *Solar Energy Materials and Solar Cells* 81 (1), 25–37, Copyright 2004, with permission from Elsevier.)

combination, orientation, and mechanical properties of most films. Generally, during the magnetron sputtering, developed ions by discharge surround the substrate through applying a potential (1 to 500 V). Although density of incident flow on the substrate film surface is low (for common magnetrons about 5% to 15% of ions per coating atom), this flow density is sufficient for most applications, but its enhancement would be effective in some cases.

In the field of preparation of hard coatings in which coating films with the lowest internal and intergrain holes and lowest defects in crystalline networks are desired for enhancing resistance and rigidity, prevention of the development of intergrain holes is possible by using increased bias voltages; however, it brings about defects within grains and tensions in films and also decreases adhesion and the quality of films. In order to cope with such limitations, it is preferable to increase ion flow density through fixing bias voltages. The energy of collided ions is so low that development of undesired defects in the film is prevented. An unbalanced magnetron system is able to go through such ionic bombardment.

The unbalanced magnetron concept was put forward in 1986 for the first time. Researchers used different magnetic arrangements (Figures 3.9 and 3.10) that have been classified into three main classes. The type I magnetron has a potent central pole and a weak external pole.

Figure 3.9 shows magnetron II. An intermediate type is almost balanced like what is widely used for all magnetrons. The usage of arrangement I of ionic bombardment is rare in the substrate, and the ratio of ion to coating atom is 25:0.1. In comparison, arrangement II makes more ionic bombardment, and the ratio of ion to coating atom in low bias potentials is 2:1. In other studies,

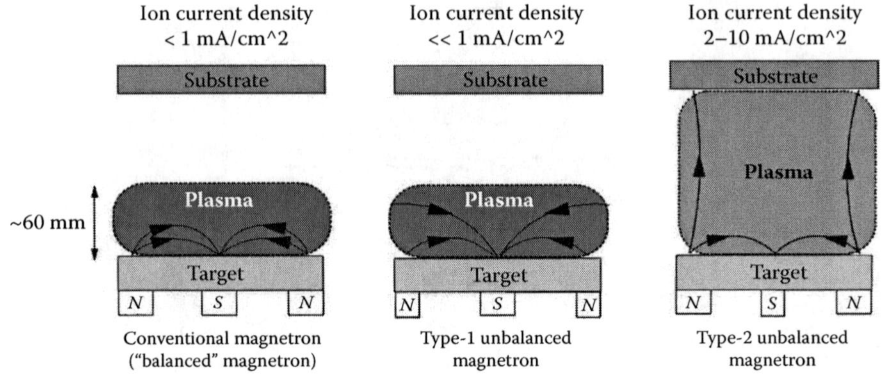

FIGURE 3.9
The plasma confinement observed in conventional and unbalanced magnetrons. (Reprinted from Kelly, P. J., and Arnell, R. D., Magnetron sputtering: A review of recent developments and applications, *Vacuum* 56 (3), 159–172, Copyright 2000, with permission from Elsevier.)

they showed that the ion flux increases linearly along with the discharge flow in the substrate area, but it does not change considerably under the influence of the whole system pressure. Figure 3.11 demonstrates some working aspects of sputtering.

The ratio of ions to coating atoms = 5:0.1, >1:2, and = 1:10 for common magnetron (CM), unbalanced magnetron (UM), and dual site sustained discharge (DSSD), respectively (comparable with what has been achieved in the ion plating process). Another advantage of UM is that the energy and

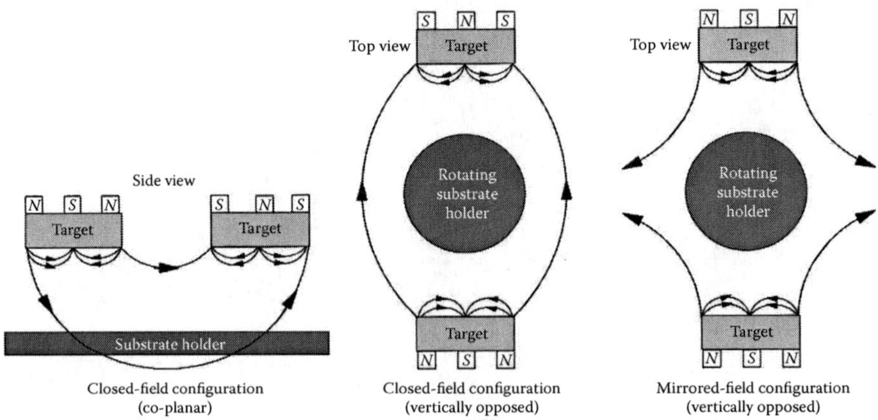

FIGURE 3.10
Dual unbalanced magnetron configurations. (Reprinted from Kelly, P. J., and Arnell, R. D., Magnetron sputtering: A review of recent developments and applications, *Vacuum* 56 (3), 159–172, Copyright 2000, with permission from Elsevier.)

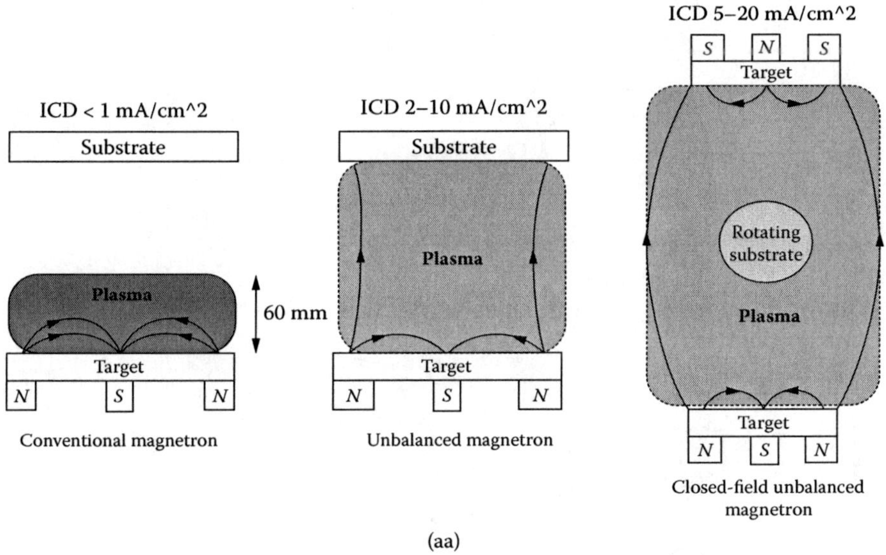

FIGURE 3.11
(aa) A comparison of the magnetic configuration and plasma confinement in conventional, unbalanced, and dual-magnetron closed-field systems. (bb) Various multiple magnetron arrangements developed to suit specific applications: (a) vertically opposed dual closed-field arrangement, (b) dual coplanar closed-field arrangement, and (c) dual magnetron barrel plater. (Reprinted from Arnell, R. D., and Kelly, P. J., Recent advances in magnetron sputtering, *Surface and Coatings Technology* 112 (1–3), 170–176, Copyright 1999, with permission from Elsevier.)

flux of ions may alter independently, so the relation between parameters of the process and microstructure of obtained films would be scrutinized comprehensively.

3.8 Chemical Vapor Deposition (CVD)

3.8.1 Introduction

Generally, CVD methods include techniques in which a flow of a gas containing volatile components of materials (which should be coated) is introduced into a vacuum chamber. The internal condition of a chamber is controlled so that it makes the chemical reactions in the vicinity or on the substrate possible, and that coating is developed at the bottom of the surface. During this process by-products of the chemical reaction leave the system. Obtained films by this method would be metal, alloy films, refractory compounds, and semiconductors.[56–60]

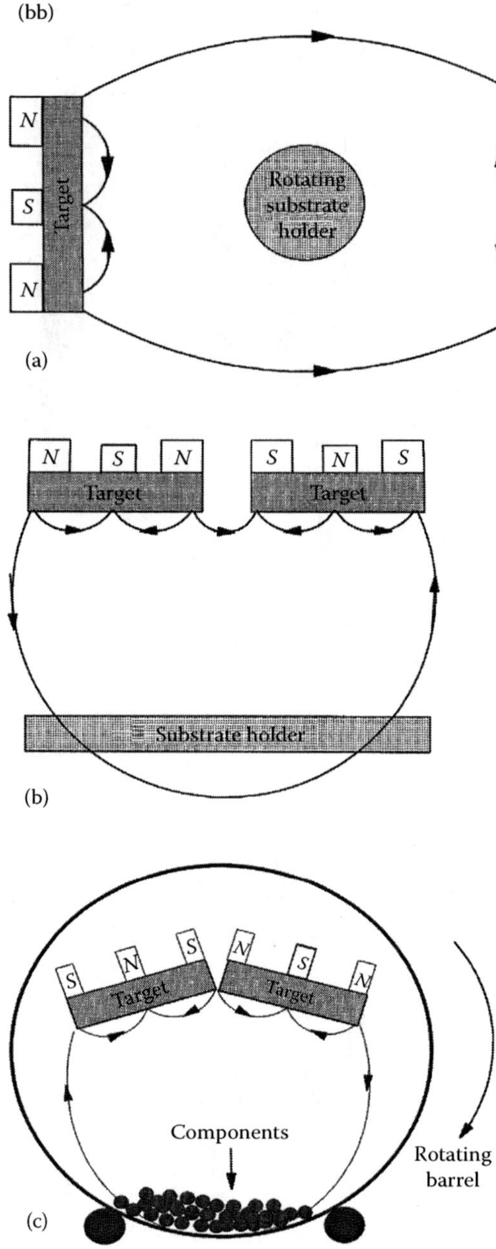

FIGURE 3.11
(Continued)

The process-relevant parameters such as the substrate temperature, the gas pressure, the concentration of reactive materials, and the gas flow rate affect properties of developed films. Certain specifications would be achieved through proper control of these parameters. Generally, reactions of CVD processes are as follows:

Decomposition reactions:

$$AB_{(g)} \rightarrow A_{(S)} + B_{(g)} \tag{3.2}$$

Oxidation and reduction reactions:

$$AB_{(g)} + C_{(g)} \rightarrow A_{(S)} + BC_{(g)} \tag{3.3}$$

Hydrolysis reactions:

$$AB_{2(g)} + 2HOH \rightarrow AO_{(S)} + 2BH_{(g)} + HOH_{(g)} \tag{3.4}$$

Polymerization reactions:

$$XA_{(g)} \rightarrow A_{X(g)} \tag{3.5}$$

Combination reactions:

$$A_{(S)} + B_{(g)} \rightarrow AB(g) \tag{3.6}$$

Figure 3.12 shows a CVD process. It is observed that initially a gas mixture is introduced into the reactor chamber. Then a chemical reaction takes place near or on the substrate; as a result, a solid material is created on the substrate.

$$2AX_{(g)} + H_2 \rightarrow A_{(s)} + 2HX_{(g)} \tag{3.7}$$

Some volatile materials are also created besides solid materials (A); they all leave the system.

In a CVD process, complicated interactions are carried out in the vapor phase, the vapor–solid interface, the substrate-coating interface, and the solid phase. Therefore, in order to improve properties of developed films on the substrate, proper conditions should be prepared for such interactions on the marked areas of Figure 3.13.

The main gas is placed on the substrate surface and film, and a boundary layer is developed in this area. Reactive materials as well as reaction by-products are transferred across these boundary layers during the process, and the precipitation rate is controlled by this process.

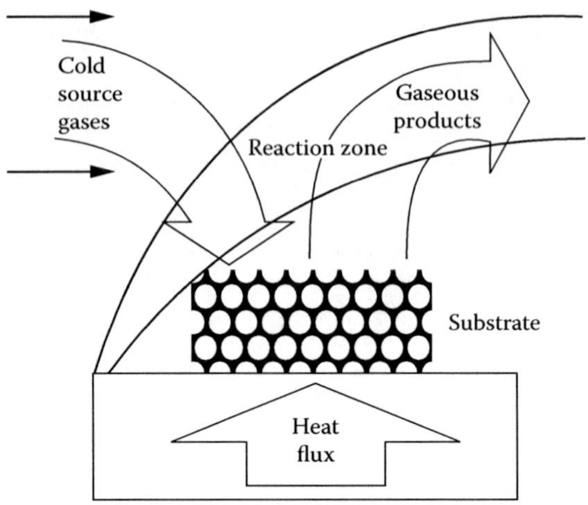

FIGURE 3.12
Foundations of chemical vapor deposition (CVD) process. (Reprinted from Ryan, G., Pandit, A., and Apatsidis, D. P., Fabrication methods of porous metals for use in orthopaedic applications, *Biomaterials* 27 (13), 2651–2670, Copyright 2006, with permission from Elsevier.)

Required heterogeneous reactions take place in the vapor phase. However, homogenous reactions are reported in most CVD processes. For preparation of a carbide layer from methane gas, for instance, formation of various carbon-containing radicals in the vapor phase that may be absorbed to the growing surface is important.

Heterogeneous reactions in section 2 of Figure 3.13 discern microstructures and hence properties of developed films. The deposition temperature varies from room temperature to 200°K (depending on the type of the process). In high temperatures, solid-state reactions (e.g., phase transformation, deposition, recrystallization, growth of grain, etc.) may take place in a CVD process.

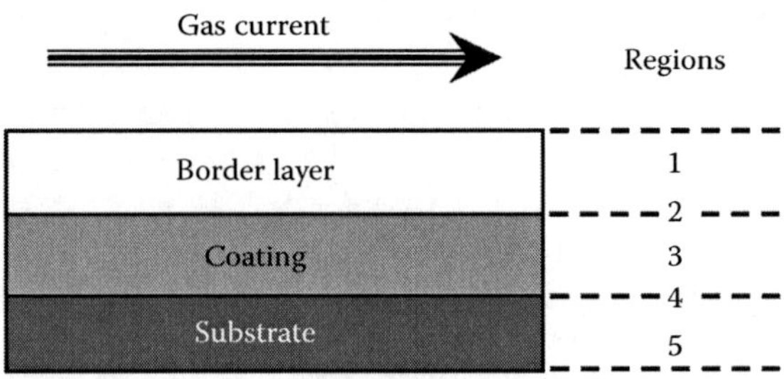

FIGURE 3.13
Interaction areas of chemical vapor deposition (CVD) process.

Components of films and substrates may permeate into each other in section 4, so intermediate phases are developed. Reactions related to section 4 play an important role in the film adhesion to substrates. CVD is used for coating a wide range of elements and compounds, some of which are summarized in Table 3.2.

3.8.2 CVD System Requirements

A CVD system is composed of three main sections: a gas distribution system, a reactor, and a discharge system (Figure 3.14).

3.8.2.1 Gas Distribution System

If inputs include gases in the room temperature, this system is completely simple and includes a pressure regulator, a regulator of the gas flow rate, gas flow throttles, and a filter to remove impure particulates from gases. If inputs include solid or liquid materials at room temperature, they enter into the system via other ways. For example an evaporator or purifier system is used to heat inputs up to a certain temperature, and then the gas containing evaporated materials is heated within the reactor.

3.8.2.2 CVD Reactors

Reactors used in CVD are classified into three classes:

1. Reactors with a cold wall
2. Reactors with a hot wall
3. Chemical reactors in a plasma-assisted chemical vapor deposition (PACVD) plasma medium

They are described more fully below.

3.8.2.2.1 Reactor with Cold Walls

Figure 3.15 shows a CVD reactor with a cold wall. The wall of such reactors is cold and no sediment is developed.

By using cold walls, the destruction line of the wall is declined via interactions between the vapor and wall. In these reactors, homogenous reactions stop, and the significance of current reactions is enhanced across the surface. A gradient of high temperature brings about severe heat transfer, so controlling uniformity of the microstructure and film thickness will be difficult. However, a reactor with a cold wall enjoys several advantages including great cleanness, fast cooling that leads to a decrease in interactions in three to five areas (Figure 3.13); likewise in this system embedding systems of the substrate are constructed easily.

TABLE 3.2

Different Reactions of Chemical Vapor Deposition (CVD) Process to Develop Sedimentary Compounds

Materials	CVD Method	°C
Nitrides BN	$BCl_3 + NH_3$	1000–2000
	Thermal decomposition $B_3N_3H_3Cl_3$	1000–2000
HfN	$HfCl_x + N_2 + H_2$	950–1300
Si_3N_4	$SiH_4 + NH_3$	950–1050
	$SiCl_4 + NH_3$	1000–1500
TaN	$TaCl_5 + N_2 + H_2$	2100–2300
TiN	$TiCl_4 + N_2 + H_2$	650–1700
VN	$VCl_2 + N_2 + H_2$	1100–1300
ZrN	$ZrCl_4 + N_2 + H_2$	2000–2500
Oxides Al_2O_3	$AlCl_3 + CO_2 + H_2$	800–1300
SiO_2	$SiH_4 + O_2$	300–450
	Thermal decomposition $Si(OEt)_4$	800–1000
Silicon-oxynitride	$SiH_4 + H_2 + CO_2 + NH_3$	900–1000
SnO_2	$SnCl_4 + VCl_4 + H_2$	
TiO_2	$TiCl_4 + O_2$ + hydrocarbon (flame)	
Silicides V_3Si	$SiCl_4 + VCl_4 + H_2$	
MoSi	Mo (substrate) + $SiCl_2$	800–1100
Borides AlB_2	$AlCl_3 + BCl_3$	~1000
HfB_x	$HfCl_4 + BX_3$ (X=Br_2Cl)	1900–2700
SiB_x	$SiCl_4 + BCl_3$	1000–1300
TiB_2	$TiCl_4 + BX_3$ (X=Br, Cl)	1000–1300
VB_2	$VCl_4 + BX_3$ (X=Br, Cl)	1900–2300
ZrB_2	$ZrCl_4 + BBr_3$	1700–2500
Carbides B_4C	$BCl_3 + CO + H_2$	1200–1800
	$B_2H_6 + CH_4$	
	Thermal decomposition of Me_2B	~550
Cr_7C_3	$CrCl_2 + H_2$	~1000
Cr_3C_2	$Cr(CO)_5 + H_2$	300–650
HfC	$HfCl_4 + H_2 + C_7H_8$	2100–2500
	$HfCl_4 + H_2 + CH_4$	1000–1300
Mo_2C	$Mo(CO)_6$	350–475
	$Mo + C_3H_{12}$	1200–1800
SiC	$SiCl_4 + C_6H_5CH_3$	1500–1800
	$MwSiCl_3 + H_2$	~1000
TiC	$TiCl_4 + H_2 + CH_4$	980–1400
V_2C	Thermal decomposition $W(CO)_6$	300–500
	$WF_6 + C_6H_6 + H_2$	400–900
VC	$VCl_2 + H_2$	~1000

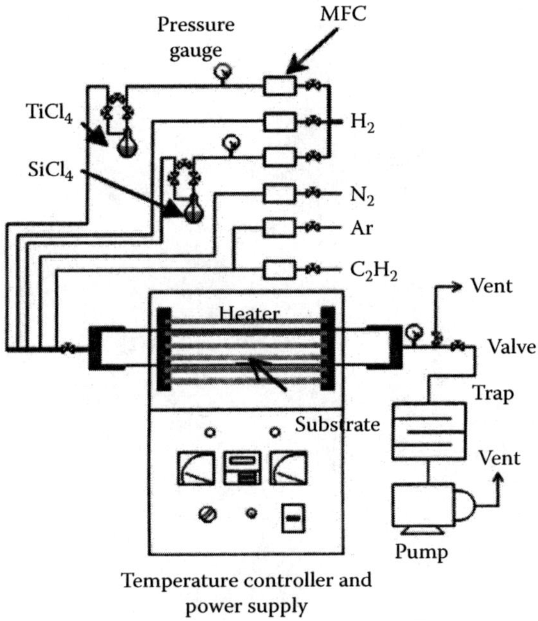

FIGURE 3.14
The chemical vapor deposition (CVD) reactor for Ti-Si-C-N depositing systems. (Reprinted from Kuo, D. H., and Huang, K. W., A new class of Ti-Si-C-N coatings obtained by chemical vapor deposition, Part 1: 1000°C Process, *Thin Solid Films* 394 (1–2), 72–80, Copyright 2001, with permission from Elsevier.)

Low-temperature reactors that work in a normal pressure are classified into three classes in terms of features of the gas flows and operation principles:

1. Horizontal tubular chambers with current dislocation
2. Discontinuous vertical rotor chambers
3. Consecutive chambers

3.8.2.2.2 Reactor with Hot Walls

In reactors with hot walls (Figure 3.16) the reactor is surrounded by a furnace, and the substrate and reactor are at the same temperature—that is, film formation not only takes place on the substrate but the internal walls of the reactor is another site for the film formation. Thicker films on the internal wall of reactors reduce speed of particulates across the wall, so they may fall in the film, which may result in creation of holes in the coating. Another destructive factor of this system is interactions between the hot walls of reactors and gases. They may result in creation of gas products and hence transportation of wall contents to the substrate and their redeposition on the substrate. Therefore, a consecutive reduction of reactive materials is observed

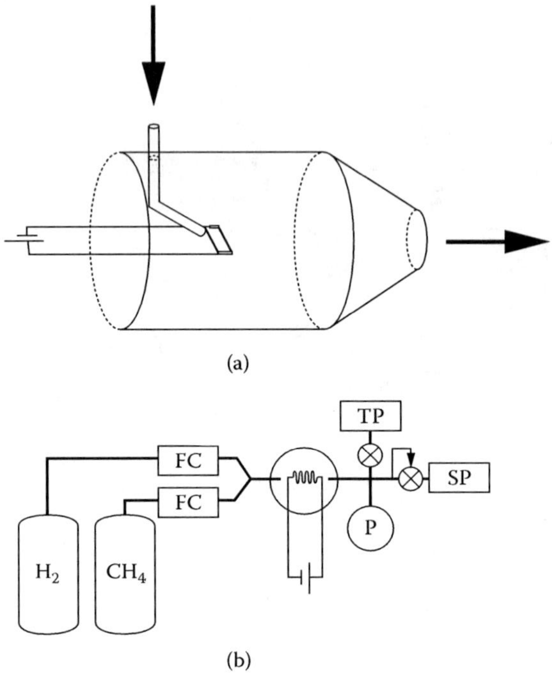

FIGURE 3.15
Cold wall chemical vapor deposition (CVD): (a) reactor schematic (approximately to scale) and (b) gas flow schematic. (FC = flow controller, TP = turbo-molecular pump, SP = scroll pump, P = Baratron gauge). (Reprinted from Finnie, P., Li-Pook-Than, A., Lefebvre, J., and Austing, D. G., Optimization of methane cold wall chemical vapor deposition for the production of single walled carbon nanotubes and devices, *Carbon* 44 (15), 3199–3206, Copyright 2006, with permission from Elsevier.)

when gas moves from the reactor to the hot wall. Different techniques may be used to compensate this reduction and to run an acceptable condition of deposition within the reactor:

1. Introduce a fresh gas in various areas of the reactor.
2. Create a temperature gradient along the reactor to enhance efficiency of the process.
3. Use geometric obstacles within the reactor to raise the linear velocity of gas flow.
4. Add some particles to the input gas to decrease deposition velocity particularly on the reactor entrance.

Reactors with hot walls usually are used in processes that are important for occurrence of homogenous reactions. Such reactors are potent and able to cover numerous substrates (about 1000) simultaneously.

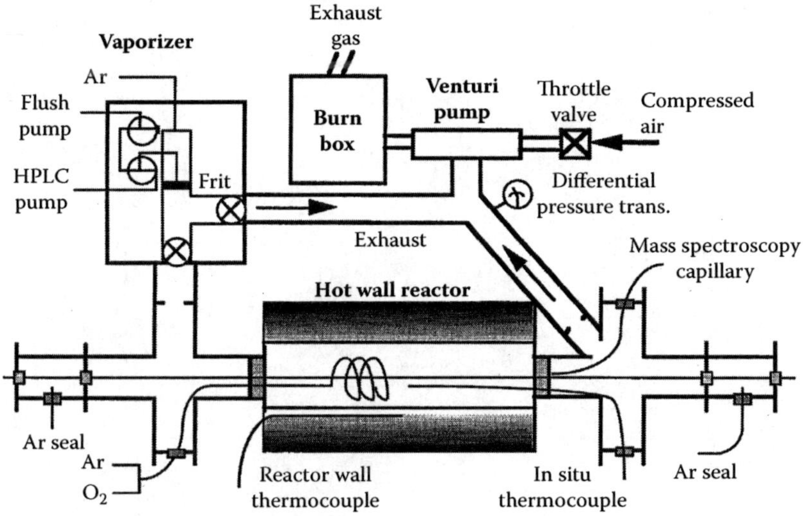

FIGURE 3.16
Hot wall chemical vapor deposition (CVD) reactor. (Reprinted from Jones, J. G., Jero, P. D., and Garrett, P. H., In-situ control of chemical vapor deposition for fiber coating, *Engineering Applications of Artificial Intelligence* 11 (5), 619–626, Copyright 1998, with permission from Elsevier.)

3.8.3 Plasma-Assisted Chemical Vapor Deposition (PACVD)

This technique is a kind of CVD process in low pressure in which the energy of gas reactive materials is enhanced via creating plasma media within reaction chambers. Its main advantage is the possibility of achieving a coating with certain properties in temperatures lower than those of typical methods of CVD. The main components of a PACVD system are similar to those of a CVD that were introduced previously; however, only a plasma generator source is added to the system. Therefore, a PACVD has the following components:

1. Initial gas or liquid source (raw material) and a device for distribution of controlled gas or liquid in the plasma medium volume
2. A plasma generator source that may be DC, radio frequence (RF), or MW
3. A discharge system to extract waste produced by chemical reactions following purification and neutralization (Figures 3.17 and 3.18 indicate some different PACVD systems using different plasma generator sources.)

3.8.3.1 Analysis of Physical and Chemical Aspects of Electrical Discharge

Free electrons constitute active particles of plasma in plasma-assisted processes that attain their energy from electric fields faster than ionic particles. Such plasmas that are created because of electrical discharge have a low

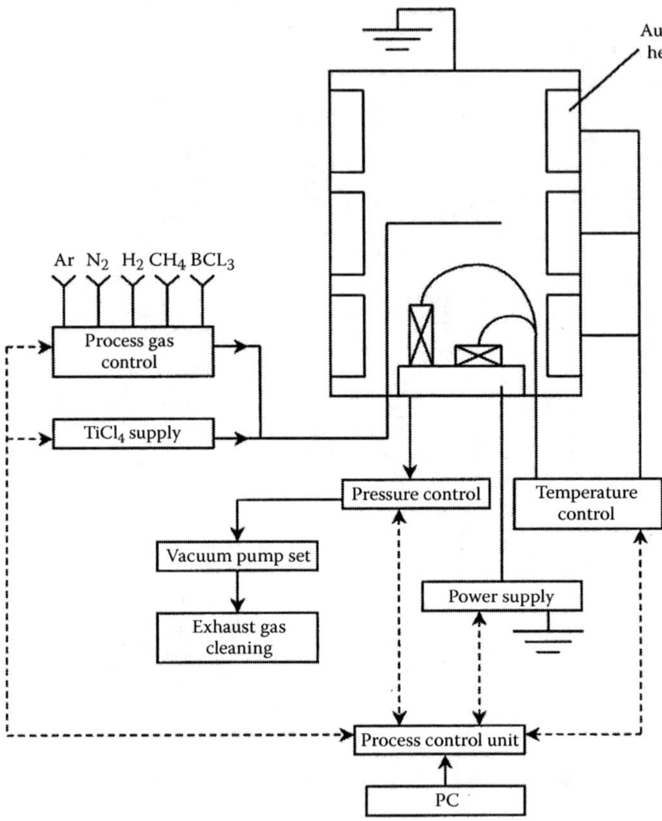

FIGURE 3.17
The plasma-assisted chemical vapor deposition (PACVD) reactor. (Reprinted from Pfohl, C., and Rie, K. T., Wear-resistant PACVD coatings of the system Ti-B-N, *Surface and Coatings Technology* 116–119, 911–915, Copyright 1999, with permission from Elsevier.)

ionization degree ($10^{-3} - 10^{-6}$) and are preserved in pressures lower than 10^2 Pa, and their electron density is $10^9 - 10^{12}$ cm^{-3}. If the density of free electrons is high enough, electron collision is carried out more than other processes. In this case, the average energy of electrons is approximately 2 to 10 eV, which generates $10^4 - 10^{5}$°K (T_e) temperature for electrons. However, the temperature of the gas (T_g) is near to the room temperature so that an unbalanced condition occurs in plasma in which $T_e \gg T_g$, and this increases active particulates and hence creates a chemical phenomenon in high temperature within a gas with low temperature. This ability to bypass the chemical balance state through producing active particles, whose flow velocity and energy are controlled considerably via monitoring chemical properties of plasma volume and electrical potential of the substrate rather than plasma, is a hallmark of PACVD. But another significant point is the control amount that may apply

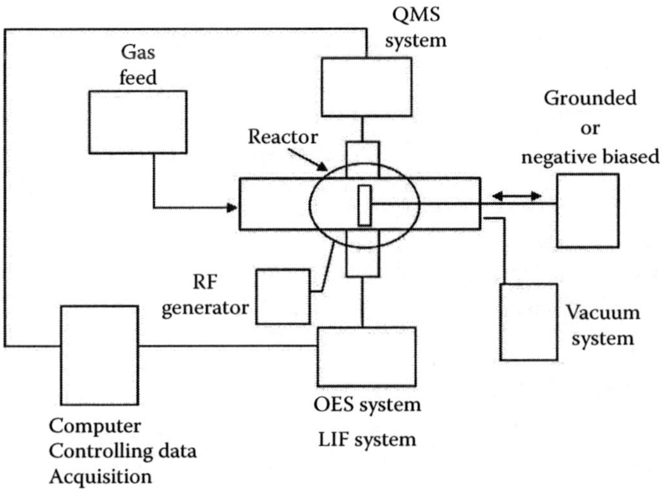

FIGURE 3.18
Experimental set-up for plasma-assisted chemical vapor deposition-radio frequency (PACVD-RF) plasma. (Reprinted from Avni, R., Fried, I., Raveh, A., and Zukerman, I., Plasma surface interaction in PACVD and PVD systems during TiAlBN nanocomposite hard thin films deposition, *Thin Solid Films* 516 (16), 5386–5392, Copyright 2008, with permission from Elsevier.)

on a wide range of interactions between the plasma and surface during plasma-assisted processes. Interactions between the plasma and surface as well as interaction in the plasma volume determine many properties of the film. Prior to analyzing the interaction between the plasma and surface and interactions within plasma volume, we deal with growth stages of a thin film and their (stages) changes due to using plasma:

First Stage: Creation of Depositor Particles

Using nonisothermal plasma makes us able to accelerate reactions through thermodynamic effects. Reactions of a PACVD happen due to a decrease in deposition temperature or via an increase in the velocity of deposition of a certain chemical reaction while they progress very slowly in typical methods.

Second Stage: Transferring Particles onto the Substrate

Controlling volume transfer in nonisothermal plasma is impossible because of the creation of energetic particle flow in the plasma.

Third Stage: Film Growth on the Substrate

Bombardment of substrates by particles with high kinetic energy annihilates inactive films over the substrate, which may have a preventive effect against chemical reactions. Similarly, it gives rise to heterogeneous reactions of absorbed molecules over the substrate surface.

TABLE 3.3

Some Reactions in the Volume of a Plasma Medium

	Process	Appearance Potential (eV)[1]
Impact ionization	$A+\overline{e}\rightarrow A^{*}+2\overline{e}$	15.75
Excitation	$A+\overline{e}\rightarrow A^{*}$	11.56
Dissociation	$AB+\overline{e}\rightarrow A+B+\overline{e}$	—
Dissociative attachment	$AB+\overline{e}\rightarrow A+B$	—
Dissociative recombination	$AB^{*}+\overline{e}\rightarrow A+B$	—
Dissociative recombination	$AB+\overline{e}\rightarrow A^{*}+B+2\overline{e}$	—

[1] Typical

Furthermore, such an effect may lead to retransmission of particles from the substrate surface. On the other hand, the microstructure of films may be controlled during the growth process because of interactions between low-energy ions and surfaces.

3.8.3.2 Main Processes of Plasma Medium

Collision processes in all plasma media are categorized into two classes: elastic and nonelastic processes. An elastic collision is a collision in which only kinetic energy interchange occurs. However, this kind of energy transmission is skipped because electrons and atoms have different masses. In these collisions only routes of the electron change, while its internal energy changes in a nonelastic collision.

Striking and ionization of electrons are the most applied processes in preserving a plasma medium in which the initial electron releases an electron from an atom and produces a positive ion and two electrons:

$$\mathrm{A} + \bar{\mathrm{e}} \rightarrow \mathrm{A} + 2\bar{\mathrm{e}} \tag{3.8}$$

The important point about decomposition reactions is that by-products of reactions are more active than initial molecules, and this chemical activity is the key point in most plasma-assisted processes. Moreover, by-products of decomposition reactions are not able to recombine with each other upon collision of two components in gas phases because of the law of conservation of energy and momentum. In conclusion, such radicals survive more in plasma media, and their density is determined by performed reactions across the surface.

In addition to the aforementioned reactions, collision between ions and neutral particles in plasma volume is possible, which may result in more ionization or facilitation of load transmission. Some significant reactions of this process are listed in Table 3.4.

Briefly, the main reactions that occur in PACVD are ionization due to electron collision, decomposition reaction, and recombination and reaction

TABLE 3.4

Reactions between Ions and Neutral Particles in Volume of Plasma Media

	Process	Appearance Potential (eV)[1]
Associative detachment	$A^- + B \rightarrow AB + \bar{e}$	15.75
Abstraction	$A + BC \rightarrow AB + C$	11.56
Atom-atom collision	$A + B \rightarrow A^* + B + \bar{e}$	
Penning dissociation	$M^* + A_2 \rightarrow 2A + M$	
Penning ionization	$M^* + B \rightarrow M + B^* + \bar{e}$	11.5 and 11.7
Electron transfer	$A + B \rightarrow A^* + B^*$	
Charge transfer	$A + B^* \rightarrow A^* + B$	
Metastable–metastable ionization	$M^* + M^* \rightarrow M + M^*$	11.55
Electron metastable ionization	$M^* + \bar{e} \rightarrow M^* + 2\bar{e}$	4.21

[1] Typical value for Argon

between ionized particles and radicals over the substrate surface. Because of numerous reactions that may take place simultaneously in plasma media, their mechanisms often remain complicated and unknown. Although knowledge of the reaction mechanism is important, the complicated or unknown nature of them has not led to the limitation of applications of plasma-assisted methods.[67–69]

3.8.3.3 Kinetic and Thermodynamic Factors

As mentioned previously, the decrease in deposition temperatures is one of the main advantages of PACVD to other typical methods. Table 3.5 indicates temperature and deposition velocity of some prepared films by plasma in low pressure.

According to some researchers such advantages are due to kinetic and thermodynamic effects. Kinetic effects, which are said to occur as a result of decreased deposition temperature or velocity and sometimes are regarded in terms of thermodynamic aspects, happen very slowly in the absence of plasma. Table 3.6 represents this kinetic effect on several reactions with negative $\Delta G^0{}_r$. Deposition velocity is zero for such reactions because their activation energy is high. However, in the presence of plasma (flow density: 10 mA.cm^{-2}) all aforementioned reactions (Table 3.5) happen quickly. This kinetic effect occurs because of the development of a new route for performing the reaction that is participated in by ions or exited particles, and they cope with kinetic obstacles against reaction performance.

Thermodynamic effects described by some researchers emerge under the influence of using severe discharge in high flow density (approximately 10 mA.cm^{-2}) and relying on chemical reactions within the plasma, which enjoy

TABLE 3.5

Deposition Temperature and Velocity of Some Prepared Thin Films by Plasma

Solid	T_1(°K)	r_4(cm s^{-1})	Reactants
Mo	≤523		$Mo(C)_4$
Ni	1073–1273		$Ni(CO)_4$
i-C	≥523	$10^{8} - 10^{-5}$	C_nH_m
C (graphite)	523–733	$\leq 10^{-5}$	$C(s)$-H_2:$C(s)$-N_2
Ovides SiO_2	673–733	$10^{8} - 10^{6}$	$Si(OC_2H_4)_4$-SiH_4-O_2N_2O
Al_2O_3	523–1273	$10^{8} - 10^{7}$	$AlCl_3$-O_2
Nitrides Si_3N_4	673–973	$10^{8} - 10^{7}$	SiH_4-N_2NH_3
TiN	523–1273	10^{-8} to 5×10^{6}	$TiCl_4$-H_2+N_2
BN	673–973		B_2H_4-NH_4
Carbides SiC	473–773	-10^{8}	SiH_4-C_nH_m
TiC	673–873	$5 \times 10^{-8} - 10^{6}$	$TiCl_4$-$CH_4(C_2H_2)$ +H_2
B_4C	673^{10}	$10^{8} - 10^{-7}$	B_2H_6-CH_4

[1] Optimum temperature.

high energy. Actually, the effective factor is that in such a medium, reversible reactions including atoms and multiatomic radicals are predominant reactions. Such reversible reactions take place in a high percentage in the plasma and allow the chemical equilibrium, as reversible chemical reactions are not observed in typical methods of thermal equilibrium. This thermodynamic

TABLE 3.6

Kinetic Effect of Plasma

Reaction (T = 500°K)	$\Delta G^0{}_r$ (500°K) (kcal mol^4)	log K_p
$SiCl_4 + O_2 \rightarrow SiO_2(s) + 2Cl_2$	–18.4	8.05
$TiCl_4 + O_2 \rightarrow TiO_2(s) + 2Cl_2$	–35.8	15.7
$AlCl_3 + NO_2 \rightarrow \frac{1}{2}Al_2O_3(s) + \frac{3}{2}Cl_2$	–47.6	20.8
$CO + \frac{1}{2}C(s) \rightarrow \frac{1}{2}CO_2$	–65.8	28.8
$SiH_4 \rightarrow Si(s) + 2H_2$	–17.3	7.54
$PH_4 \rightarrow P(s) + \frac{1}{2}H_2$	–6.8	2.99
$\frac{1}{2}B_2H_6 \rightarrow B(s) = \frac{3}{2}H_2$	–15.4	6.73
$SiH_4 + \frac{4}{2}NH_3 \rightarrow \frac{1}{3}Si_3N_4(s) + 4H_2$	–65.0	28.4
$\frac{1}{2}B_2H_4 + NH_3 \rightarrow BN(s) + 3H_2$	–70.2	30.7

effect is used for chemical transfer of solids under plasma conditions for creating amorphous and crystalline thin films.

3.8.3.4 Interactions between Plasma and Surface

We analyzed current reactions in the plasma volume that promote the PACVD process along with changing first and second stages of the film growth. Interactions between plasma and surface constitute the second application of plasma to develop thin films that will lead to the development of the third and final stages of the film growth process.

A layered area near the substrate surface in contact with plasma is created because of different mobility of electrons and ions whose electrical load is negative in contrast to plasma potential (*V*p) and its energy is 5 to 30 eV. For PACVD, the field potential (*V*s) may be used to improve morphology of the film growth through changing DC or RF voltage in the 30 to 300 eV range, and it may change (up to several keV) for improving adhesion properties of films.

Actually, bombardment of substrates by ions, which is done either prior to the film growth or during the film growth, makes it possible to control both chemical structure and layered microstructure that has been deposited directly. Usually, such controls are not possible in typical CVD methods. In Figure 3.19, represented interactions change along with changes in

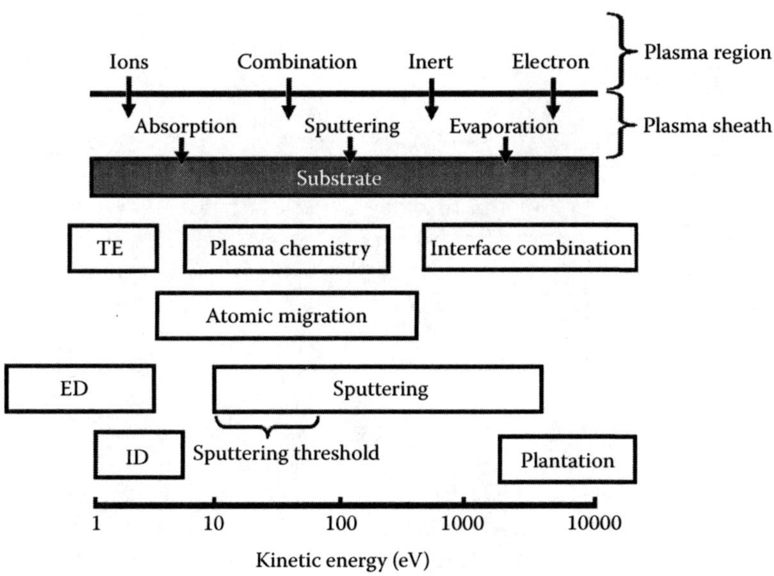

FIGURE 3.19
Current interaction within plasma: TE, thermal electrodes; ED, empty data; and ID, intermediate data.

applied voltages on substrates. The key point is that the applied energy on the substrate may be equal to or highly more than the characteristic energy of a solid.

Solid characteristic energy may refer to the bonding energy of atoms across the surface or sublimation energy (Us) (3 to 10 eV) or sputtering threshold energy (approximately 4 Us); hence, changing many chemical, physical, and mechanical properties of coatings under ionic bombardment of raising films will be feasible. Such effects are divided into two classes:

1. Bombardment by energetic neutral ions or particles
2. Bombardment by low-energy neutral ions or particles

3.8.3.4.1 Bombardment by Energetic Neutral Ions or Particles

Ions or energetic particles are able to penetrate into substrates or growing films. They may be imprisoned within substrates or growing films or to make atoms of substrates move back to films or vice versa. And, they are able to increase adhesion between films and substrates. The possibility of penetration or imprisonment of an ion is increased through decreasing throwing mass and increasing energy. Also it is possible to increase the quantity of current reactions across the surface through breaking chemical bonds and creating adsorption areas. Likewise for preparing surfaces, a preliminary operation is carried out by using a sputtering process that is developed by bombardment with energetic ions; such preparation enhances adhesion of film to substrate.

Using plasma-assisted methods makes us able to control the interface area more than typical coating methods.

3.8.3.4.2 Bombardment by Low-Energy Neutral Ions or Particles

Bombardment by low-energy ions and particles (lower than 100 eV) is performed for controlling the quality of film growth and hence controlling physical properties of the film produced by PACVD. By controlling the applied voltage on substrates, we can control the possibility of exposing film to the bombardment during the growth process which in turn affects density, grain size, and texture and chemical structure of films produced via PACVD. In typical CVD methods, the microstructure of produced films is determined mainly by deposition temperatures. Increased temperature due to application of CVD increases superficial penetration of adding atoms to the surface, which in turn will result in removing atomic shadowing effect and creation of an aggregated film; this effect is found in PACVD via bombardment of growing film with low-energy ions, which gives rise to creation of an aggregated and hole-free film. Indeed in PACVD, atoms move via bombardment by low-energy ions instead of increasing temperature. Similarly, bombardment of the surface with low-energy ions makes us able to resputter atoms from the growing film and redistribute them across different areas of substrates, which will lead to creation of a uniform film. In addition to improved microstructure, sputtering by low-energy ions removes impurities from growing films and alters the chemical composition of films.

It is worth mentioning that the energy of ions is not only the determinant factor of chemical structure of produced films, but concentration of gas phases and chemical composition of plasma are very important in this case, too. However, bombardment by low-energy ions leads to an increase in internal stresses of produced films that may break the coating suddenly, or it may alter physical properties of films through imposing point defects due to radiation harms. These alterations usually are not desirable, but bombardment by low-energy ions makes a condition in which producing certain coatings is easier than other typical methods.

3.8.4 Discharge System

Discharge systems usually include vacuum pumps, total pressure control systems, and a filter. The selection of a vacuum pump depends on the type of processes (discharge capacity, required pressure range, and the gas that should be discharged). Due to the activity of gases used in CVD, selecting vacuum pumps containing oil is important because of their capability to control total pressure. There are several tools to control total pressure that are selected based on the process. Finally, toxic, corrosive, and explosive gases that may be produced during the process are filtered prior to discharging. A combination of different filters is used for some combined processes in order to minimize concentration of hazardous gases.

3.8.5 Advantages of the CVD Method

1. The thickness of the deposited film is uniform and is independent from substrate forms/shapes.
2. Deposition velocity is relatively high and may reach 10 to 100 nm/sec.
3. The composition of the coating is easily controllable and stoichiometric compositions may be prepared through controlling the process parameters.
4. Adhesion and cohesion in CVD are more than PVD methods. High temperatures and their effects on acceleration of distribution as a factor to control reactions are of great importance.
5. Numerous parts may be coated in an individual process.
6. CVD is very suitable for multiple and mixed films of tools in mass productions.
7. Deposition temperatures of refractory and hard materials are lower than their sintering and melting points.
8. Sedimentation of films in reference directions of gradation is possible.

9. Grain dimensions are controllable in films.
10. Porous and dented parts are coated by this method very well.
11. The process is implemented easily in typical pressures.
12. Grains that may grow over microscopic structures of another grain can rise in this method.
13. Doping films with controlled amounts of impurities is very simple.
14. Uniform epitaxy films with minimum impurities are produced.[70–72]

3.8.6 Limitations and Disadvantages of CVD

1. The necessity of high temperatures for reaction of coating is the limitation of this method that affects substrates negatively.
2. Because of using H_2 in most of CVD processes as the reductive factor for coating materials and also for preparing evaporation and reaction, with regard to high temperatures of chambers, hydrogen explosion in presence of oxygen is possible. So, a watering system is essential, and required cautions must be considered.
3. The coatings consist of volatile metallic chlorides and fluorides as their main materials. The presence of chloride and other gases may be hazardous, poisonous, and problematic because of the high temperature of the chambers. So, a great deal of skill is required to use gases and to control pores. Chloride gas is a suffocating and detrimental gas for the body, respiratory systems, and the environment.
4. Some reactive materials are expensive and rare, such as volatile chlorides of Ti, Zr, Al, and Si.
5. There are limited available reactions in CVD in low temperatures.
6. There are numerous limitations for coating materials because of reactions and the necessity of evaporation of coating materials in atmospheric pressure.
7. Substrates may be damaged because of the corrosive nature of current vapors in deposition processes. Also, gas feeding lines and other parts of the machine may be corroded. Thus, volatile products will be produced that take part in film growth as an impurity.
8. In high temperatures, some undesired processes may be induced during deposition; however, it is worth mentioning that selection of substrates will be limited in high temperatures.
9. Controlling uniformity of deposition is very difficult.
10. Thermodynamic and kinetic details of reactions of this process are intricate.
11. By-products of the reaction may be deposited in undesired areas of deposition systems, and sometimes removing them will be impossible.[73,74]

3.9 Molecular Beam Epitaxy (MBE)

Molecular beam epitaxy (MBE) is used to construct parts of microwaves and optoelectronics. MBE has many capabilities. Any MBE system requires certain equipment. This section describes this technique and reviews requirements of MBE.

3.9.1 Ultrahigh (UHV) System

The MBE process is an oblique growth in an ultrahigh vacuum (UHV). Several molecular beams with different flux density make chemical bonds with heated monocrystalline substrates. Figure 3.20 shows a schematic sample of MBE growth.

There are chemical elements or their compounds that are necessary for growth in cells. The temperature of each cell is high enough for generating thermal energy. Epitaxy with desired chemical composition is generated through choosing proper evaporative cells and also appropriate temperature

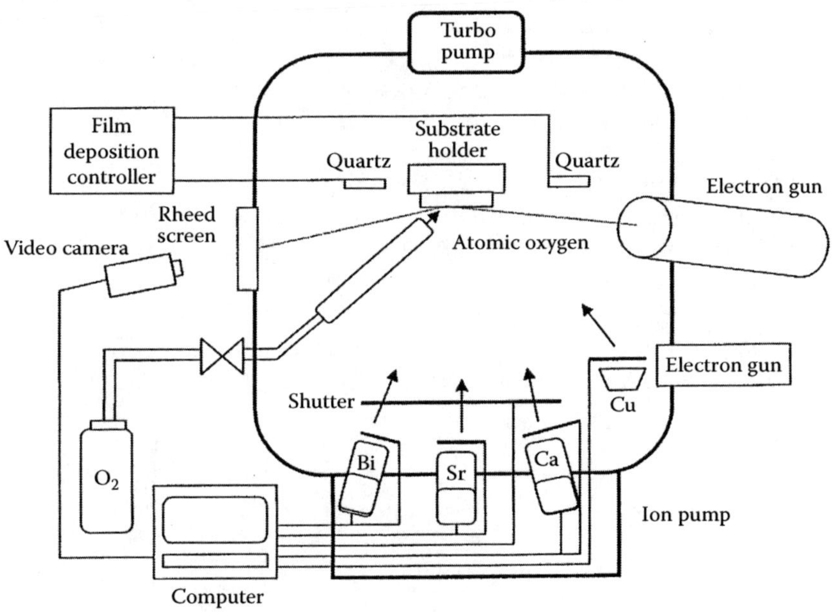

FIGURE 3.20
Molecular beam epitaxy (MBE) growth chamber. (Reprinted from Moussy, J. B., Laval, J. Y., Xu, X. Z., Beuran, F. C., Deville Cavellin, C., and Laguës, M., Percolation behaviour in intergrowth BiSrCaCuO structures grown by molecular beam epitaxy, *Physica C: Superconductivity and Its Applications* 329 (4), 231–242, Copyright 2000, with permission from Elsevier.)

TABLE 3.7

Specifications of Analyzed Kinetic Theory in the Air

Pressure (torr)	Mean Free Path (cm)	Collision Rate (s^{-1})	Impingement Rate ($s^{-1}.cm^{-2}$)	Number of Monolayer (s^{-1})
1.0E-02	5.0E-01	9.0E+04	3.8E+18	4.4E+03
1.0E-04	5.1E+01	9.0E+02	3.8E+16	4.4E+01
1.0E-05	5.1E+02	9.0E+01	3.8E+15	4.4E+00
1.0E-07	5.1E+04	9.0E-01	3.8E+13	4.4E-02
1.0E-09	5.1E+06	9.0E-03	3.8E+11	4.4E-04

for the substrates of films. Moreover, deposition with a proper rate begins via a shutter placed between cells and substrates of films. Deposition with a relative low growth rate (1 μm/h) is one of the specifications of MBE. Therefore, a shutter makes a trivial deposition (which is <1 film/sec) possible. There is no bulk penetration in MBE up to $5A$, and the surface is completely flat. In MBE, the molecular beam is with a neutral electrical load and its energy is equal to 0.1 to 1 eV.

In MBE systems, cryopumps containing liquid nitrogen are used, and a lower pressure may be accessible by using cryopumps containing helium.

MBE pressure is about 10^{-13} to 10^{-11} Torr, and its pressure in working time is 10^{-9} torr. Regarding the information resulting from Table 3.7, it is observed that there is almost no collision with remained atoms or molecules within chambers.

For different evaporation rates with changing pressure of chambers, the numbers of molecules of remained gases, which lie over the surface, are different from the numbers of molecules of evaporated materials that lie over the surface. Table 3.8 shows these alterations.

$$K = \text{number of molecules of remained } N \text{ gas that lie over the surface}/N \text{ evaporated molecules that lie over the surface} \quad (3.9)$$

TABLE 3.8

Alterations of Evaporation Rate through Changing Chamber Pressure, Number of Gas Molecules in Fixed K

	R(nm/s)			
P(torr)	1.0E-01	1.0E+00	1.0E+01	1.0E+02
1.0E-09	1.0E-03	1.0E-04	1.0E-05	1.0E-06
1.0E-07	1.0E-01	1.0E-02	1.0E-03	1.0E-04
1.0E-05	1.0E+01	1.0E+00	1.0E-01	1.0E-02
1.0E-03	1.0E+03	1.0E+02	1.0E+01	1.0E+00

3.9.2 Substrate Heater and Preserver

Preserving the substrate to the surface that supplies its necessary temperature is very important to ensure uniformity and iteration of properties of films. The maximum allowed temperature alteration for MBE is about ±10°C. Also, the substrate must be preserved in a way that is not pressurized and does not make impurity on the film. For III-V compounds it is installed on the heater with an indium wafer.

Refractory materials with high conductivity including Mo are used to bring about temperature uniformity. Sometimes, using Ga has been reported instead of In. However, Ga makes corrosion on the Mo. For IV-IV compounds Tin is used to preserve substrates. In growth temperatures, which are lower than 600°C, vapor pressure of such liquids is very low. If heating is not applied on the substrate uniformly it creates dislocation in the raised film. According to some researchers, to preserve Si, graphite was used to preserve the substrate that not only creates proper thermal contact but also prevents from doping.

3.9.3 Source Cell for MBE

MBE is a method for deposition of physical evaporation that takes place preferably in an ultrahigh vacuum. In this method the molecular beam flux of one or more elements or their combination are placed in the thermal source and are evaporated toward a warm substrate. Atoms or molecules are partially or completely condensed on the substrate and make a chemical bond with the substrate and form deposition films. Film quality depends on some factors including source structures, temperature of substrate surfaces, beam flux values, and growth rates that should be determined experimentally. Using MBE makes us able to control film composition, crystallization, and crystal purity; formation of a film through injection; and also flatness of morphology of surface up to an atomic layer. During the MBE process, a beam flux of an evaporative source should be controlled precisely. Vapor, liquid, and solid states are in an equilibrium state in MBE cells. Given to the kinetic theory, molecular flux is as

$$J = \frac{a}{2\pi\sqrt{WKT}} = 3.51\times10^{22}\frac{P}{\sqrt{WT}} \tag{3.10}$$

where J is vapor flux (molecule/cm^2/sec.), P is pressure (Torr), W is the molecular weight of vapor, and T is the cell temperature (Kelvin).

Molecular beam flux output of the MBE cell valve is $V = JA$, where A is the valve area, J is the vapor pressure, and V is the output vapor pressure from the valve. All II-IV compounds are evaporated as

$$MX_{(s)} \rightarrow M_{(g)} + \frac{1}{2}X_2(g) \tag{3.11}$$

that is, vapor for II and VI groups will be single or dual. The equilibrium constant for the reaction in Equation (3.10) depends on the vapor pressure.

$$K_P = P_m^2 .P_{X2} \tag{3.12}$$

For II-IV compounds, major elements are very volatile compared with their combinations. Their evaporation is much closer to each other, so we have

$$J_{Mx} = J_M = 2J_{X2} \tag{3.13}$$

Combining the above-mentioned equations we will have protect it better.

$$J_{Mx} = 4.42 \times 10^{22} Kp^3 / T^{\frac{1}{2}} w_M^{\frac{1}{3}} w_{X_2}^{\frac{1}{6}} \tag{3.14}$$

Given aforementioned notes, the key factors necessary for designing MBE cells are cleanness, isothermality, and easy application. A cylindrical cell body made of spectroscopic graphite has a cleavage on the cylinder that regulates output rates and concentrates output vapor on the substrate. The tungsten heater (W) crosses the alumina insulator, and it is preserved by Ta, which is a thermal reflector. A nitrogen cooler preserver that consists of copper planed coated covered by gold reflects heat while condenses dopings of output gases.

3.10 Ion Beam Assisted Film Deposition

3.10.1 Partially Ionized Molecular Beam Epitaxy (PI-MBE)

Figure 3.21 shows a schematic picture of the MBE method. This unique technique prepares doped films with good controllability. Here an atom is partially ionized in a molecular beam of film materials (M) and became M^+ and is implanted in the doped sample as

$$M^+ + X \rightarrow M + X^+ \tag{3.15}$$

By applying a negative voltage on the substrate, M^+ and X^+ are accelerated and fall on the substrate along with neutral beams.

In this process, because of implanted ions an extra energy is generated in the superficial area where the deposition process promotes. This extra energy enhances migration of atoms and declines temperature of epitaxy. In line with the development of electronic industries and manufacturing microstructures, controlling the development procedure has gained more importance and attention.

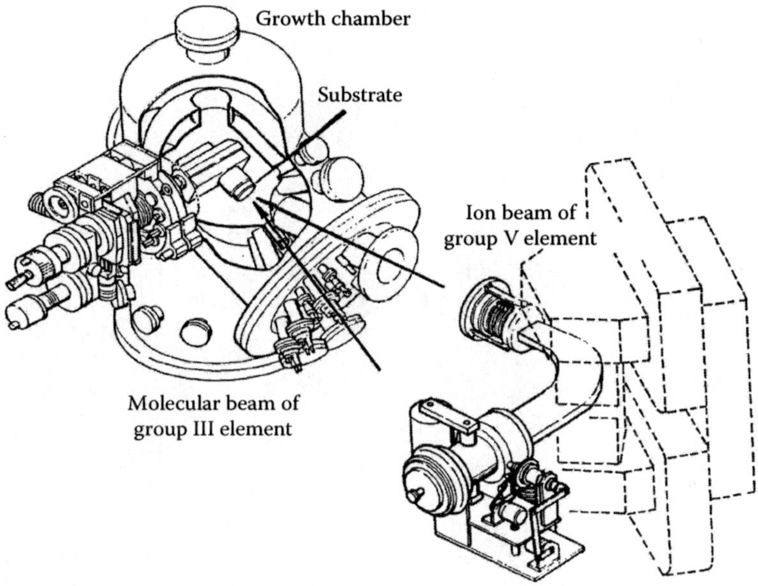

FIGURE 3.21
Molecular beam epitaxy (MBE) growth system. (Reprinted from Shimizu, S., Tsukakoshi, O., Komiya, S., and Makita, Y., Molecular beam epitaxy of InP using low energy P^+ ion beam, *Japanese Journal of Applied Physics, Part 2: Letters* 24 (2), 115–118, Copyright 1985, with permission from Japan Society of Applied Physics.)

For manufacturing advanced microelectronic parts, typical methods are not able to develop certain epitaxy films. PI-MBE is a modern method that meets the requirements of today's technology. A Si epitaxy film with good crystalline quality, for example, may be developed in a low-temperature substrate. Film thickness is controllable by this technique. Similarly, findings show that it is possible to control condensers through the crucible temperature of impure samples. Ion implanting and MBE have succeeded in adding dopers and crystal growth, respectively, but ion implanting in association with MBE is very complicated.

3.10.2 Implant-Epitaxy Rig

Si implant-epitaxy has been carried out by PI-MBE systems. Ionic pumps are used as vacuum pumps in this method, which declines pressure by 2×10^{-8} Torr and basic pressure reaches 5×10^{-7} during the film growth process. Si is evaporated from the source by using an electronic gun and produces a Si beam; its ionization part includes a ring-shaped tungsten fiber, an anodic lattice (W), and a repeller (Mo). A part of a Si molecular beam is ionized upon crossing this area through thermal bombardment of electrons, which emit from W filament and then are accelerated by the anode potential. The number of ions within Si beams varies according to changes in the emitted flow from the filament.

The furnace temperature is 760 to 880°C when its vapor pressure is about $5 \times 10^{-4} - 5 \times 10^{-2}$ Torr. The substrate connects to a graphite sensitometer (diameter: 70 mm, thickness: 6 mm) and a negative potential (10 KV), and is heated with a RF source (power: 2KW, frequency: 1.2 MHz).

A quartz plane (thickness: 3 mm) has been interposed between the sensitometer and helicoidal RF to make sure about their insulation. Induction heating or RF is a significant technique used in epitaxy phases. Moreover it allows us to heat the substrate uniformly in a large time interval. For example, when the temperature of a substrate is 850°C, changes in temperatures over the substrate that are about ±5°C and are measured in 2 min relate to the Si wafer.

There is a plane-formed radiative shield (Mo) wrapping around the thermal system. Applying negative voltage on the shield leads to emission of secondary electrons from the substrate. In this case, the distance between the substrate and Si source is 20 cm.

Number and density ratio of colliding ions (η) with deposited atoms are calculated by measuring ion flow density in the substrate. If ions are univalent, the ratio is defined as

$$\eta = \frac{i}{RNq} \times 100 \tag{3.16}$$

where i, R, N, and q are measured and substrate ion flow density, deposition rate of film, atomic density of film, and electron load, respectively.

For Si, $N = 5 \times 10^{22}$ atom/cm^2, and i and R are in terms of mA/cm^2 and $\mathring{A}$/sec that are measured experimentally, so we have

$$\eta = 1.25\, i/R \tag{3.17}$$

Implant-ion is used to manufacture p-n junctions of Si solar cells and ohmic electrodes.

3.10.3 IVD Technique

Figure 3.22 shows this technique schematically. The ionic vapor deposition (IVD) technique is used to deposit films. It is employed for both ion implanting and vapor deposition simultaneously, and also is used to produce combined coating films such as metallic nitrides.

To produce such films, metal is deposited through evaporation from a vacuum, which is simultaneously bombarded by molecular ions (e.g., nitrogen) at 25 to 40 kV. The advantage of IVD is formation of a coating (1 μm) that provides a high adhesion between the film and substrate because of

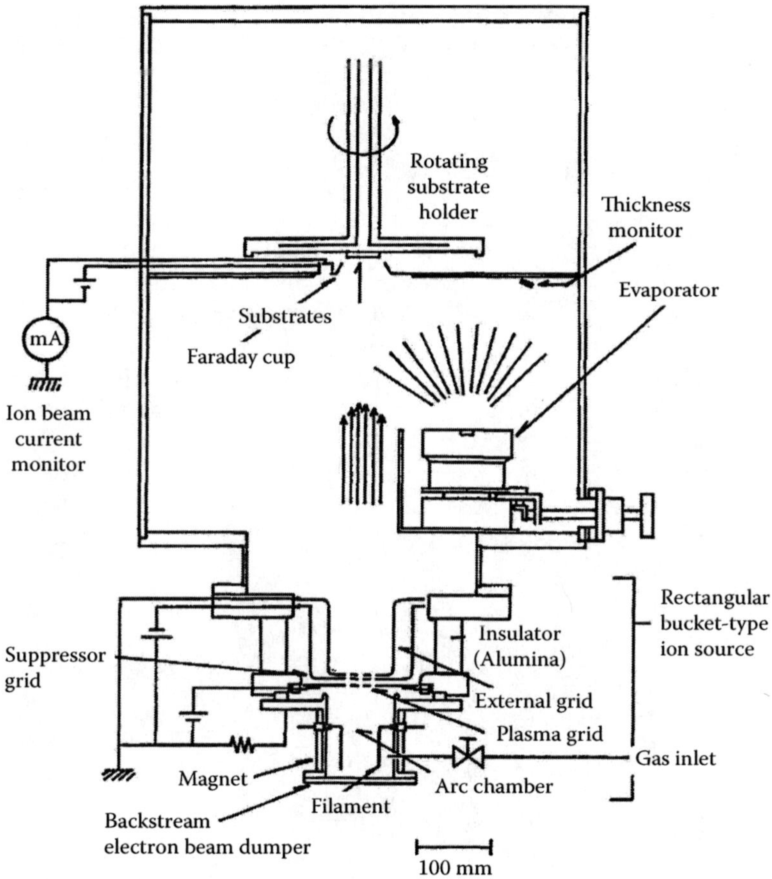

FIGURE 3.22
An ionic vapor deposition (IVD) machine. (Reprinted from Satou, M., Andoh, Y., Ogata, K., Suzuki, Y., Matsuda, K., and Fujimoto, F., Coating films of titanium nitride prepared by ion and vapor deposition method, *Japanese Journal of Applied Physics, Part 1: Regular Papers and Short Notes* 24 (6), 656–660, Copyright 1985, with permission from Japan Society of Applied Physics.)

the development of an intermixed layer in the boundary area between films and substrates. Similarly, IVD lends a hand to form III-V combined semiconductor films, and the problem of controlling stoichiometry would be solved by enhancing the adhesion coefficient of V group elements. In this technique, elements of III group and V group are used as molecular beams and a mass segregator. The ionic beam is pure and is done in order to increase the implant adhesion coefficient in low energy, about 100 eV. Using an ionic implant, with low energy during the film growth process, introduces many intricate phenomena such as radiation-induced, surface roughening, and recoil. Radiation operates sputtering through energetic ions across the film. Thus, in order to achieve thin films, the condition must be selected so that the sputtering rate is not lower than the deposition rate.

3.10.4 Double Ionic Beam (DIB) Technique

In previous literature, using this technique to raise combined films was not reported; however, a refined version of this method was developed and used for the growth of combined semiconductive films. However, there is no difference between them principally. A reactive vessel and a RF coil including discharging plasma have been located within the chamber. Ga and Sb are preserved in crucibles that are resistant against heating. The substrate preservers are also insulators. The density of carriers and dynamicity are measured via the van der Pauw method.

Growth of Sb and Ga epitaxy films with a substrate temperature of 340 to 440°C in hydrogen plasma is possible by this technique. A schematic view of the plasma-assisted epitaxy (PAE) method can be seen in Figure 3.23.

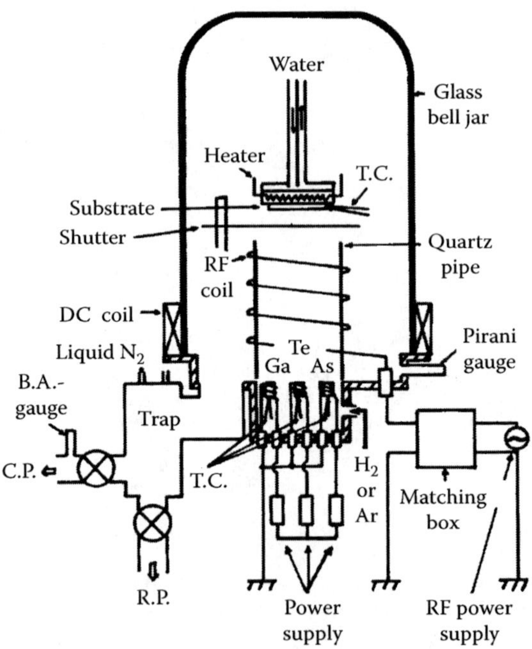

FIGURE 3.23
Plasma-assisted epitaxy (PAE) apparatus. (Reprinted from Matsushita, K., Sugiyama, Y., Igarashi, S., Hariu, T., and Shibata, Y., Heavily Te-doped GaAs layers by plasma-assisted epitaxy, *Japanese Journal of Applied Physics, Part 2: Letters* 22 (9), 602–604, Copyright 1983, with permission from Japan Society of Applied Physics.)

3.10.5 Ionized Cluster Beam (ICB)

This relatively new technique was offered by some researchers, and then researchers began working on improving and extending its applications and advantages.

3.10.5.1 Physical Process

In this technique, initially the material, which should be deposited, is put in the crucible, which usually is in a cylindrical form. And then it will be evaporated by bombardment or direct heating. It is done by a wrapped fiber (W, Re) around the crucible. The transferred current is about 4A. Vapor of atoms builds up in the crucible and the leaving vapor of the nozzles will reach the saturated state due to adiabatic expansion whenever the pressure ratio of the crucible to its peripheral environment is about 10^4. Atoms lose their energy because of frequent collisions and become cold, so they begin to congregate around each other and finally develop nuclei. Nuclei smaller than the critical size are broken into smaller parts, and nuclei bigger than critical size begin to grow and form clusters.

Clusters produced by this technique usually have 500 to 2000 atoms. The number may be obtained by using various methods such as time-of-flight (TOF) and brake electric field methods. It has been demonstrated that few ionized clusters affect developed film features. Thus, in the next phase, left clusters of the crucible are ionized. Electron emission from fiber is used for ionizing clusters (Figure 3.24).

Subsequent to ionization of clusters, they should be accelerated, so applying high voltage (0 to 10 kV) between the substrate and an electrode accelerates them toward the substrate.

3.10.5.2 ICB Rig

Figure 3.25 shows a schematic figure of an ICB rig.

There are controllable parameters in all stages, each of which affects structures and specifications of films. According to the presented classification by some researchers, the ICB technique has two versions: standard and reactive. The reactive version is represented as R-ICB. A reactive gas (e.g., oxygen) is introduced to the system from the point the vapor leaves. Some atoms or molecules of oxygen are ionized and accelerated to substrates in association with ionized clusters. They make a reaction with cluster atoms in the substrate and then lie over the substrate as an oxide layer of that material. We can use one or more crucibles for each version. Several crucibles are used for compounds with vapor pressures of their elements that are not similar.

3.10.5.3 ICB Advantages and Specifications

Recently some semiconductive parts with high-quality crystalline structures have been manufactured by the ICB that are comparable with the MBE

FIGURE 3.24
The ionized cluster ion beam source (two crucible system; 1, acceleration electrode; 2, focusing electrode; 3, electron gun assembly; 4, crucible chamber; and 5, holder). (Reprinted from Cho, S. J., Park, D. K., Kwon, T. W., Yoo, D. S., and Kim, I. G., The role of PI interlayer deposited by ionized cluster beam on the electroluminescence efficiency, *Thin Solid Films* 417 (1–2), 175–179, Copyright 2002, with permission from Elsevier.)

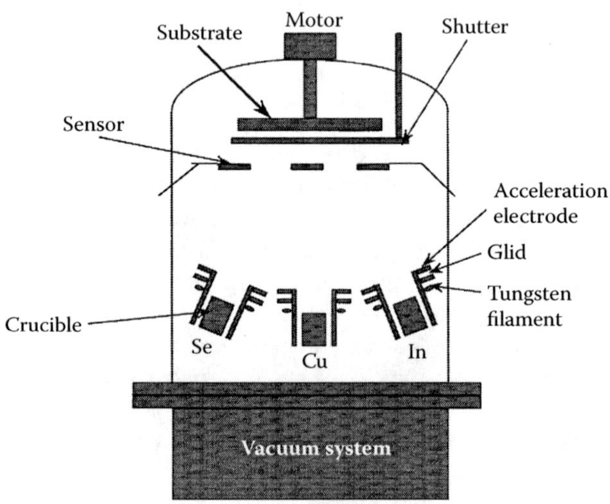

FIGURE 3.25
An ionized cluster beam deposition. (Reprinted from Kondo, K., Sano, H., and Sato, K., Nozzle diameter effects on $CuInSe_2$ films grown by ionized cluster beam deposition, *Thin Solid Films* 326 (1–2), 83–87, Copyright 1998, with permission from Elsevier.)

technique. Even after considering the increased growth rate of them, lower substrate temperature, and greater substrate area, we will learn that the ICB has more advantages in terms of industrial production than MBE. It has made the construction of epitaxy films in low temperatures possible. The following features are the reasons why this method has more advantages over other typical methods:

1. Self-cleaning of surface: Bombardment of ionized clusters on the substrate surface results in cleanliness of the surface, which is done without falling vacuum pressure.
2. Growth of epitaxy is possible in low temperatures.
3. Chemical reactions increase because of increased atomic migration, so laminar growth is possible.
4. The spatial load problem is solved because of the low ratio of e/m in the ionized clustered beam.
5. The metallization problem for the preparation of micro parts is solved through controlling film structures by ICB parameters.

These advantages have made the ICB technique a good option for growth of epitaxy compared with other techniques including MBE, VPE, and LPE.

3.11 Pulsed Laser Deposition (PLD)

It is possible to raise the temperature of the surface if a laser is radiated to a target with sufficient intensity. This method is used not only for evaporation purposes but also to recool crystals and to form epitaxy layers. Refrigeration by pulse laser was implemented in 1975 for the first time to repair harm due to ion substitution. This technique enjoys marked advantages for warming all samples, because it is limited to shallow films and prevents distribution of impurities inside the solid. The depth of laser absorption is proportionate to α^{-1}, where α is the absorption coefficient for the wavelength used in a laser and usually is smaller than the distance the thermal energy penetrates $L = \sqrt{2D\tau}$.

During the radiation period, temperature changes are calculated as $\Delta T = I\tau/C_v\rho L$, where ρ is density, C_v is special thermal capacity, and I is energy intensity (W/cm^2). The refrigeration rate is very large. Beams of consecutive waves of the laser are very appropriate for 10^{-4} S $< \tau < 10^{-1}$ S, but for 10^{-8} S $< \tau < 10^{-7}$ S, which has been achieved from Q-switched lasers, only a multimicron layer of the target was warmed and its refrigeration rate was 10^{10} C/sec. This range is suitable for microelectric applications. Iteration of this process for several times will result in the formation of a desired film. Density of the impurity is minimized, and in this case both penetration and excitation of flow will be dominant. Most of the time, the electrical properties of thin films remain stable by heating after deposition. By increasing the temperature, the rate of the recrystallization process increases, while at room temperature this process proceeds slowly. Protection will be achieved against other reactions via coating the whole surface by this penetrable layer, and the electrical resistance of film remains stable.

Figure 3.26 indicates the main components of a PLD system. A concentration of laser energy on the target, which might be a metal, a semiconductor, a polymer, or a ceramic, evaporates it (the target). It makes deposition of films on the substrates possible.

3.12 Chemical Bath Deposition (CBD)

Because of its simple and cost-effective nature, the application of the CBD has long been under consideration. The CBD may be used to develop uniform thin films of semiconductors, oxide films, and also alloys. It is also able to deposit on large surfaces. In this method, solutions that produce combined thin films

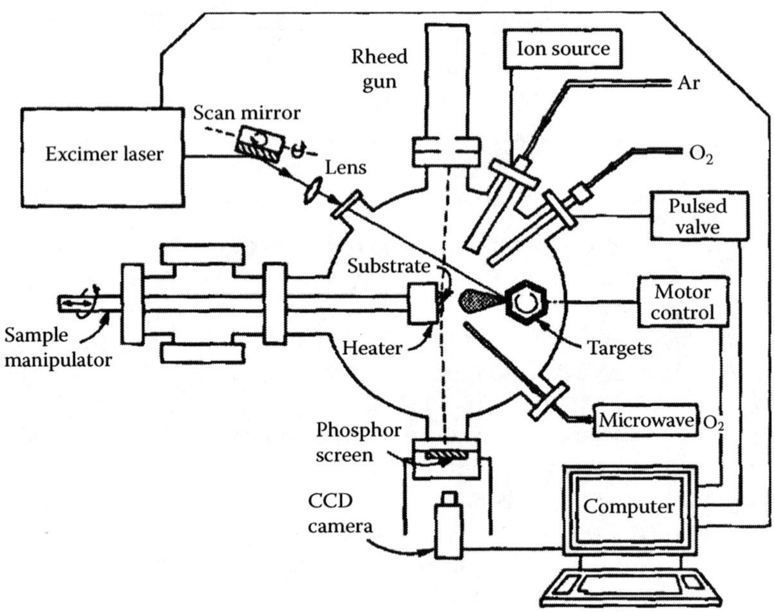

FIGURE 3.26
The pulsed laser deposition system. (Reprinted from Chern, M. Y., Gupta, A., Hussey, B. W., and Shaw, T. M., Reflection high-energy electron diffraction intensity monitored homoepitaxial growth of $SrTiO_3$ buffer layer by pulsed laser deposition, *Journal of Vacuum Science and Technology A: Vacuum, Surfaces, and Films* 11 (3), 637–641, Copyright 1993, with permission from American Vacuum Society.)

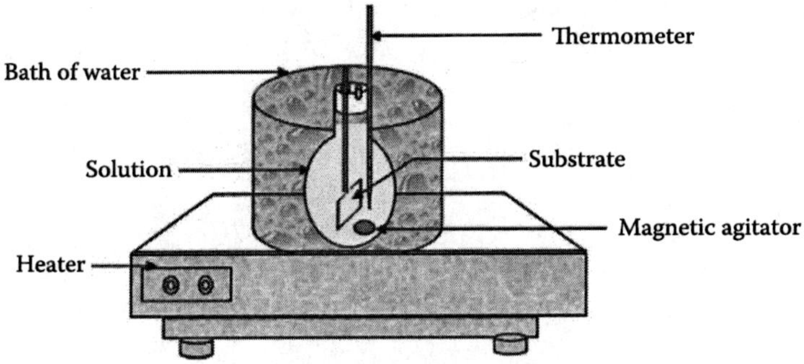

FIGURE 3.27
Chemical bath deposition (CBD) system. (Reprinted from Kamoun Allouche, N., Ben Nasr, T., Turki Kamoun, N., and Guasch, C., Synthesis and properties of chemical bath deposited ZnS multilayer films, *Materials Chemistry and Physics* 123 (2–3), 620–624, Copyright 2010, with permission from Elsevier.)

are poured into a container, and then a substrate is immersed vertically into the container, so the material lies on the substrate uniformly (Figure 3.27).

The CBD depends on controlling sedimentation from solutions of combined materials on the proper substrate. The solution is preserved in basic media. Film thickness and deposition velocity are controllable through changing pH, the solution temperature, and concentration. Being familiar with issues pertaining to the chemistry of the problem is essential to control parameters of thin films in the CBD technique.

For example, for the development of PbS films, concentrations of S^{2-} ions and metallic ions of Pb^{2+} should be controlled so that a good film is obtained. Metallic sulfide as a result of the reaction of the mentioned ions is developed when the solubility product principle already exists. For a saturated solution with low-soluble compounds, the product of molar concentration of its ions is constant for the given temperature. Based on the solubility product principle, there is a certain numerical relation among concentration of ions in a saturated solution, as an electrolyte that is relevant to its solid phase. If this relation does not exist, the equilibrium state will not appear. Therefore, the dissolved material precipitates or the solid will go on to be dissolved by the time the concentration of ions can be applied in the solubility product principle. Sedimentation takes place when the ionic product (IP) is dominant compared with the solution product (SP); hence, the solution will contain more ions than the required rate for the saturated solution. When IP < SP, sedimentation does not occur.

Instantaneous precipitation should be removed in order to form thin films through an ion-by-ion reaction. This is achievable in the presence of a semistable complex of metallic ions, as it controls the number of ions based on an equilibrium reaction as

$$M(A)^{2+} \Leftrightarrow M^{2+} + A \tag{3.18}$$

where M^{2+} is for metallic ions, and A stands for stabilizer.

The concentration of free metallic ions in a given temperature is measured as

$$K_1 = \frac{(M^{+2}).(A)}{(M(A)^{+2})} \tag{3.19}$$

where, K_I is defined as the constant of instability. There is a reverse relation between this constant and stability of metallic complex, so concentration of metallic ions will decline in the solution. The concentration of metallic ions is controlled approximately through controlling the concentration and temperature of complex factors.

Local instantaneous precipitation of a sulfide is possible if there is a high concentration of S^{2-} ions. In this case, SP will be dominant. It may be solved with the first chalcogen ions. It occurs for example by having thio-urea $(NH_2)_2CS$ in a basic solution based on the following reactions:

$$(NH_2)_2CS + OH^- \Leftrightarrow CH_2N_2 + H_2O + Hs^- \tag{3.20}$$

$$Hs^- + OH^- \Leftrightarrow H_2O + S^{2-} \tag{3.21}$$

When the IPs of metallic ions and chalcogen are dominant ions compared with SP, metallic chalcogen (e.g., PbS) is formed by the ion combination process on the surface that has been immersed under the substrate in the solution in order to create nucleation centers. Substrate is immersed vertically in the solution, and then solution is stirred by a magnetic stirrer.

Regarding a description of the solution growth process, the following items are expected:

1. Under the given condition, film should reach a predefined final thickness by all metal ions and chalcogen (i.e., IP < SP).
2. Either deposition rate or thickness of layers depends on the chemical nature of solutions, the complex factor, the concentration, and the temperature of substrates.
3. Though crystallographic structures and microstructures of films depend on created energy by the deposition process, stoichiometry of films does not depend on deposition condition.
4. If the IP of each impure insoluble compound of solution is not more than its SP, the placement of solution impurity inside the film cannot be expected.

In the kinetics of film growth in the CBD, initially embryonic sites by ions are created on the clean substrate, and then the film growth process starts. It is expected that growth kinetics depends on concentration of ions, their velocity and nucleation processes, and growing on the substrate surfaces. Different conditions of deposition that effectively affect these parameters are

1. Type of salt: Growth kinetic depends on salts combinations used for producing chalcogen and metal ions. Although there has not been any principal study in this regard, there has been some research on CdSe and CdS. Generally, it can be said that deposition rate and final thickness for sulfide films are more than those of Se films under similar deposition conditions.
2. Complex factor: Concentration of M^{2+} ions (metallic ion) decreases along with an increased concentration of complex ions. Thus, reaction velocity and hence the precipitation rate decline and create more thickness for the film.
3. pH of solution: If OH^- are placed within the complex form (like $Pb(OH)C_6H_5O_7^{2-}$), by adding OH^- (i.e., increasing pH) the complex becomes more stable and thus concentration of free ions of M^{2+} decreases. As a result, by raising the pH of the solution, the deposition rates decrease, and the final thickness increases.
4. Substrate: When the lattice and parameters of deposited material are adapted with the substrate layer, free energy of nucleation is changed rarely, so nucleation is promoted easily. As a result, a high deposition rate and a high final thickness are observed for such substrates. Figure 3.28 shows this effect for $Pb_{0.86}Hg_{0.14}S$ thin films deposited on different substrates.

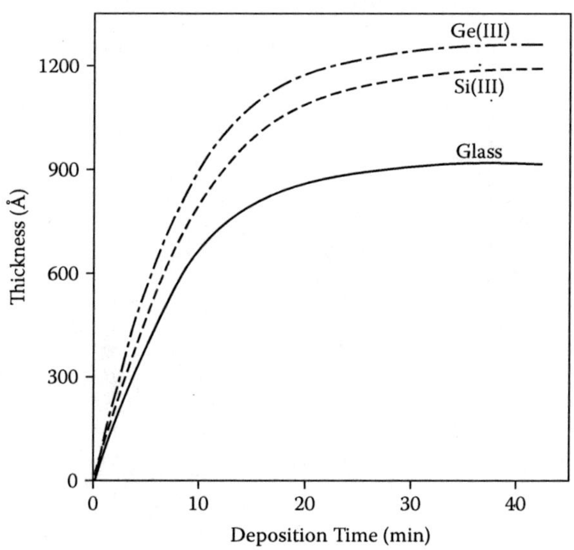

FIGURE 3.28
The thickness as a function of deposition time for $Pb_{0.86}Hg_{0.14}S$ thin films grown on different substrates. (Reprinted from Sharma, N. C., Pandya, D. K., Sehgal, H. K., and Chopra, K. L., Electroless deposition of epitaxial $Pb_{1-x}Hg_xS$ films, *Thin Solid Films* 59 (2), 157–164, Copyright 1979, with permission from Elsevier.)

5. Bath—substrate temperature: Decomposition of complex and chalcogenic compounds (including sixth group elements) speeds up by enhancing the solution temperature. Increasing the concentration of metallic ions and chalcogen along with high kinetic energy of ions raises the interaction between ions and, in turn, the deposition rate. On the other hand, the final thickness, depending on the saturation degree of the solution, rises or declines with increasing or decreasing temperatures.

The final thickness increases initially with the increase in saturation (due to increased concentration of ions), and then in high saturation, in which precipitation and sedimentation are dominant, decreases.

Figure 3.29 represents effects of temperatures on the final thickness of $Pb_{0.86}Hg_{0.14}S$ films.

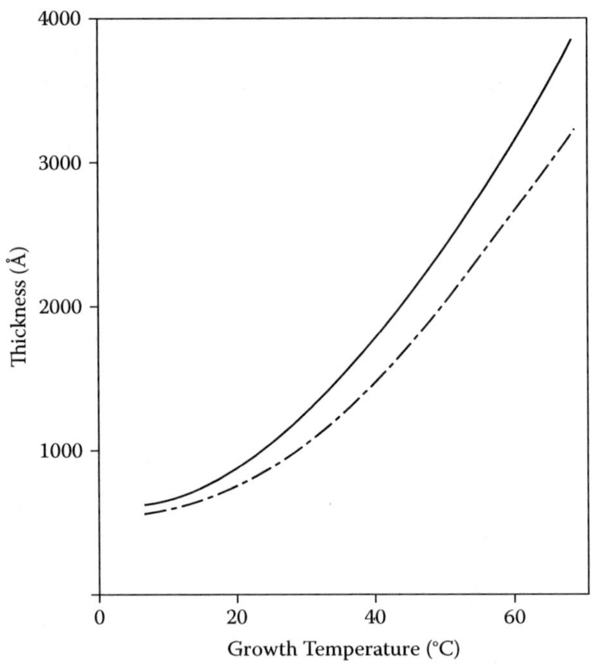

FIGURE 3.29
Final thickness of $Pb_{0.86}Hg_{0.14}S$ films as a function of growth temperature grown on glass substrate. (Reprinted from Sharma, N. C., Pandya, D. K., Sehgal, H. K., and Chopra, K. L., Electroless deposition of epitaxial $Pb_{1-x}Hg_xS$ films, *Thin Solid Films* 59 (2), 157–164, Copyright 1979, with permission from Elsevier.)

References

1. Hajati, S., Zaporojtchenko, V., Faupel, F., and Tougaard, S., Characterization of Au nano-cluster formation on and diffusion in polystyrene using XPS peak shape analysis, *Surface Science* 601 (15), 3261–3267, 2007.
2. Jensen, H., Soloviev, A., Li, Z., and Søgaard, E. G., XPS and FTIR investigation of the surface properties of different prepared titania nano-powders, *Applied Surface Science* 246 (1–3), 239–249, 2005.
3. Maffes, T. G. G., Owen, G. T., Penny, M. W., Starke, T. K. H., Clark, S. A., Ferkel, H., and Wilks, S. P., Nano-crystalline SnO_2 gas sensor response to O_2 and CH_4 at elevated temperature investigated by XPS, *Surface Science* 520 (1–2), 29–34, 2002.
4. Tang, F., Yu, L., Huang, X., and Guo, J., Characterization of adsorption and distribution of polyelectrolyte on stability of nano-zirconia suspensions by Auger Electron Spectroscopy, *Nanostructured Materials* 11 (4), 441–450, 1999.
5. Tougaard, S., Quantitative XPS: Non-destructive analysis of surface nanostructures, *Applied Surface Science* 100–101, 1–10, 1996.
6. Wannaparhun, S., Seal, S., and Desai, V., Surface chemistry of Nextel-720, alumina and Nextel-720/alumina ceramic matrix composite (CMC) using XPS—A tool for nano-spectroscopy, *Applied Surface Science* 185 (3–4), 183–196, 2002.
7. Jungblut, H., Wille, D., and Lewerenz, H. J., Nano-oxidation of H-terminated p-Si(100): Influence of the humidity on growth and surface properties of oxide islands, *Applied Physics Letters* 78 (2), 168–170, 2001.
8. Hill, N. A., Pokrant, S., and Hill, A. J., Optical properties of Si-Ge semiconductor nano-onions, *Journal of Physical Chemistry B* 103 (16), 3156–3161, 1999.
9. Gracin, D., Juraic, K., Dubcek, P., Gajovic, A., and Bernstorff, S., The nanostructural properties of hydrogenated a-Si and Si-C thin films alloys by GISAXS and vibrational spectroscopy, *Applied Surface Science* 252 (15), 5598–5601, 2006.
10. Stauffer, L., Van, S., Bolmont, D., Koulmann, J. J., and Minot, C., Non-equivalence of the adatoms in the DAS model of the Si(111) 7×7 surface, an extended Hückel model calculation, *Solid State Communications* 85 (11), 935–940, 1993.
11. Mari, F., Hideo, N., and Akio, Y., Electronic structure of the DAS model for the Si(111)7×7 reconstructed surface by energy band calculations, *Surface Science* 242 (1–3), 229–232, 1991.
12. Yanagisawa, J., and Yoshimori, A., Analysis of high energy ion scattering for the Si(111)7×7 DAS model by computer simulation, *Surface Science* 231 (3), 297–303, 1990.
13. Fujita, M., Nagayoshi, H., and Yoshimori, A., Electronic structure of the DAS model for Si(111)7×7 by energy band calculations—simplified models, *Surface Science* 208 (1–2), 155–163, 1989.
14. Kanamori, J., Lattice gas model analysis of the (111) surface structures of Si, Ge and related systems. II. On the DAS model, *Journal of the Physical Society of Japan* 55 (8), 2723–2734, 1986.
15. Kawabata, T., Okuyama, F., and Tanemura, M., Fundamental and practical aspects of reactive N^{+}_{2-} ion sputtering in Auger in-depth analysis, *Journal of Applied Physics* 69 (6), 3723–3728, 1991.

16. Levenson, L. L., Fundamentals of auger electron spectroscopy, *Scanning Electron Microscopy* (pt 4), 1643–1653, 1983.
17. Kondratenko, A. V., and Mazalov, L. N., Theoretical fundamentals of high resolution Auger electron spectroscopy, *Chemical Physics* 64 (1), 139–142, 1982.
18. Holloway, P. H., Fundamentals and applications of auger electron spectroscopy, in *Advances in Electronics and Electron Physics* 241–298, 1980.
19. Bauer, E., Low energy electron microscopy, *Reports on Progress in Physics* 57 (9), 895–938, 1994.
20. Libuda, J., Frank, M., Sandell, A., Andersson, S., Brühwiler, P. A., Bäumer, M., Mårtensson, N., and Freund, H. J., Interaction of rhodium with hydroxylated alumina model substrates, *Surface Science* 384 (1–3), 106–119, 1997.
21. Starke, U., Bram, C., Steiner, P. R., Hartner, W., Hammer, L., Heinz, K., and Müller, K., The (0001)-surface of 6HSiC: Morphology, composition and structure, *Applied Surface Science* 89 (2), 175–185, 1995.
22. Kleinle, G., Moritz, W., and Ertl, G., An efficient method for LEED crystallography, *Surface Science* 238 (1–3), 119–131, 1990.
23. Knall, J., Sundgren, J. E., Hansson, G. V., and Greene, J. E., Indium overlayers on clean Si(100) 2×1: Surface structure, nucleation, and growth, *Surface Science* 166 (2–3), 512–538, 1986.
24. Bain, C. D., Troughton, E. B., Tao, Y. T., Evall, J., Whitesides, G. M., and Nuzzo, R. G., Formation of monolayer films by the spontaneous assembly of organic thiols from solution onto gold, *Journal of the American Chemical Society* 111 (1), 321–335, 1989.
25. Gelius, U., Wannberg, B., Baltzer, P., Fellner-Feldegg, H., Carlsson, G., Johansson, C. G., Larsson, J., Münger, P., and Vegerfors, G., A new ESCA instrument with improved surface sensitivity, fast imaging properties and excellent energy resolution, *Journal of Electron Spectroscopy and Related Phenomena* 52 (C), 747–785, 1990.
26. Anpo, M., and Che, M., Applications of photoluminescence techniques to the characterization of solid surfaces in relation to adsorption, catalysis, and photocatalysis, in *Advances in Catalysis* 119–257, 1999.
27. Takeda, S., Suzuki, S., Odaka, H., and Hosono, H., Photocatalytic TiO_2 thin film deposited onto glass by DC magnetron sputtering, *Thin Solid Films* 392 (2), 338–344, 2001.
28. Ingall, M. D. K., Honeyman, C. H., Mercure, J. V., Bianconi, P. A., and Kunz, R. R., Surface functionalization and imaging using monolayers and surface-grafted polymer layers, *Journal of the American Chemical Society* 121 (15), 3607–3613, 1999.
29. Kong, Y. C., Yu, D. P., Zhang, B., Fang, W., and Feng, S. Q., Ultraviolet-emitting ZnO nanowires synthesized by a physical vapor deposition approach, *Applied Physics Letters* 78 (4), 407–409, 2001.
30. Lyu, S. C., Zhang, Y., Lee, C. J., Ruh, H., and Lee, H. J., Low-temperature growth of ZnO nanowire array by a simple physical vapor-deposition method, *Chemistry of Materials* 15 (17), 3294–3299, 2003.
31. Helmersson, U., Lattemann, M., Bohlmark, J., Ehiasarian, A. P., and Gudmundsson, J. T., Ionized physical vapor deposition (IPVD): A review of technology and applications, *Thin Solid Films* 513 (1–2), 1–24, 2006.
32. Narayan, J., Tiwari, P., Chen, X., Singh, J., Chowdhury, R., and Zheleva, T., Epitaxial growth of TiN films on (100) silicon substrates by laser physical vapor deposition, *Applied Physics Letters* 61 (11), 1290–1292, 1992.

33. Rossnagel, S. M., Directional and ionized physical vapor deposition for microelectronics applications, *Journal of Vacuum Science and Technology B: Microelectronics and Nanometer Structures* 16 (5), 2585–2608, 1998.
34. Wang, L., Zhang, X., Zhao, S., Zhou, G., Zhou, Y., and Qi, J., Synthesis of well-aligned ZnO nanowires by simple physical vapor deposition on c-oriented ZnO thin films without catalysts or additives, *Applied Physics Letters* 86 (2), 024108-1–024108-3, 2005.
35. Johnson, C. A., Ruud, J. A., Bruce, R., and Wortman, D., Relationships between residual stress, microstructure and mechanical properties of electron beam-physical vapor deposition thermal barrier coatings, *Surface and Coatings Technology* 108–109 (1–3), 80–85, 1998.
36. Sung, M. F., Kuan, Y. D., Chen, B. X., and Lee, S. M., Design and fabrication of light weight current collectors for direct methanol fuel cells using the micro-electro mechanical system technique, *Journal of Power Sources* 196 (14), 5897–5902, 2011.
37. Ho K. T., Sebastian, P., and Quirante, T., Multiobjective optimization of the design of two-stage flash evaporators: Part 1. Process modeling, *International Journal of Thermal Sciences* 49 (12), 2453–2458, 2010.
38. Harris, M. L., Doraiswamy, A., Narayan, R. J., Patz, T. M., and Chrisey, D. B., Recent progress in CAD/CAM laser direct-writing of biomaterials, *Materials Science and Engineering C* 28 (3), 359–365, 2008.
39. Karlsson, L., Hultman, L., Johansson, M. P., Sundgren, J. E., and Ljungcrantz, H., Growth, microstructure, and mechanical properties of arc evaporated TiC_xN_{1-x} ($0 \le x \le 1$) films, *Surface and Coatings Technology* 126 (1), 1–14, 2000.
40. Puckett, S. D., Lee, P. P., Ciombor, D. M., Aaron, R. K., and Webster, T. J., Nanotextured titanium surfaces for enhancing skin growth on transcutaneous osseointegrated devices, *Acta Biomaterialia* 6 (6), 2352–2362, 2010.
41. Lim, J. H., Kong, C. K., Kim, K. K., Park, I. K., Hwang, D. K., and Park, S. J., UV electroluminescence emission from ZnO light-emitting diodes grown by high-temperature radiofrequency sputtering, *Advanced Materials* 18 (20), 2720–2724, 2006.
42. Coburn, J. W., and Kay, E., Positive-ion bombardment of substrates in RF diode glow discharge sputtering, *Journal of Applied Physics* 43 (12), 4965–4971, 1972.
43. Kim, H. K., Kim, D. G., Lee, K. S., Huh, M. S., Jeong, S. H., Kim, K. I., and Seong, T. Y., Plasma damage-free sputtering of indium tin oxide cathode layers for top-emitting organic light-emitting diodes, *Applied Physics Letters* 86 (18), 1–3, 2005.
44. Aita, C. R., Basal orientation aluminum nitride grown at low temperature by RF diode sputtering, *Journal of Applied Physics* 53 (3), 1807–1808, 1982.
45. Coburn, J. W., A system for determining the mass and energy of particles incident on a substrate in a planar diode sputtering system, *Review of Scientific Instruments* 41 (8), 1219–1223, 1970.
46. Carcia, P. F., McLean, R. S., Reilly, M. H., and Nunes Jr, G., Transparent ZnO thin-film transistor fabricated by RF magnetron sputtering, *Applied Physics Letters* 82 (7), 1117–1119, 2003.
47. Minami, T., Sato, H., Nanto, H., and Takata, S., Group III impurity doped zinc oxide thin films prepared by RF magnetron sputtering, *Japanese Journal of Applied Physics, Part 2: Letters* 24 (10), 781–784, 1985.
48. Yabuta, H., Sano, M., Abe, K., Aiba, T., Den, T., Kumomi, H., Nomura, K., Kamiya, T., and Hosono, H., High-mobility thin-film transistor with amorphous

$InGaZnO_4$ channel fabricated by room temperature RF-magnetron sputtering, *Applied Physics Letters* 89 (11), 2006.
49. Torng, C. J., Sivertsen, J. M., Judy, J. H., and Chang, C., Structure and bonding studies of the C:N thin films produced by rf sputtering method, *Journal of Materials Research* 5 (11), 2490–2496, 1990.
50. Axelevitch, A., Gorenstein, B., Darawshe, H., and Golan, G., Investigation of thin solid ZnO films prepared by sputtering, *Thin Solid Films* 518 (16), 4520–4524, 2010.
51. Svadkovski, I. V., Golosov, D. A., and Zavatskiy, S. M., Characterisation parameters for unbalanced magnetron sputtering systems, *Vacuum* 68 (4), 283–290, 2002.
52. Bugaev, S. P., and Sochugov, N. S., Production of large-area coatings on glasses and plastics, *Surface and Coatings Technology* 131 (1–3), 474–480, 2000.
53. Zhang, Q. C., and Shen, Y. G., High performance W-AlN cermet solar coatings designed by modelling calculations and deposited by DC magnetron sputtering, *Solar Energy Materials and Solar Cells* 81 (1), 25–37, 2004.
54. Kelly, P. J., and Arnell, R. D., Magnetron sputtering: A review of recent developments and applications, *Vacuum* 56 (3), 159–172, 2000.
55. Arnell, R. D., and Kelly, P. J., Recent advances in magnetron sputtering, *Surface and Coatings Technology* 112 (1–3), 170–176, 1999.
56. Wu, J. J., and Liu, S. C., Low-temperature growth of well-aligned ZnO nanorods by chemical vapor deposition, *Advanced Materials* 14 (3), 215–218, 2002.
57. Chhowalla, M., Teo, K. B. K., Ducati, C., Rupesinghe, N. L., Amaratunga, G. A. J., Ferrari, A. C., Roy, D., Robertson, J., and Milne, W. I., Growth process conditions of vertically aligned carbon nanotubes using plasma enhanced chemical vapor deposition, *Journal of Applied Physics* 90 (10), 5308–5317, 2001.
58. Reina, A., Jia, X., Ho, J., Nezich, D., Son, H., Bulovic, V., Dresselhaus, M. S., and Jing, K., Large area, few-layer graphene films on arbitrary substrates by chemical vapor deposition, *Nano Letters* 9 (1), 30–35, 2009.
59. Kong, J., Cassell, A. M., and Dai, H., Chemical vapor deposition of methane for single-walled carbon nanotubes, *Chemical Physics Letters* 292 (4–6), 567–574, 1998.
60. Gorla, C. R., Emanetoglu, N. W., Liang, S., Mayo, W. E., Lu, Y., Wraback, M., and Shen, H., Structural, optical, and surface acoustic wave properties of epitaxial ZnO films grown on (0112) sapphire by metalorganic chemical vapor deposition, *Journal of Applied Physics* 85 (5), 2595–2602, 1999.
61. Ryan, G., Pandit, A., and Apatsidis, D. P., Fabrication methods of porous metals for use in orthopaedic applications, *Biomaterials* 27 (13), 2651–2670, 2006.
62. Kuo, D. H., and Huang, K. W., A new class of Ti-Si-C-N coatings obtained by chemical vapor deposition, Part 1: 1000°C Process, *Thin Solid Films* 394 (1–2), 72–80, 2001.
63. Finnie, P., Li-Pook-Than, A., Lefebvre, J., and Austing, D. G., Optimization of methane cold wall chemical vapor deposition for the production of single walled carbon nanotubes and devices, *Carbon* 44 (15), 3199–3206, 2006.
64. Jones, J. G., Jero, P. D., and Garrett, P. H., In-situ control of chemical vapor deposition for fiber coating, *Engineering Applications of Artificial Intelligence* 11 (5), 619–626, 1998.
65. Pfohl, C., and Rie, K. T., Wear-resistant PACVD coatings of the system Ti-B-N, *Surface and Coatings Technology* 116–119, 911–915, 1999.

66. Avni, R., Fried, I., Raveh, A., and Zukerman, I., Plasma surface interaction in PACVD and PVD systems during TiAlBN nanocomposite hard thin films deposition, *Thin Solid Films* 516 (16), 5386–5392, 2008.
67. Aliofkhazraei, M., and Sabour Rouhaghdam, A., Study of anodic voltage on properties of complex nanocrystalline carbonitrided titanium fabricated by duplex treatments, *Materials Research Innovations* 14 (2), 177–182, 2010.
68. Aliofkhazraei, M., Rouhaghdam, A. S., Ghobadi, E., and Mohsenian, E., Study of shape and distribution of TiO_2 nanorods produced by atmospheric pressure plasma, *Plasma Processes and Polymers* 6 (Suppl. 1), S214–S217, 2009.
69. Mofidi, S. H. H., Aliofkhazraei, M., Rouhaghdam, A. S., Ghobadi, E., and Mohsenian, E., Improvement of surface characteristics by electroplating hard chromium coating post treated by nanocrystalline plasma electrolytic carbonitriding, *Plasma Processes and Polymers* 6 (Suppl. 1), S297–S301, 2009.
70. Matsumura, H., Hasegawa, T., Nishizaki, S., and Ohdaira, K., Advantage of plasma-less deposition in Cat-CVD to the performance of electronic devices, *Thin Solid Films* 519 (14), 4568–4570, 2011.
71. Matsumura, H., Ohdaira, K., and Nishizaki, S., Advantage of plasma-less deposition: Cat-CVD fabrication of a-Si TFT with current drivability equivalent to poly-Si TFT, *Physica Status Solidi (C) Current Topics in Solid State Physics* 7 (3–4), 1132–1135, 2010.
72. Ruige, J. B., Dekker, J. M., Bakker, S. J. L., and Heine, R. J., Insulin and risk of CVD: Advantages and limitations using meta-analysis, *Cardiovascular Reviews and Reports* 21 (1), 32–34, 2000.
73. Ouazzani, J., Chiu, K. C., and Rosenberger, F., On the 2D modelling of horizontal CVD reactors and its limitations, *Journal of Crystal Growth* 91 (4), 497–508, 1988.
74. Schintlmeister, W., Metallwerk Plansee GmbH, R., Kanz, J., and Wallgram, W., Possibilities and limitations for the application of CVD-coated cemented carbide tools and steel tools, *International Journal of Refractory Metals and Hard Materials* 2 (1), 41–43, 1983.
75. Moussy, J. B., Laval, J. Y., Xu, X. Z., Beuran, F. C., Deville Cavellin, C., and Laguës, M., Percolation behaviour in intergrowth BiSrCaCuO structures grown by molecular beam epitaxy, *Physica C: Superconductivity and Its Applications* 329 (4), 231–242, 2000.
76. Shimizu, S., Tsukakoshi, O., Komiya, S., and Makita, Y., Molecular beam epitaxy of InP using low energy P^+ ion beam, *Japanese Journal of Applied Physics, Part 2: Letters* 24 (2), 115–118, 1985.
77. Satou, M., Andoh, Y., Ogata, K., Suzuki, Y., Matsuda, K., and Fujimoto, F., Coating films of titanium nitride prepared by ion and vapor deposition method, *Japanese Journal of Applied Physics, Part 1: Regular Papers and Short Notes* 24 (6), 656–660, 1985.
78. Matsushita, K., Sugiyama, Y., Igarashi, S., Hariu, T., and Shibata, Y., Heavily Te-doped GaAs layers by plasma-assisted epitaxy, *Japanese Journal of Applied Physics, Part 2: Letters* 22 (9), 602–604, 1983.
79. Cho, S. J., Park, D. K., Kwon, T. W., Yoo, D. S., and Kim, I. G., The role of PI interlayer deposited by ionized cluster beam on the electroluminescence efficiency, *Thin Solid Films* 417 (1–2), 175–179, 2002.
80. Kondo, K., Sano, H., and Sato, K., Nozzle diameter effects on $CuInSe_2$ films grown by ionized cluster beam deposition, *Thin Solid Films* 326 (1–2), 83–87, 1998.

81. Chern, M. Y., Gupta, A., Hussey, B. W., and Shaw, T. M., Reflection high-energy electron diffraction intensity monitored homoepitaxial growth of $SrTiO_3$ buffer layer by pulsed laser deposition, *Journal of Vacuum Science and Technology A: Vacuum, Surfaces, and Films* 11 (3), 637–641, 1993.
82. Kamoun Allouche, N., Ben Nasr, T., Turki Kamoun, N., and Guasch, C., Synthesis and properties of chemical bath deposited ZnS multilayer films, *Materials Chemistry and Physics* 123 (2–3), 620–624, 2010.
83. Sharma, N. C., Pandya, D. K., Sehgal, H. K., and Chopra, K. L., Electroless deposition of epitaxial $Pb_{1-x}Hg_xS$ films, *Thin Solid Films* 59 (2), 157–164, 1979.

4

Mechanical Fabrication/Properties of Two-Dimensional Nanostructures

4.1 Introduction

Properties of thin films and tiny structures, with typical sizes in the range of a few microns or below, cannot just be guessed from the properties of bulk samples. This is due to two main effects: (1) materials for bulk mechanical testing usually have sizes that are much larger than the microstructural features, like grains or particles, while in thin films the geometrical and microstructural sizes are usually on the same order of scale. (2) Mechanical behavior is restricted by certain fundamental length scales. Some examples for length scale effects on mechanical properties are shown in Figure 4.1. Elastic properties are recognized by the atomic bonds, with lengths in the range of 1 Å. Plasticity in metals consists of dislocation movements, which are delayed when they try to pass among barriers more closely spaced than about 100 nm. Through fatigue in metals, complex dislocations structures are formed, with typical sizes of a few micrometers. In brittle materials, fracture is started at defects with a critical dimension of several tens of micrometers. As a result, it is predictable that the mechanical properties of a material will basically change as the sample sizes become smaller than these various intrinsic lengths.[1–14]

Figure 4.1 also shows the usual geometrical sizes of materials used in different technological purposes. In particular, it can be noted that microelectromechanical systems (MEMS) and microelectronics fall precisely in the range where fundamental changes in material properties are expected to happen. It is then essential to measure the mechanical properties at a length scale similar to the feature dimensions used in these technological applications. As a consequence, several specialized testing methods have been developed, mainly in the last 15 years, to study mechanical properties in small sizes. The aim of mechanical testing is to characterize properties such as Young's modulus, yield, fracture strength, creep, and fatigue resistance. In some cases, the basics of macroscopic testing can be shifted directly to small dimensions, as for instance in microtensile or nanoindentation testing.

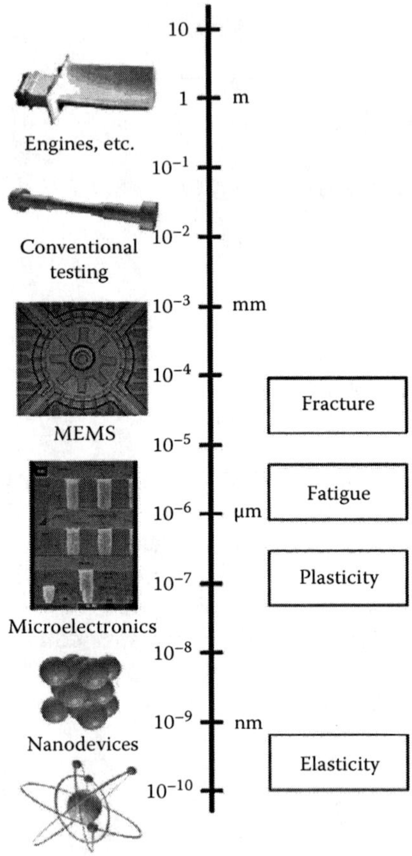

FIGURE 4.1
Length scale effects on mechanical properties of materials. (From Kraft, O., and Volkert, C. A., Mechanical testing of thin films and small structures, *Advanced Engineering Materials* 3 (3), 99–110, Copyright 2001, Wiley-VCH. Reproduced with permission.)

However, new ways of measuring forces and strains in small volumes such as two-dimensional nanostructures are often required.

4.2 Multiple-Layer Coatings

Multiple-layer coatings are made of alternative layers of metals, various alloys, or both. For example, multiple-layer coating of A/B* is composed of alternative layers of A and B, which is schematically shown in Figure 4.2.

* In order to distinguish alloy coatings of A and B, which is shown as A-B, it is decided to use A/B as a sign for multilayer coating.

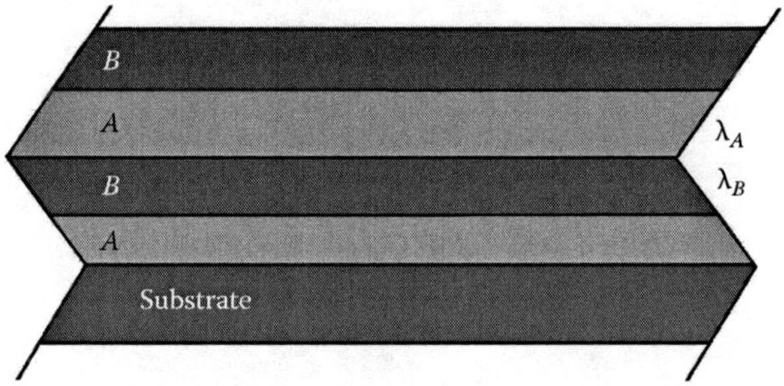

FIGURE 4.2
A multiple-layer coating made of alternative layers of A and B. The thickness of layers A and B and wavelength of the coating are λ_a, λ_b, and $\lambda_a + \lambda_b$, respectively.

Each layer of sample A (or sample B) is of fairly equal thickness of λ_a (or λ_b) in the film. Two layers of these coatings ($\lambda_a + \lambda_b$) can have a thickness of several angstroms to micrometers. Regarding the alternative frequency of the layers in multiple-layer coatings, the thickness gained from two alternative layers is known as coatings wavelength. Generally, there are no clear boundaries between the layers, and the combination of the film is in the form of alternative waves in the coating's length. These coatings are referred to in scientific papers with various names such as multiple-layer thin films, superlattice alloys, or composition-modulated alloys (CMAs).

Nanostructured materials have far better properties than polycrystalline materials. Thus, several studies have been performed in this area. In general, nanostructured materials are categorized into three classes, including co-axis (or three-dimensional [3D]) structures, filament nanostructures (or two-dimensional [2D] nanostructures), and layered nanostructures (one-dimensional [1D] nanostructures). These structures are shown in Figure 4.3. Co-axis nanostructures have nanometric range in all three dimensions, while in layered nanocrystals just the thickness and in filaments nanostructures just the width is in nanometric range.

For their very fine grains sizes, nanostructured materials are mainly unstable; for instance there was reported a significant growth in grain sizes (doubling the crystal size within 24 hours) for single-phase nanostructured crystallites in ambient temperatures or even lower (once the balance melt temperature of the material is lower than 600°C). However, multiple-layer nanostructured materials are somehow more stable than the other nanostructures. Besides, in multiple-layer nanocoatings some unusual, considerable properties such as optical, electrical, mechanical, and magnetic ones are present, which are absent in the other nanostructures. Figure 4.4 illustrates measured changes in atomic force microscopy (AFM) topography and friction force on graphene layers as well as different thicknesses friction behavior for graphene thin sheets with different numbers of atomic layers.[16–23]

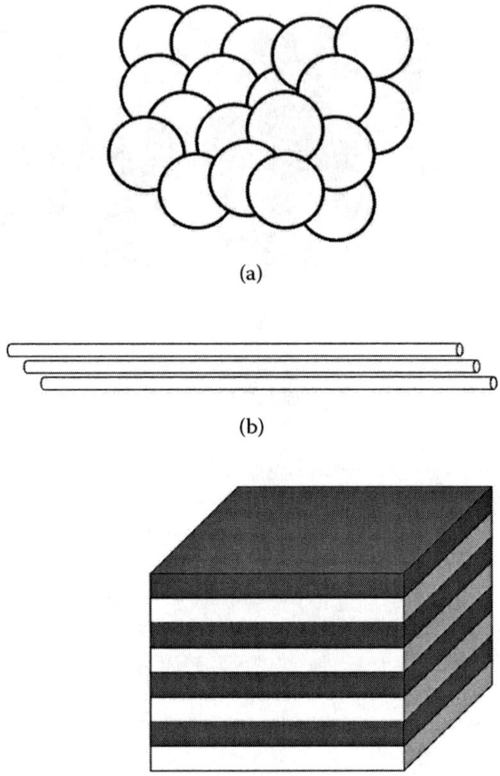

(a) (b) (c)

FIGURE 4.3
Nanostructures: (a) co-axis, (b) filament, and (c) layered.

4.3 Fabrication Methods of Multiple-Layer Coatings

Among the methods for producing multiple-layer structures one can name deposition from vapor phase, electrochemical deposition, sputtering, and chemical vapor deposition (CVD) processes. The rolling of metallic foils or mechanical techniques can also be used to produce layers with thicknesses less than 1 μm. Some researchers have developed the semi-industrial process of vapor deposition with thermal source of electron radiation (Figure 4.5a) and were able to produce alternative layers of Al (with thicknesses of 20 to 1600 nm) and other metals such as Fe, Mg, Mn, Ni, or Ti (with thicknesses of 0.1 to 20 nm).

These coatings are mainly produced using the dry processes and in gas phases. Although in these methods there is not much control in film growth and the material is prepared without any impurities, they need to increase

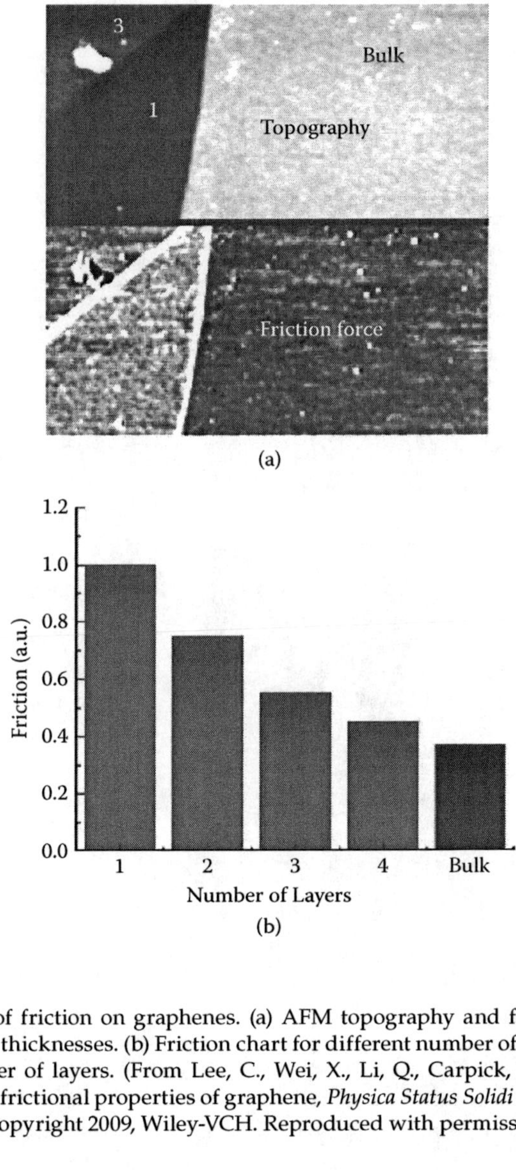

FIGURE 4.4
Layer dependence of friction on graphenes. (a) AFM topography and friction force on graphenes for different thicknesses. (b) Friction chart for different number of layers. The numbers in (a) denote number of layers. (From Lee, C., Wei, X., Li, Q., Carpick, R., Kysar, J. W., and Hone, J., Elastic and frictional properties of graphene, *Physica Status Solidi (B) Basic Research* 246 (11–12), 2562–2567, Copyright 2009, Wiley-VCH. Reproduced with permission.)

temperatures and the vacuum. In addition, the high costs of equipment and the wastes produced by raw materials are among other drawbacks. In contrast, electrochemical deposition is a rather cheap process and leads to less environmental problems compared to gas methods. Besides, this method needs a quite lower temperature, conducting to a lower rate of diffusion between layers and a higher distinction of the layers' boundary. Some other advantages of this method are quick coating, high operational capability, easy control of the composition and thickness of the coating, and simple

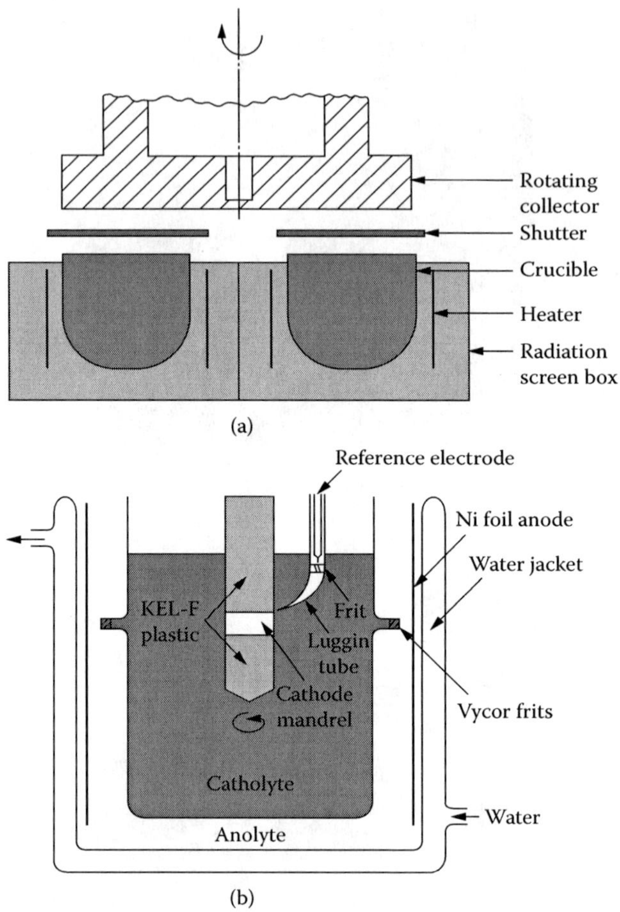

FIGURE 4.5
(a) Vapor deposition rig with spinning collector, and (b) the electrodeposition cell.

equipment. Considering the above-mentioned advantages, the electrochemical method has been widely used to produce nanostructured materials for various applications including magnetic substances, noble gas catalyzers, and so forth.

The electrochemical deposition of the multiple-layer metals can be performed using two separated electrolytes or even, as shown in Figure 4.5b, in a simple process from an electrolyte with efficient control of agitation and electrical conditions (particularly the voltage).

It is obvious that the great deal of time needed to produce multiple layers with significant nano wavelengths and total coating thickness is a barrier to simultaneously using baths. Today this problem is solved through the mechanical cleaning of the samples during the displacement between two solutions. In these systems, the cathode is in disk form and is constantly in rotation. Using this method, the alloying speed can be 2 to 4 μm per hour. The creation of oxide layers during the displacement between two baths is another drawback of using the two-bath method.

In the single-bath method the quality of the produced layer is higher, and there is no need for mechanical equipment for moving samples. However, it must be noticed that all electrolytes or all oxidized or reduced ions in the solution are not compatible with each other, and for some coatings it is necessary to use two baths.[25–30]

4.3.1 Multiple-Layer Electrodeposition Using a Single Bath

In order to perform multiple-layered electrodeposition by one bath, it is required to have 100 mV deposition voltage differences between each of the samples. In the single-bath multiple-layered electrodeposition method, it is necessary to have precise control of each ion's concentration for production of fairly pure samples. The bath commonly starts with preparation of the electrolytes, which have a high speed for deposition of the active species B, and then a small amount of noble species A (about 1% of the concentration of the species B) is added to the bath. Using this method causes lessening of the species A deposition during the B deposition and the development of multiple coatings of A/B with a layer purity higher than 95%.

Generally, in more noble species, displacement of the mass is the controller factor of the electrodeposition rate and in the more active part the kinetic conditions control the electrodeposition rate. Figure 4.6 presents

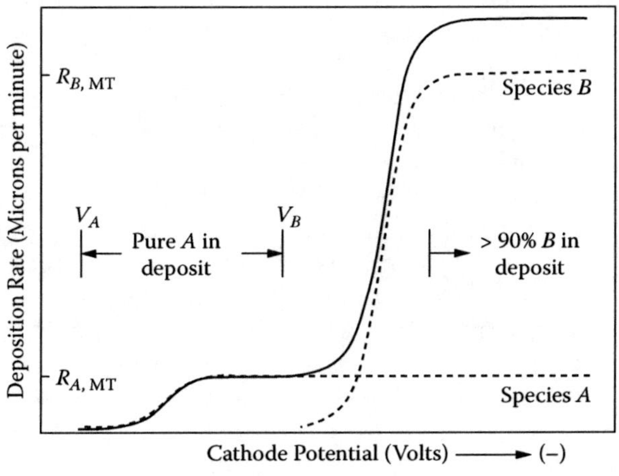

FIGURE 4.6
Ideal deposition rate with applied cathode potential for electrochemical deposition of the multiple-layer coating of A/B. Dashed lines show the deposition rate of each sample, while the solid lines present total deposition rate.

the deposition speed with applied voltage for multiple-layer electrodeposition baths in ideal states. The sample A (which is more noble) and sample B, respectively, deposits in potentials of V_A and V_B. In the voltages among the V_A and V_B, it is possible to perform pure A electrodeposition. Figure 4.6 shows that the sample A deposits with sample B with a limitation of its rate for mass transfer (R_{AMT}).

Once the low current density is used during the electrodeposition, the ratio of sample A is higher, while the ratio of sample B is more in high current density. In the average current density there would be an intermediate state.

The molar percentage of sample B in alloy A-B is controlled by coating potential (or current), agitation intensity of the bath, and concentration of samples A and B in the electrolyte. Stirring during the noble species' electrodeposition process increases deposition rate in the given layer. In contrast, a decrease in the agitation rate of the bath during the sample B deposition would result in an increase of species A's deposition rate (R_{AMT}) and in turn the accumulation rate of the film decreases and layers with higher degrees of purity would be obtained. Furthermore, an increase in the concentration ratio of B to A in the electrolyte directly affects the R_{AMT}-to-R_{BMT} ratio.

Constant fluctuation of the current or the voltage within the area where the pure sample A and A/B alloy deposits leads to the development of a multiple-layer coating of A/A-B. The thickness of each layer is obtained using the Faraday rule by measuring the passing charge (where the current output is also known). However, it must be noted that in multiple-layer electrodeposition of the coating from a single bath, it may be possible for resolution of the active species during the electrodeposition of the noble element.

In production of the multiple-layered coatings, pulse current electrodeposition is of a higher quality in contrast to that of pulse voltage electrodeposition. Yet, the best result can be achieved by a triple pulse current, where a short pulse of high current is followed for active element deposition with a short pulse of zero, and at the end a lengthy pulse with low current is applied for the noble element. The amount and time of each pulse depend on the agitation method, bath composition, and ideal wavelength.

In Table 4.1, two types of triple pulse programs are presented for the production of a multiple-layer coating of Cu/Ni with coherent nanometric layers (in both states the sulfamate bath is used). Due to the application of high current density, type 1 electrodeposition is of a high coating rate and the developed alloy has higher contents of nickel. The high rate of electrodeposition is somehow achieved by stirring the solution using the cylindrical cathode in the solution. Yet, the use of high electrodeposition speed leads to the development of an Ni-rich alloy.

Type 2 pulses are a way in which the electrodeposition rate is low and the alloy has a higher ratio of Cu in contrast to type 1. In the type 1 program the agitation rate of the solution in each current pulse is modulated. In this condition, it is necessary to precisely control the coincidence of the agitation times with current pulses during the multiple-layer electrodeposition.[31–38]

TABLE 4.1

Condition of Electrodeposition Pulses for Making Multiple-Layer Coatings Cu/Ni Using the Sulfamate Bath

Condition	Bath Agitation	Growth Rate (Microns per Hour)	$\lambda_{Ni}/\lambda_{Cu}$
Nickel pulse: 90 mA/cm^2 for ~0.7 s	60 rpm		
"Rest" pulse: 0 mA/cm^2 for 0.25 to 1 s		~7	~10
Copper pulse: 1.5 mA/cm^2 for ~4 s	600 rpm		
Nickel pulse: 12–20 mA/cm^2 for ~0.5 s	None		
"Rest" pulse: 0 mA/cm^2 for 0.5 s		~1	~2
Copper pulse: 0.3 mA/cm^2 for ~11 s	None		

4.3.2 Mechanical Cleavage for Synthesis of Graphene Layers

There is a burning need to expand a large-scale method to produce graphene two-dimensional nanostructures reliably for various hopeful applications being developed. These usages rely largely on the sole properties of graphene, and the properties are mainly affected by the technique of synthesis. Even though several laboratory techniques to create graphene have been created and reported, the suitability of these methods to large-scale manufacturing remains to be proven. These techniques can be generally classified as epitaxial growth, colloidal suspension, unconventional methods, and exfoliation. Some researchers created a new technique to produce few-layer graphene from bulk graphite by mechanical cleavage. The technique involves the use of an ultrasharp single crystal diamond wedge to cleave a highly ordered pyrolytic graphite sample to create the graphene layers. Cleaving is done by employing ultrasonic oscillations beside the wedge. Characterization of the layers demonstrates that the method is able to create graphene layers with an area of a few micrometers. Usage of oscillation improves the quality of the layers formed with the layers having a lower crystallite size as obtained from the Raman spectrum.

They used highly ordered pyrolytic graphite (HOPG) as the starting substrate material. The HOPG is first cut into small pieces and then embedded into an epofix embedding medium. Then they trimmed it as shown in Figure 4.7a into a pyramid shape. The ultrasharp wedge employed for sectioning is made of a single crystal diamond with a sharpness less than 20 Å and has an included angle of 35°. The diamond wedge is mounted on an ultrasonic oscillation arrangement capable of providing tunable frequencies with a range of vibration of a few tens of nanometers. The diamond wedge mounted on the oscillation arrangement is aligned carefully with respect to the HOPG mount (Figure 4.7b). The HOPG and the diamond wedge system are mounted on two dissimilar high-precision slide arrangements (Figure 4.7c).

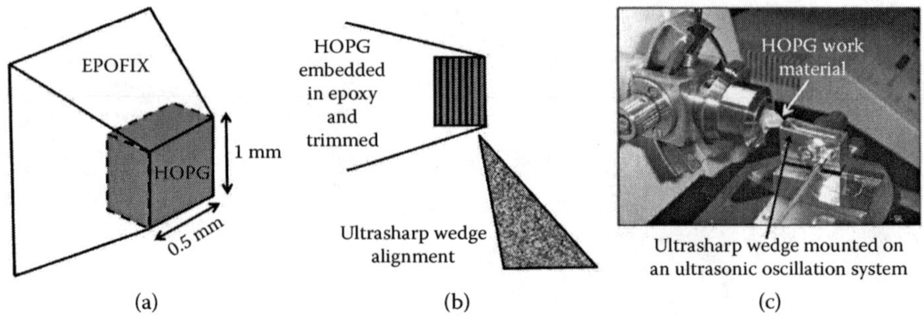

FIGURE 4.7
Highly ordered pyrolytic graphite (HOPG), SPI grade ZYH. (a) HOPG mounted in epofix and trimmed to pyramid shape. (b) Setup showing wedge alignment with HOPG layers. (c) Actual experimental setup. (Reprinted based on general permission from SpringerOpen. From Jayasena, B., and Subbiah, S., A novel mechanical cleavage method for synthesizing few-layer graphenes, *Nanoscale Research Letters* 6 (1), 1–7, Copyright 2011.)

They observed two-dimensional nanostructures using transmission electron microscopy (TEM) for the few-layer graphenes obtained with and without oscillations (Figures 4.8 and 4.9, respectively). In the images of layers without use of oscillations, the folded graphene sheet is obviously visible (marked as 1). In addition, several grain boundaries (marked as 2) are observed.

Figure 4.8 illustrates an area where the sheet appears to be heavily crumpled. In the images of layers obtained with the use of oscillations, grain boundaries, folded graphene sheets, and smooth areas of the sheets are also obviously observed. No heavily crumpled regions were observed, but some structures that resemble nanohorns can be observed (marked as 3 in the enlarged area). Nanohorns are considered as structures resulting from crushing a single sheet of graphene. The large surface area of the nanohorns is reported to be functional in various usages such as hydrogen gas storage. AFM measurements show that the proposed mechanical cleaving technique is able to produce thin-layer graphene with a thickness of tens of nanometers. It has been revealed that there is a large amount of attention required to know the edge formation with ultrasonic oscillation usage because structures that seem to resemble nanohorns were observed. Application of ultrasonic vibrations along the tool edge is seen to considerably reduce the relative ratios observed in a Raman spectrum. Hence, the applied oscillations may have potential to decrease the defects in cleaved layers. The use of ultrasonic vibration also reduces the crystallite dimension.

4.3.3 Chemomechanical Synthesis of Thin Films

4.3.3.1 $Hg_xCd_{1-x}Se$ Thin Films

Some kinds of thin films such as mercury-cadmium chalcogenides attract attention because of their potential capability in a wide spectrum of the

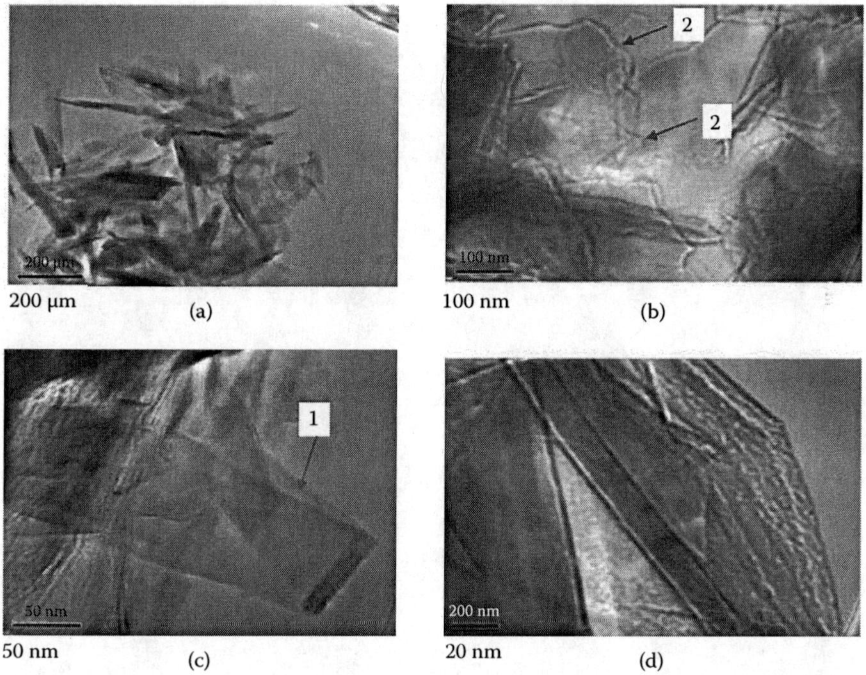

FIGURE 4.8
Transmission electron microscopy (TEM) images without ultrasonic oscillation: (a) large few-layer graphene (FLG) edges, (b,d) folded layered graphene (LG), and (c) large graphene sheet with rolled edge. (Reprinted based on general permission from SpringerOpen. From Jayasena, B., and Subbiah, S., A novel mechanical cleavage method for synthesizing few-layer graphenes, *Nanoscale Research Letters* 6 (1), 1–7, Copyright 2011.

opto-electronic devices. $Hg_xCd_{1-x}Se$ is one of such ideal ternary materials for use in visible and infrared (IR) detection. The band structures, optical characteristics, and crystal structures of both CdSe and HgSe are very comparable, and consequently the system $Hg_xCd_{1-x}Se$ would not only result in the possibility of a graded energy gap of a broad spectral sensitivity but many more material characteristics could be changed and outstandingly controlled by controlling the arrangement composition (x). Depending on the work and the conditions of research, both CdSe and HgSe are known to be in hexagonal wurtzite or cubic zinc blend crystal forms. Thin films of these materials have been frequently manufactured via the vacuum techniques. However, creation of that growth by a solution bath method is very simple, suitable, and viable compared to other cost-intensive methods. One of the beautiful features of the chemosynthesis way is the easiness with which the alloys can be produced without the use of any complicated instrumentation and method control.

Recently chemomechanical synthesis of thin films has attracted the attention of researchers. $Hg_xCd_{1-x}Se$ thin films with a changeable mercury

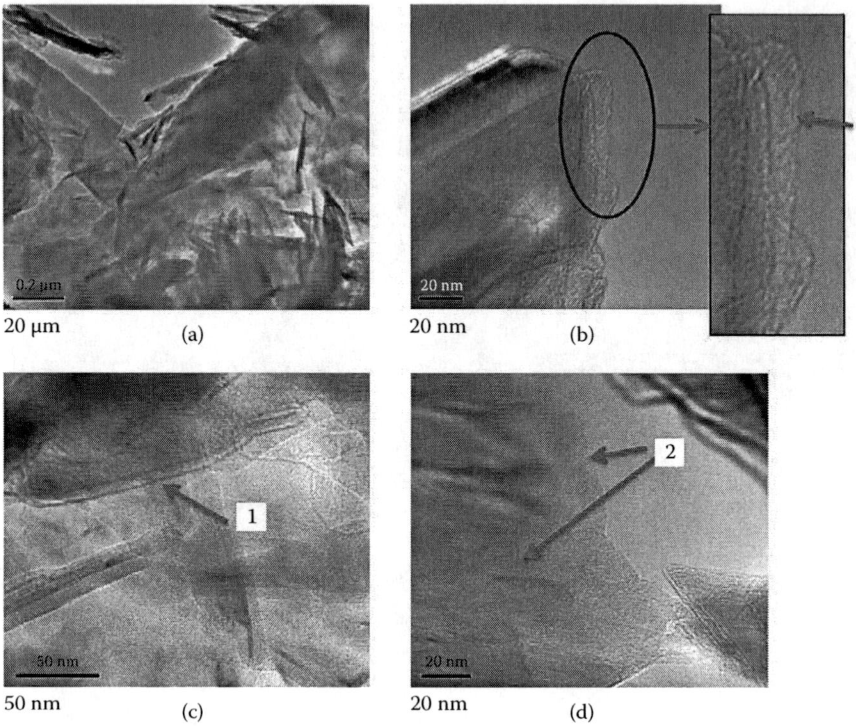

FIGURE 4.9
Transmission electron microscopy (TEM) images with ultrasonic oscillation: (a) FLG, (b) edge of graphene sheet, (c) and (d) folded FLG. (Reprinted based on general permission from SpringerOpen. From Jayasena, B., and Subbiah, S., A novel mechanical cleavage method for synthesizing few-layer graphenes, *Nanoscale Research Letters* 6 (1), 1–7, Copyright 2011.)

concentration has been done with the aim to enhance the physical, microscopic, compositional, and structural characteristics sourced due to the incorporation of Hg^{2+} in CdSe. The various effective parameters and the deposition conditions such as deposition time, temperature and concentration of the chemical species, pH, speed of the mechanical churning, and so forth, were optimized to receive high-quality films. The compositional analysis demonstrated a nonlinear behavior between the Hg:Cd ratio of the bath and fabricated film. The films are polycrystalline over the entire range studied with a main wurtzite structure in addition to the cubic zinc blend. The crystallite dimension determined from different methods is observed to be rising with an increase in the film composition (x) up to 0.05 and then stays more or less similar for upper values of x. The anticipated optical gaps illustrated a monotonic nonlinear reduction in the band gap with rising Hg content in the CdSe.[40–46]

Thin films of different Hg:Cd ion concentration ratio were attained on the optically plane amorphous glass substrates by a chemical growth method. For deposition of the films, equimolar solutions of the relative chemical components were mixed in exact volume stoichiometric proportion. Triethanolamine was used as a complexing agent, and pH of the reaction mixture was adjusted to its relative level. Glass substrates were mounted on a particularly designed substrate holder and kept turned with a 60 rpm speed by means of a stable speed gear motor. This offers a constant and uniform mechanical stirring of the reaction mixture. To obtain high-quality samples, time and temperature of deposition and speed of the substrate revolution were optimized. The film composition was then determined by an energy-dispersive x-ray spectroscopy (EDS) method. The x-ray diffractograms were achieved on these thin films to get the structural/crystallographic data. The surface texture of the films was observed through a scanning electron microscope.

These researchers reported that as the mole percentage of Hg in solution was raised, the Hg percentage in the film increased and the consequent Cd content in the film decreased constantly. This is demonstrated as the variation of x_{film} versus x_{bath} as shown in Figure 4.10. The obtained diagram is nonlinear.

It was found that there existed a cubic phase whose amount is more or less the same without any considerable change in its *d*-value throughout the range studied. Thus, there must be a solid solution arrangement of the kind $Hg_xCd_{1-x}Se$ in the range between 0 and 0.05 for only the hexagonal phase. The *d*-values and the intensity levels of both cubic CdSe and HgSe are alike.

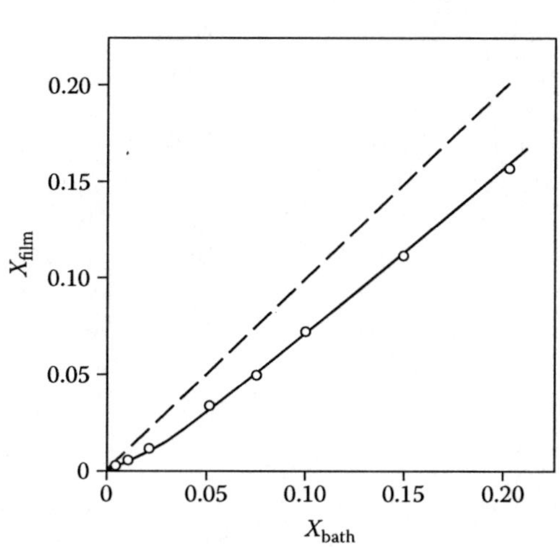

FIGURE 4.10
Variation of the x_{film} with x_{bath} in chemomechanical synthesis method. (Reprinted from Pujari, V. B., Mane, S. H., Karande, V. S., Dargad, J. S., and Deshmukh, L. P., Mercury-cadmium-selenide thin film layers: Structural, microscopic and spectral response studies, *Materials Chemistry and Physics* 83 (1), 10–15, Copyright 2004, with permission from Elsevier.)

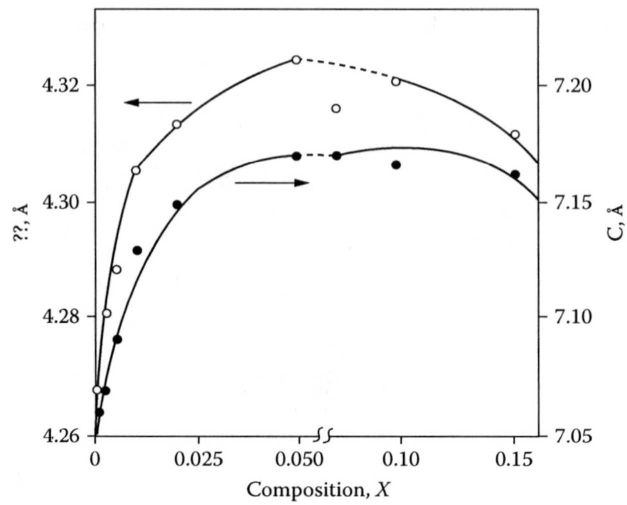

FIGURE 4.11
Variation of lattice parameters *a* and *c* with composition parameter, *x*. (Reprinted from Pujari, V. B., Mane, S. H., Karande, V. S., Dargad, J. S., and Deshmukh, L. P., Mercury-cadmium-selenide thin film layers: Structural, microscopic and spectral response studies, *Materials Chemistry and Physics* 83 (1), 10–15, Copyright 2004, with permission from Elsevier.)

Further, for the range $0.05 \leq x \leq 0.2$, separate phase arrangement of both HgSe and CdSe resulted. The variations in the lattice parameters *a* and *c* have also been observed (Figure 4.11) for the solid solution range. It is noticed that there is a weak difference of the lattice parameters with the composition parameter (x). An average crystallite dimension was also noticed for all the ranges. The different dimensions are cited in Table 4.2.

These researchers synthesized thin films that are polycrystalline composed of the hexagonal wurtzite and cubic zinc blend phase structures. The surface observations demonstrated that the crystalline nature of all the samples went on rising with Hg content in CdSe. The average crystallite dimension increased up to a level and then remained more or less steady for higher values of x. The spectral studies illustrated a high absorption coefficient with the optical gap reduced nonlinearly.

4.3.3.2 $Cd_{1-x}Zn_xSe$ Thin Films

Pseudo binaries of different group compounds are attracting attention due to their potential abilities in a wide spectrum of the optoelectronic devices. $Cd_{1-x}Zn_xSe$ is one of such ideal ternary materials for use in electroluminescent, photoluminescent, photoconductive, and photovoltaic device usages. The band structures, optical characteristics, and crystal structures of both CdSe and ZnSe are comparable, and thus the system $Cd_{1-x}Zn_xSe$ would not only result in the feasibility of a graded energy gap of a broad spectral

TABLE 4.2

Some Parameters of $Hg_xCd_{1-x}Se$ Thin Films Synthesized by the Chemomechanical Method

Composition Parameter, *x*	Crystallite Size (XRD) (angstrom)	Crystallite Size (SEM) (angstrom)	Power Factor (m)
0	177	463	0.41
0.0025	183	474	0.43
0.005	189	496	0.42
0.0075	196	512	0.5
0.01	203	532	0.4
0.02	218	567	0.48
0.05	258	670	0.45
0.075	253	664	0.46
0.10	247	656	0.42
0.15	249	661	0.49
0.20	253	668	0.48

Source: Reprinted from Pujari, V. B., et al. Mercury-cadmium-selenide thin film layers: Structural, microscopic, and spectral response studies, *Materials Chemistry and Physics* 83(1), 10–15, copyright 2004, with permission from Elsevier.

sensitivity, but many more material characteristics would be enhanced and outstandingly controlled by controlling the system composition (x). Thin films of these materials have been mostly synthesized via vacuum methods. However, this research that already created that growth by a solution bath method is very simple and viable compared to the other cost-intensive techniques. Especially, one of the beautiful characteristics of chemomechanical synthesis is the simplicity with which the alloys can be fabricated without the use of any sophisticated instrumentation and method control.

Therefore, these researchers used the chemomechanical synthesis method for fabrication of (Cd, Zn)Se thin films with a changeable composition with the aim to enhance the deposition behavior, growth kinetics, structural changes, and optical properties. They studied the effect of different process parameters on the growth and quality of the films. The films were crystalline over the same ranges studied with a predominant wurtzite structure and a zinc blend structure (solid solution). The crystal dimension calculated from the x-ray diffraction (XRD) and scanning electron micrography (SEM) is observed to reduce with rising x.

The decomposition of sodium selenosulfate is achievable in an aqueous alkaline electrolyte containing the cadmium and zinc sulfates and a triethanolamine (TEA) complexing agent that allows control of the Cd^{2+} and Zn^{+2} ion amounts and soluble types of Cd^{2+} and Zn^{2+} in the reaction bath. The deposition procedure is based on the slow release of Cd^{2+}, Zn^{2+}, and Se^{2-} ions in solution, which then condense, on an ion-by-ion basis, onto the substrates that are vertically mounted in the reaction solution. Due to the fact that the solubility products of

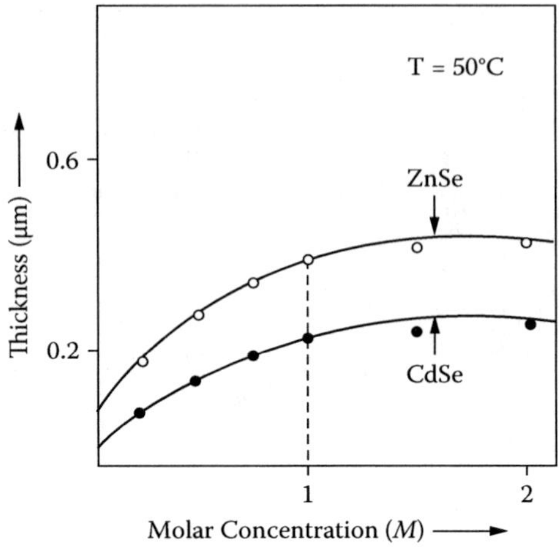

FIGURE 4.12
Variation of the film thickness with concentration of the Cd^{2+} and Zn^{2+} species (T = 50°C). (Reprinted from Sutrave, D. S., Shahane, G. S., Patil, V. B., and Deshmukh, L. P., Microcrystallographic and optical studies on $Cd_{1-x}Zn_xSe$ thin films, *Materials Chemistry and Physics* 65 (3), 298–305, Copyright 2000, with permission from Elsevier.)

metal selenides are very minute, control of Cd^{2+} and Zn^{2+} ions in a reaction bath leads to the control of the rate of precipitation and hence the rate of deposition.

It appears that an increase in growth rate is probable when the amounts of the basic ingredients in the bath are increased. However, researchers have observed that the growth rate initially is almost linear with the molar concentration (Figure 4.12) and then saturates for a further increase in the amount of the species in the bath. The optimum values of the concentrations of Cd^{2+} and Zn^{2+} ions were kept steady. The second important feature in physical studies is the influence of the deposition time and temperature on film creation. The film growth on the glass strip surface has been studied as a function of bath temperature and the deposition time (Figures 4.13a and 4.13b). The initial growth rate can be designed only in the quasi-linear stage. The second stage is known by means of a saturation phenomenon that appears to be mainly due to the depletion of Cd^{2+}, Zn^{2+}, and Se^{2-} species in the reaction bath. At higher temperature, this saturation stage shows itself earlier and is due to the two competitive procedures: the film formation and the homogeneous precipitation. A raise in the deposition temperature leads to the homogeneous precipitation rather than the film creation, which causes saturation to occur.

It is notable that the physical, structural, and optical characteristics observed for chemical bath synthesized thin films have a close relation with what had been observed for other semiconductor solid solution arrangements that shows their common characteristics. It has also been shown that the variations in the deposition behavior lead to composition scenarios ranging from a physical mixing of CdSe and ZnSe to $Cd_{1-x}Zn_xSe$ solid solution

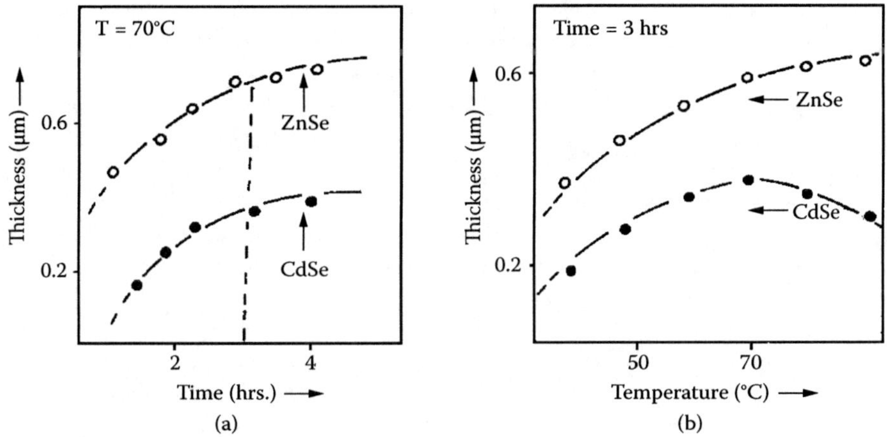

FIGURE 4.13
Film growth studies as a function of (a) time and (b) temperature (at optimum concentration). (Reprinted from Sutrave, D. S., Shahane, G. S., Patil, V. B., and Deshmukh, L. P., Microcrystallographic and optical studies on $Cd_{1-x}Zn_xSe$ thin films, *Materials Chemistry and Physics* 65 (3), 298–305, Copyright 2000, with permission from Elsevier.)

and also the superstructures with alternating stacks of CdSe and ZnSe layers that can be built. These characteristics make the chemomechanical bath procedure a simple, inexpensive, and beautiful means of obtaining (Cd, Zn) Se thin films for a range of usages.

4.3.4 Achieving Enhanced Mechanical Properties

Researchers have studied the mechanical properties of electroplated thin films. They fabricated nanocrystalline Cu-Ni films by electroplating galvanostatically onto Cu/Ti/Si substrates. The excellent mechanical properties of these metallic films can be recognized by their nanocrystalline nature and with the presence of stacking faults and the related formation of intragranular nanotwins during film growth. Due to their nanocrystalline character, these films also demonstrate a very low surface roughness (near 2 nm). The combination of desired properties, together with the ease of the fabrication technique, makes this system attractive for widespread technical applications, including hard metallic layers or magnetic micro-/nano-electromechanical devices.

Nanocrystalline (nc) metals are drawing huge attention within the scientific population as they possess novel and improved mechanical properties compared with coarse-grained materials. In particular, advantages include high yield, fracture strength, better wear resistance, and superplasticity at low temperatures and high strain rates. Apart from the obvious technological significance of bulk nanostructured specimens, production of nc-metals and metallic alloys is also of importance in thin-film knowledge, mainly for the implementation of components in micro-/nano-electromechanical systems (MEMS/NEMS) or to obtain mechanically hard, corrosion-resistant coatings. MEMS/NEMS are usually produced from either pure silicon or silicon-based compounds, such as silicon nitride or silicon carbide. However, these materials are not appropriate for magnetic applications or tools requiring high electrical conductivity. So, metallic alloys are emerging as suitable candidates to replace Si compounds in certain MEMS/NEMS applications, mainly in magnetic MEMS/NEMS, which have the extra value in that they can be remotely actuated. In fact, metallic films illustrate more ductility and fracture toughness than silicon.

Several techniques are traditionally used to produce nc-materials: inert gas condensation, ball milling, electrodeposition, crystallization from amorphous materials, severe plastic deformation, surface mechanical attrition, and friction stir processing; they are either the bottom-up or the top-down approaches. From all these techniques, electrodeposition stands out from the rest as it allows the creation of fully dense nc-metals, as well as thin films, with grain dimensions well below 100 nm. For many years, electrodeposition of metallic alloys was considered for decorative uses and protection against corrosion. There has been renewed attention in electrodeposition as it can fabricate nanostructured materials for a range of functions.

Researchers used the sulfate-based bath that contains citrate as the complexing agent. The x-ray diffraction patterns of the $Cu_{1-x}Ni_x$ films contain wide diffraction peaks that match those of copper (seed-layer) and the face-centered cubic Cu-Ni phase. The 38° to 56° 2θ range, covering the (111) and (200) reflections, is illustrated in Figure 4.14a. A progressive shift in the Cu-Ni peak positions, toward larger 2θ angles, is noticed as the Ni percentage is raised, indicating reduction in the cell parameter (see Figure 4.14b). The cell parameter relation with alloy content obeys Vegard's law.

Figure 4.15a illustrates three usual nanoindentation curves regarding $Cu_{1-x}Ni_x$ films. The penetration depth received at the end of the loading section is lower for larger Ni percentages, showing that alloys with larger Ni content are mechanically harder. The overall dependence of the hardness, H, on the Ni percentage, shown in Figure 4.15b, indicates that H increases from 6.4 GPa (for x=0.45) to 8.2 GPa (for x=0.87). It should be said that care has to be taken by comparing hardness measured using different methods and different applied loads. In fact, the hardness values achieved from nanoindentation at low loads (e.g., 10 mN) are typically larger than those evaluated from Vickers microhardness, that the maximum applied load is of the order of 500 mN or more.

Even if an indentation size effect (ISE) of 10% to 30% is considered, the hardness values obtained in the $Cu_{1-x}Ni_x$ thin films investigated here are clearly much larger than those of microcrystalline or nanocrystalline Cu thin films, in concordance with the larger hardness values of pure Ni with respect to Cu. The hardness of obtained films is also significantly larger than in electroplated micrometer-sized pure Ni; coarse-grained Co-Ni alloys; as-deposited ultrananocrystalline Ni-P, Ni-W-P, or Ni-W-Cu-P films; or Ni-P-W multilayered

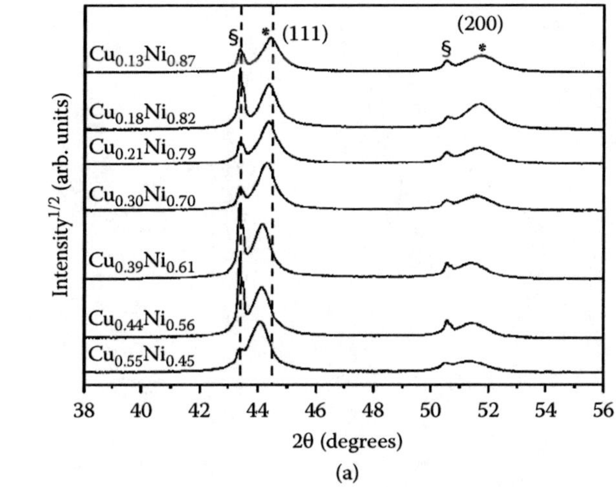

FIGURE 4.14
(a) X-ray diffraction (XRD) patterns of $Cu_{1-x}Ni_x$ thin films in the 38° to 56° 2θ region. The dashed lines indicate the position of (111) fcc reflection for pure copper and pure nickel. Peaks denoted by § and * belong to the Cu seed-layer and the Cu–Ni phase, respectively. (b) Cell parameter variation with Ni amount in the alloy. The line is a fit of the data using Vegard's law (linear correlation coefficient, *R*, of 0.996). (From. Pellicer, E., Varea, A., Pané, S., Nelson, B. J., Menéndez, E., Estrader, M., Suriñach, S., Baró, M. D., Nogués, J., and Sort, J., Nanocrystalline electroplated Cu-Ni: Metallic thin films with enhanced mechanical properties and tunable magnetic behavior, *Advanced Functional Materials* 20 (6), 983–991, Copyright 2010, Wiley-VCH. Reprinted with permission.)

structures. The main cause for the high hardness displayed by the electrodeposited alloys is the saccharine-assisted crystallite dimension refinement through thin film growth. For small crystallite dimensions, the role of grain boundaries in delaying dislocation motion is improved, resulting in bigger stress concentration at grain boundaries due to dislocation pileup.

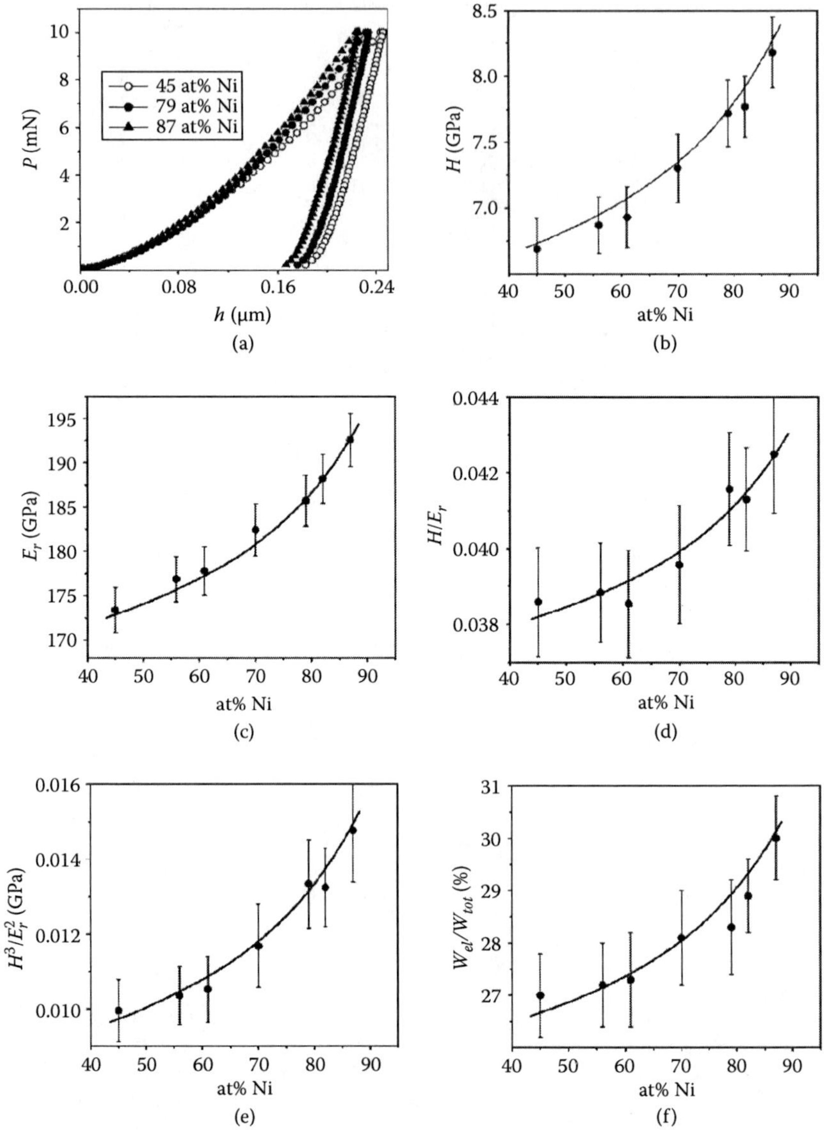

FIGURE 4.15
(a) Load–unload nanoindentation curves (applied load, P, versus penetration depth, h) corresponding to $Cu_{0.55}Ni_{0.45}$, $Cu_{0.21}Ni_{0.79}$, and $Cu_{0.13}Ni_{0.87}$ films. The following panels show the dependence of (b) the nanoindentation hardness H, (c) reduced elastic modulus E_r, (d) $\frac{H}{E_r}$, (e) $\frac{H^3}{E_r^2}$, and (f) elastic recovery (i.e., $\frac{W_{el}}{W_{tot}}$ ratio, where W_{el} and W_{tot} denote, respectively, the elastic and total energies during nanoindentation), on the thin film composition. (From Pellicer, E., Varea, A., Pané, S., Nelson, B. J., Menéndez, E., Estrader, M., Suriñach, S., Baró, M. D., Nogués, J., and Sort, J., Nanocrystalline electroplated Cu-Ni: Metallic thin films with enhanced mechanical properties and tunable magnetic behavior, *Advanced Functional Materials* 20 (6), 983–991, Copyright 2010, Wiley-VCH. Reprinted with permission.)

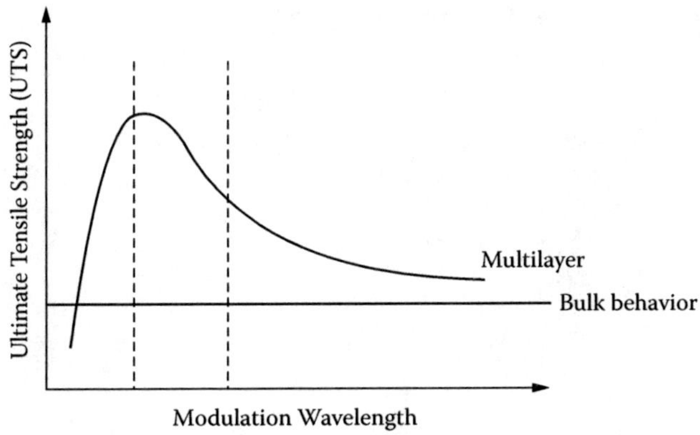

FIGURE 4.16
Changes of ultimate tensional strength with modulation wavelength for metallic multilayers. (Reprinted from Foecke, T., and Lashmore, D. S., Mechanical behavior of compositionally modulated alloys, *Scripta Metallurgica et Materiala* 27 (6), 651–656, Copyright 1992, with permission from Elsevier.)

4.4 Examining the Characteristics of the Multiple-Layer Coatings

4.4.1 Mechanical Properties of the Multiple-Layer Nanostructures

Several researchers have reported the high rate of yield strength and failure strength for multiple-layer coatings compared with those for their constituent materials. Multiple-layer structures' tensional strength highly depends on their pulse wavelength* and increases with a decrease of wavelength. The empirically similar results of this relationship are shown in Figure 4.16. Figure 4.16 shows that the tensional strength in wavelengths of 10 to 200 nm increases up to its peak amount. It seems that the position of this peak is significantly influenced by the multiple-layered process and constituent materials. The mechanical properties of the multiple-layered coatings depend on fine structures as well as inter-layer structures.

The results of some studies show that tensional strength of the multiple-layer foil of Cu/Ni ($\lambda_{Cu} \approx 2$ nm, $\lambda_{Ni} \approx 18$ nm) developed by the electrochemical deposition method is about three times more than that of homogenous Cu or Ni. In addition, coating the steel by multiple-layer foil of Cu/Ni leads to the improvement of sliding wear resistance. With respect to all mentioned points, one can say these coatings are an efficient choice in the increase of industrial parts such as a printer's lifetime, which is subjected to wear.

* The pulse wavelength is defined as the thickness of its one couple of layers.

4.4.2 Strengthening Theories in Multiple-Layer Systems

In general, the theoretical methods used for predicting the yield strength of the multiple layers are classified into three groups. The first group is based on the strengthening mechanism of Orowan, the second group of models is based on the Hall–Petch reinforcing method, and the third group is founded by the virtual forces. In the third model, the enhancement force is created by the difference in an elastic modulus between two substances and the layer boundary.

4.4.2.1 Orowan Mechanism

One can use Equation (4.1) between the coatings' strength and layers' thickness, where σ and d are strength yield and the thickness, respectively:

$$\sigma_y = \sigma_0 + kd^{-n} \tag{4.1}$$

Here, n is a constant value that is –0.5 and –1 for mechanisms of Hall–Petch and Orowan, respectively.

In an Orowan model, reinforcing occurs by stopping the development and movement of dislocations within the layers. Also, in this model it is supposed that plastic deformation just takes place in the softest layers and the strength is measured from the needed stress for diffusion of the dislocation circle within one of the layers. In this model, the yield strength is inversely related with the thickness of the soft layer, and dislocations are limited by the stronger barrier between the layers in the soft phase.[51–53]

4.4.2.2 Hall–Petch Mechanism

In this model, reinforcement occurs by resistance against movement of dislocations from layer boundaries throughout the coating. In the Hall–Petch model it is assumed that the yielding occurs immediately after dislocations overcome the barrier of the interlayer boundary—once dislocations are released from sticking to the boundaries. In this model the critical factor for defining the yield strength of the multiple-layer system is the facility of dislocations diffusion in the width of the layer boundary. In both models, Orowan and Hall–Petch, informing from the movement of first dislocation is of great importance.

One may use the value of 0.5 for n, which is common in many bulk materials for low-temperature deformation, and Equation (4.1) is suitable for predicting the behavior of multiple-layer systems with layer distances higher than 100 nm. However, in wavelengths of 25 to 50 nm, the Hall–Petch model is not appropriate, and the Orowan mechanism is more convenient for these coatings.[54–58]

4.4.2.3 Reinforcing by Virtual Forces

Kohler introduced another reinforcing method for multiple layers developed by epitaxy growth and coherent growth of two materials with the fairly same network parameters. Using this model requires that the elastic modulus of one layer not be more than twice the other, so the virtual part of the mutual connection of the dislocations would cause rejection of the boundary dislocations. The main mechanism of reinforcement in this formulation is resistance against the movement of dislocations in the interface of the material with a low modulus to material with a higher modulus. This researcher has offered a relationship for the least flexural strength needed for movement of dislocations as Equation (4.2):

$$\mu_{\min} = \frac{\mu_b}{8\pi} \frac{\mu_a \mu_b}{\mu_a + \mu_b} \tag{4.2}$$

where μ_a and μ_b are the flexural modulus of the materials *a* and *b*, respectively. Inserting the proper values to this equation, it is possible to create very strong multiple-layer structures. The behavior described by Kohler was empirically approved by some researchers for Al/Cu films from the evaporative process. These researchers have shown that Equation (4.2) works with good preciseness in a layer thickness of less than 50 nm for the strength surface of multiple-layer coatings. Yet, in higher thicknesses the strength is inversely related to the layer's thickness.

In a model of reinforcing with virtual forces this fact has been neglected; however, in the strength of materials with very low thicknesses (such as monocrystal whiskers), development of dislocations is more important than their movement. Though high boundary energy is a good barrier for dislocations' movement, layer boundary is a good source for development of dislocations. This may increase the negative aspect of raise in layer boundaries.

Some researchers have shown that in some Cu/Ni layers, a decrease in strength for samples whose thickness in two layers is less than 20 nm is accompanied by a decline in electrical resistance of the coating. Development of noncoherent Cu layers leads to an increase of conductivity in the Ni matrix and negates the increase of strength that occurred by a decrease of two layers' thickness. Then a decrease of coatings' strength by a decrease of wavelength (more than 20 nm) may be developed by the creation of noncoherent layers of Cu.

Deposition circumstances that lead to the development of twins result in a decrease of strength due to a decrease of coherency of the faces. In addition, development of needle-form grains decreases the load bear capacity, as well as apparent strength and malleability.

In addition to layers' thickness, two species' ratio of layer thicknesses in multiple-layer structures is of great importance in their strength. Another parameter effective in the strength of multiple-layer coatings is the concentration of the noble species in electrolyte electrodeposition. For instance, in multiple-layer Cu/Ni for constant thicknesses, the strength is significantly influenced by Cu concentration in the electrolyte and has a maximum that can be changed according to the agitation rate during the Cu deposition. The results of experiments have shown that in a multiple-layer coating of Cu/Ni, where the thickness of the Ni layer and Cu layer is 90 and 10 nm, respectively, changing the concentration of the electrolyte from 8 to 7 millimolar increases the coatings tensional strength from 700 to 1150 MPa. Strength of this coating is the maximum in Cu concentration of 7 millimolar and with further decrease of its concentration (up to 4.5 millimolar) the strength again begins to decrease and reaches 850 MPa.

A change in coatings' strength by the effect of Cu concentration changes in electrolyte occurs due to competition of two factors. Once the concentration of Cu increases in the electrolyte, the coherency increases between Cu and Ni layers, and then strength increases through this phenomenon, although increase in coherency leads to a decrease of boundary dislocation density, an increase in coherency stresses makes it difficult for dislocations to pass from the boundaries. But, on the other hand, as the Cu concentration increases in the electrolyte, Ni concentration would also increase. This makes it difficult to recognize the boundary between the layers and so it would be in a diffusive state, which results in a decrease of strength.

It is predicted that in multiple-layer coatings with semicoherent layers, an increase in mismatch strains (i.e., increase of coherency stresses and decrease of dislocations distances in semicoherent boundary) leads to an increase of the strength. Yet, in systems with high mismatch strains, such as Cu/Ag multiple layer, the number of boundary dislocation centers is higher, and then their motions are more than that of multiple layers with lower mismatch strains, such as Cu/Ni multiple layers. This point leads to less strength of boundary dislocations against the pass of a dislocation from their interface. Hence, multiple layers of Cu/Ni, in contrast to multiple layers of Cu/Ag, have a higher hardness peak in the same condition. Figure 4.17 presents a comparison between the strength of mentioned multiple layers.

However, both mentioned references[19,28] are considered for discussions about the strength of multiple-layer coatings, and resistance against dislocations passing from the boundary. However, as previously mentioned, in layers with a thickness of lower than 100 nm, Orowan and Kohler mechanisms are generally used for the prediction of multiple-layer systems' behavior. It must be added, in Orowan model resistance against internal dislocations and in Kohler model differences between two layers flexural modulus, which depends mainly on layers' material, is a reinforcing factor. Thus, for analyzing the observed behaviors the above points also have to be considered.

4.4.3 An Example from Cubic Boron Nitride Thin Films

Some researchers have studied the mechanical and tribological properties of epitaxial cubic boron nitride thin films that grow on diamond substrates.

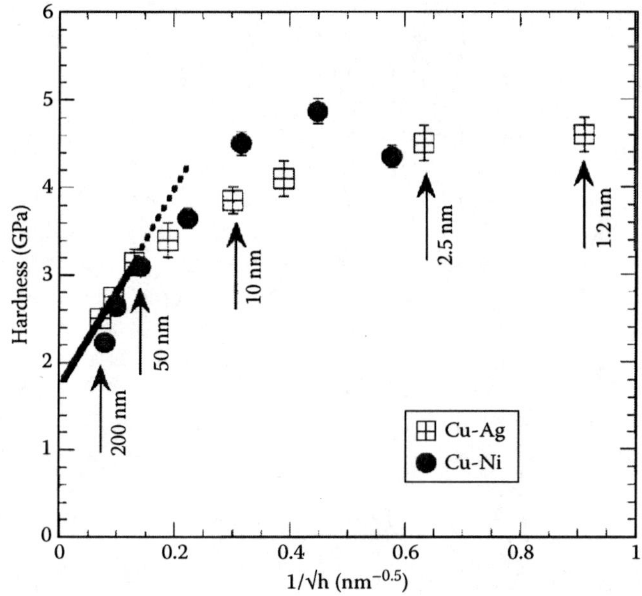

FIGURE 4.17
Hardness of Cu/Ag and Cu/Ni multilayers as a function of $1/\sqrt{h}$. (Reprinted from McKeown, J., Misra, A., Kung, H., Hoagland, R. G., and Nastasi, M., Microstructures and strength of nanoscale Cu-Ag multilayers, *Scripta Materialia* 46 (8), 593–598, Copyright 2002, with permission from Elsevier.)

Cubic boron nitride (c-BN) is an ideal material based on its applications as a hard coating for cutting tools due to its many attractive properties such as hardness, oxidation resistance, and being inert against iron even at elevated temperatures, with the latter property resulting in the ability to machine ferrous metals. However, these potential usages are hindered by the weak adhesion of c-BN films to most underlayer materials. This trend of c-BN films to delaminate also puts important obstacles to precise measurements of their mechanical properties. The weak adhesion of such samples is known as being due to the high compressive stress caused by energetic ion bombardment during growth and to a mechanically soft turbostratic boron nitride (t-BN) interlayer between c-BN and substrates. Most c-BN films consist of nanodimension crystallites and have the above-mentioned layered arrangement. Therefore, the elastic and mechanical properties that have been reported with different methods are surely not representative for c-BN alone. In most cases sample polishing is necessary before hardness measurements, as these c-BN films are significantly rough. Recently, researchers reported that thick hetero-epitaxial c-BN films without any intermediate turbostratic boron nitride (t-BN) layer can be arranged on CVD diamond films and single-crystal diamond materials using ion beam assisted deposition (IBAD).

Scratch tests have been extensively used as a suitable method for estimating adhesion of thin hard wear-resistant layers such as titanium nitride (TiN), diamond-like carbon (DLC), and c-BN to substrates. In the conventional scratch test method a diamond tip is drawn across the coating surface under a progressively rising vertical load until the coating becomes detached or fractured. The minimum load at which any identifiable failure occurs is called the "critical load" L_c. However, the connection between this critical load and the adhesive strength of the substrate/coating interface is still vague. In addition to the coating/substrate bond strength, a broad range of factors are recognized to affect the critical load value obtained from the scratch test. Some researchers showed the relative enhancement of the mechanical and tribological characteristics of the c-BN films on diamond as compared to the c-BN films on silicon substrates.

Figure 4.18 stands for the scratch tracks on the c-BN films on diamond. For the nonirradiated c-BN film a thin groove is observed in Figure 4.18a followed by an unexpected increase of the width of the scratch track, showing fracture and delamination of the coating. The critical load and normal displacement at which failure of the coating occurred were near 4 mN and 82 nm, correspondingly (Point A), while the opening of the scratch track occurs at the normal load of 8.6 mN and resultant normal displacement of 150 nm (Point B denoting also film/substrate interface). For this film a compressive stress of 2 GPa was expected. It appears that this compressive stress in the film tends to close the crack up to Point B where the film–substrate interface is contacted. From Point B to the end of the scratch, the substrate becomes locally uncovered by the film while deforming elastically. This results in a scratch groove that is flat at its base.

For the irradiated c-BN film on diamond, the AFM image captured without delay after scratching (Figure 4.18b) shows obviously the smooth scratch track. This track starts directly at the Point A*/B* when the indenter gets to the film–substrate interface (normal displacement of 100 nm), and it broadens and deepens with increasing the load from 2.9 mN up to 10 mN. In this case, beyond Point A*/B*, the sliding procedure produces film fracture near the deforming area ahead of the tip, but the film stays adherent to the substrate. This points out that the interfacial adhesion of the irradiated c-BN film on diamond is higher than the interfacial adhesion of the nonirradiated c-BN film on diamond. The internal stress in the irradiated c-BN film on diamond was indicated to be less than («1 GPa) nonirradiated c-BN film (2 GPa). Thus, the difference in normal displacement in nonirradiated and irradiated c-BN films (seen in Figures 4.18c and 4.18d) can be clarified by different stresses. The compressive stress pushes the penetrating indenter tip up resulting in the higher normal load ($L = 5$ mN) required to reach the normal displacement of 100 nm (Point C in Figure 4.18c) for the nonirradiated c-BN film on diamond as compared to that for the irradiated c-BN film ($L = 2.9$ mN, Point C* in Figure 4.18d).

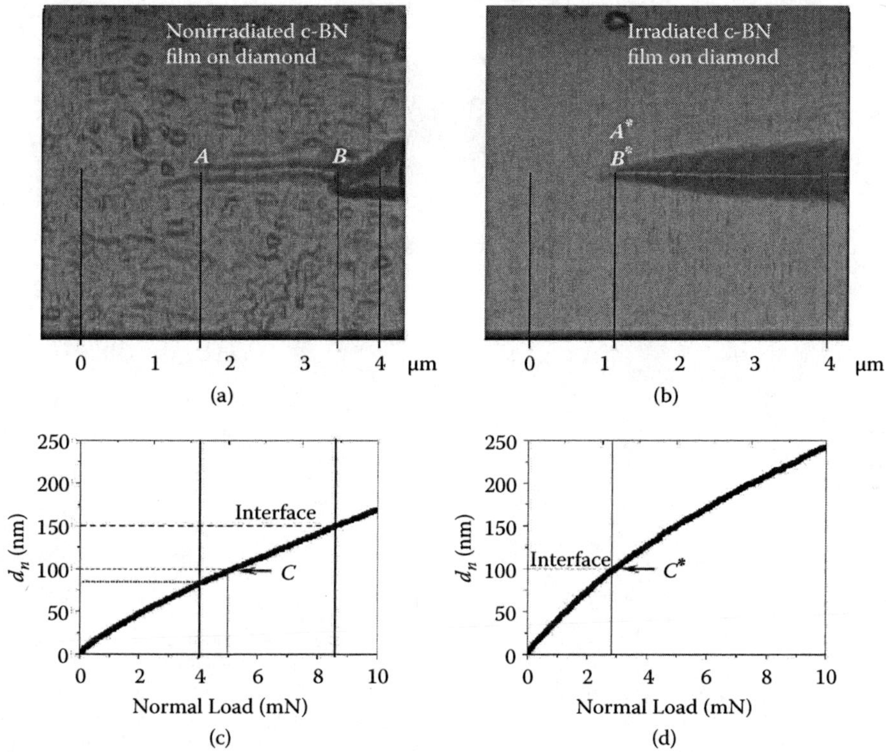

FIGURE 4.18
AFM images after the progressive load scratch test over load range from 0 to 10 mN (plotted in a three-dimensional view with a vertical scale of 730 nm) with corresponding normal displacement d_n plotted against the normal load for (a), (c) the 150 nm thick nonirradiated c-BN film on diamond and (b,d) the 100 nm thick irradiated c-BN film on diamond substrate. Solid lines indicate the beginning and the end of the scratch test as well as characteristic points of the scratch: A, A*, crack onset; B, B*, film/substrate interface. The point at d_n = 100 nm on the normal load-displacement curve is marked by C and C*. (From Deyneka-Dupriez, N., Herr, U., Fecht, H. J., Zhang, X. W., Yin, H., Boyen, H. G., and Ziemann, P., Mechanical and tribological properties of epitaxial cubic boron nitride thin films grown on diamond, *Advanced Engineering Materials* 10 (5), 482–487, Copyright 2008, Wiley-VCH. Reprinted with permission.)

Figure 4.19 illustrates the results of three scratch tests on an irradiated c-BN film on diamond. They show fairly repeatable friction behavior. Combining coefficient of friction and normal displacement data (in Figure 4.18d or schematically in Figure 4.19) and the depth of the resulting scratch track (see three-dimensional AFM image in Figure 4.19), it seems that the rise of μ is caused by different stages of deformation and fracture. In stage I probably only the c-BN film is elastically deformed. The coefficient of friction remains almost constant at this stage except for the beginning of the scratch

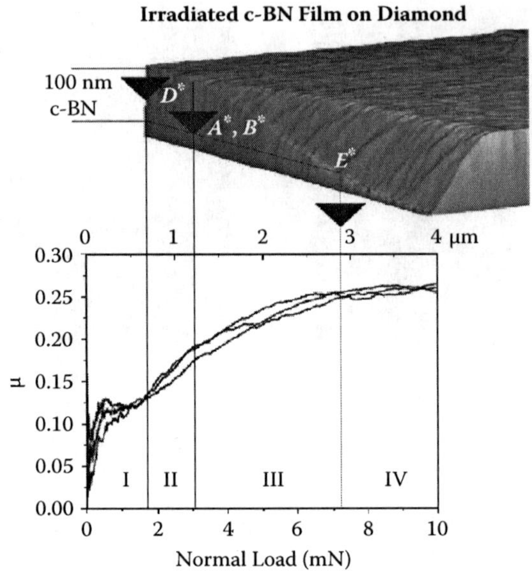

FIGURE 4.19
The coefficient of friction *l* plotted against the normal load that was obtained from the three scratch tests on the 100 nm thick irradiated c-BN film on diamond together with one of the sequential three-dimensional AFM images cross-sectioned at the center of the scratch track. Solid lines relate the coefficient of friction and the position of the indenter (black triangle) in the sample and scratch track topography. They also indicate the borders of the different regions on the *l*-curve between the characteristic points of the scratch: D*, beginning of the substrate deformation; A*, B*, beginning of the film fracture; E*, beginning of the substrate fracture. (From Deyneka-Dupriez, N., Herr, U., Fecht, H. J., Zhang, X. W., Yin, H., Boyen, H. G., and Ziemann, P., Mechanical and tribological properties of epitaxial cubic boron nitride thin films grown on diamond, *Advanced Engineering Materials* 10 (5), 482–487, Copyright 2008, Wiley-VCH. Reprinted with permission.)

at very low loads. Because no fracture takes place at stage I, the μ value in this part of the curve up to Point D*, may be interpreted as the coefficient of friction of c-BN. At stage II, starting from Point D*, both film and substrate are elastically deformed. Here, the coefficient of friction rises even though at this stage the indenter scratches only the film and hence goes through resistance caused by elastic deformation of the film similar to adhesion to c-BN. The energy dissipation due to fracture of the film makes further increase of the coefficient of friction at stage III. In this case the coefficient of friction is mostly influenced by the fracture toughness of c-BN and deformation of the diamond. At Point E*, substrate also starts to fracture, and the coefficient of friction arrives at its saturation level. Finally, at stage IV, the coefficient of friction is known by the fracture of both film and substrate.

These researchers analyzed the epitaxially grown c-BN films on diamond for different thicknesses and characterized their mechanical properties by using nanoindentation and scratch methods. In comparison with similar values for films on Si substrates, a great improved value of the hardness up to 73 GPa was observed and understood as being due to the considerably enhanced sample quality. During the scratch test the c-BN film on Si was detached suddenly at very low loads as a result of high internal stresses. The associated mechanical failure mode is delamination at the interface due to a low adhesive strength. The load dependence of coefficient of friction of irradiated epitaxial c-BN films on diamond can be explained by a superposition of the coefficients of friction for the c-BN film and the diamond substrate, bearing in mind the c-BN film thickness.

4.4.4 The Effect of the Supermodule in Multilayer Systems

It was found that the elastic modulus of the layer structures mainly rests upon the layer's thickness. In the critical thickness of about 2 nm, the modulus sharply increases with a factor of 4 (Figure 4.20), showing the effect of the supermodule. This effect can just be found is some states, such as Cu-Ni and Cu-Pd and Au-Ni and Ag-Pd, as well as multiple-layer films with high content of [111] texture. Table 4.3 presents the available data on the effect of the supermodulus on multiple-layer depositions.

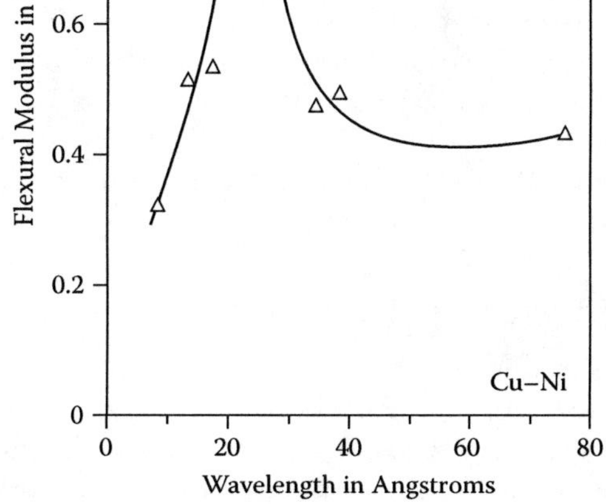

FIGURE 4.20
Changes of flexural modulus with a wavelength of modulation for Cu-Ni foil with constant composition of 50 atomic percent of Cu and [111] texture. (Reprinted from Baral, D., Ketterson, J. B., and Hilliard, J. E., Mechanical properties of composition modulated Cu-Ni foils, *Journal of Applied Physics* 57 (4), 1076–1083, Copyright 1985, with permission from American Institute of Physics.)

TABLE 4.3

Effect of the Supermodulus on Layered Nanocrystals

Lattice Structures	Component Couple	Mutual Solubility	Pct Lattice Mismatch	Supermodulus Effect
Fcc-fcc	Au-Ni	Miscible	13.5	Yes
	Cu-Pd	Miscible	7.5	Yes
	Ag-Pd	Miscible	4.9	Yes
	Cu-Ni	Miscible	2.5	Yes
	Cu-Au	Miscible	12.7	No
	Pt-Ni	Miscible		No
Bcc-fcc	Nb-Cu	Immiscible	8.9	No
	Mo-Ni	Immiscible	10.8	No
Hcp-fcc	Ti-Ni	Immiscible	19.0	No

The theoretical models for describing the supermodulus effect are categorized into two groups. The first group has a purely electrical interpretation. The presence of a high deal of coherency strains, produced in layers with high compatibility, is the base of the second important theory to explain the supermodulus effect.

Because the coherency strains can reach several percent, they have the ability to move the atoms out of the Hook zone of interatomic potential, leading to an increase of the modulus. Although in Cu-Ni films that have low coherency strains the effect of supermodulus has been reported, in multiple-layer films of Cu-Au where the coherency strains are large, the supermodulus has not been observed. Then, both theories are deficient for describing all observations, and further empirical data are needed for a clear account of the effect of the supermodulus. Recently, some researchers have suggested that the effect of the supermodulus in multilayer thin films is partly due to nonlinear elastic effects created by large biaxial strains that are produced by surface stresses of noncoherent interfaces throughout the layer. This theory can explain dependency of the supermodulus effect on wavelength and average magnitude of this effect.

4.4.5 Tribological Behavior of the Multilayer Coatings

One of the main factors in the deterioration of automobile pieces is fretting or wear with two-way movements with low pulse amplitude. Fretting by two-way movements of the pieces can occur in the range of less than several hundred nanometers and presents not only during the work, but also during the maintenance or transition period. One important aspect of fretting is that as well as occurrence of mechanical destruction, significant chemical reactions such as rapid oxidization and coherency may also happen. The studies show that multiple layers with wavelengths of less than 60 nm have a high resistance against the fretting wear.[18,62–70]

TABLE 4.4

Measured Wear Coefficient for Multiple-Layer Coatings

Specimen	Below Critical Load	Above Critical Load
Copper	—	5.0×10^{-4}
Nickel	—	2.4×10^{-4}
100 nm Ni/Cu	0.46×10^{-4}	2.2×10^{-4}
10 nm Ni/Cu	0.38×10^{-4}	5.5×10^{-4}
3.8 nm Ni/Cu	0.52×10^{-4}	5.3×10^{-4}

Note: Wear coefficient = (wear volume × hardness)/(load × distance).

Also, the result of some other studies implies that multiple-layer coatings of Cu/Ni have higher strength during the wear with steel in comparison with their constituents. Wear coefficients for multiple layers with different layer thicknesses are shown in Table 4.4. Table 4.5 demonstrates that there are different critical forces for each layer thickness, where passing that force the wear coefficient would be as much as 10 times. The critical force can be developed by an increase of resistance to plastic current and brittle failure.

Increase in mechanical strength and wear strength is created by

- Barriers for dislocations slide in the interface of two neighboring layers of Cu and Ni
- Increase of current stress for plastic deformation due to small dimension of the Cu and Ni layers

The first effect depends on both materials properties but not on layers distances of the inserted force during the experiments. The second effect increases with the decrease of layers distance and depends on the experiment force; as the large enough force overcomes this effect. This is the reason for the observed critical force in wear experiments. The critical forces of the multiple layers are listed in Table 4.5. It is concluded from the table that as the layers' distance decreases, critical force increases.

TABLE 4.5

Critical Forces for Multiple-Layer Coatings

Specimen	Observed Critical Load (*N*)	Increase in Flow Stress[a]	Calculated Critical Load (*N*)
Cu, Ni	~1	×1	~1
100 nm Ni/Cu	4.5	×3.3	3.3
10 nm Ni/Cu	14	×10	10
3.8 nm Ni/Cu	31	×16	16

[a] Based on (layer size)$^{-0.5}$.

4.5 Examples of Mechanical Affected Properties of Two-Dimensional Nanostructures

4.5.1 TiN/CrN Nanomultilayer Thin Film

Thin films with periodic depositions of two (or more) resources have recently garnered extensive attention because coating mechanical properties can be considerably enhanced. Multilayer thin films usually show improved mechanical properties and thermal stability compared to their single-layer counterparts. Some researchers showed that nanostructured TiAlN/CrN multilayers with a bilayer thickness ranging from 6 to 12 nm have high hardness values of near 36 GPa. Multilayered TiN/SiN_x coatings exhibited improved thermal stability compared to the single-layer counterpart. The researchers showed that the multilayers can be preserved after annealing at 1000°C when the SiN_x layer thickness is near 0.8 nm of multilayered TiN/SiN_x coatings. It was concluded that the lesser the bilayer period thickness is, the superior these properties are before a critical level is reached.

Researchers deposited TiN and CrN single-layer thin films along with a TiN/CrN multilayer, which has a typical layer thickness of 4 nm on tungsten carbide substrates by a commercially used unbalanced DC magnetron sputtering. They investigated the microstructure and mechanical characteristics of the as-deposited films. The TiN/CrN multilayer has a single-phase arrangement with the preferential orientation of (2 0 0), a finer surface morphology, and a finer columnar crystal arrangement. Furthermore, the TiN/CrN multilayer thin film exhibits better adhesion on substrate and hardness than the single-layer thin films. The hardness of the annealed multilayer thin film is slightly increased after annealing at temperatures of 600 to 800°C. The mechanical characteristics of the samples were assessed by the nanoindentation scratch test and hardness test. Figure 4.21 illustrates the acoustic emission signal as a function of indenter loading of the TiN single layer, CrN single layer, and TiN/CrN multilayer thin films. The curve illustrates that critical load (L_C) is not where the acoustic emission signal started. This is done by some small fractures on the two sides of scratch before thin film failure happened and therefore interfered with the acoustic emission signals. It can be seen that the critical loads are 24.7N, 34.5N, and 45.1N for a TiN single layer, CrN single layer, and TiN/CrN multilayer thin films, correspondingly.

The variation of hardness with indentation depth of the selected TiN and CrN single-layer thin films (fabricated with nitrogen flow rate of 30 sccm) and TiN/CrN multilayer are illustrated in Figure 4.22. The greatest hardness values are like the two single-layer thin films (23.9 ± 0.4 GPa in TiN and 21.4 ± 0.9 GPa in CrN). Comparable to the single-layer thin films, the hardness of the TiN/CrN multilayer thin film increases with a indentation depth near 200 nm but followed by slight hardening to an indentation depth

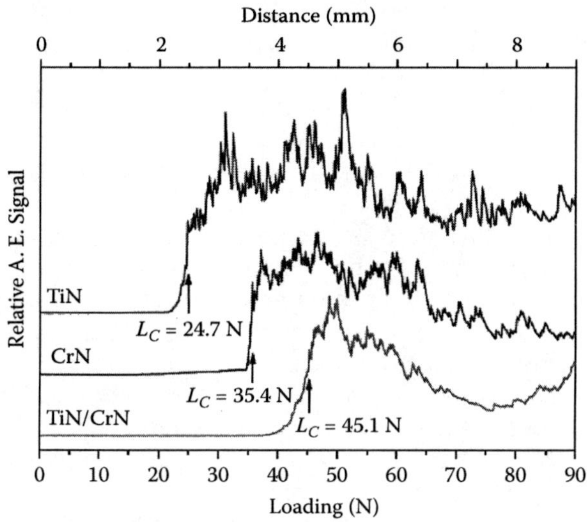

FIGURE 4.21
Scratch test results of the TiN single layer, CrN single layer, and TiN/CrN multilayer thin films. (Reprinted from Sun, P. L., Su, C. Y., Liou, T. P., Hsu, C. H., and Lin, C. K., Mechanical behavior of TiN/CrN nano-multilayer thin film deposited by unbalanced magnetron sputter process, *Journal of Alloys and Compounds* 509 (6), 3197–3201, Copyright 2011, with permission from Elsevier.)

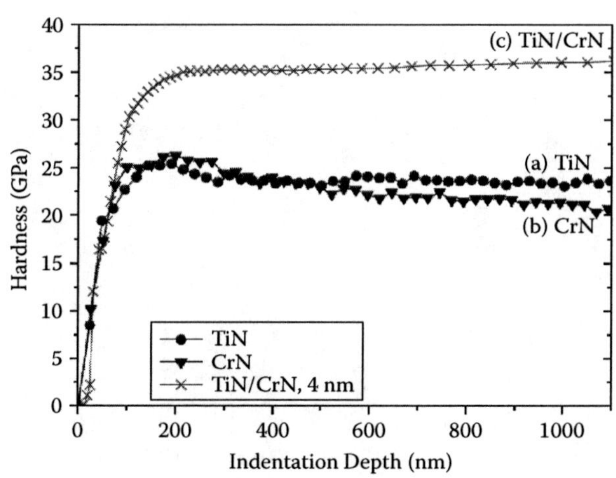

FIGURE 4.22
Hardness versus indentation depth in the TiN single layer, CrN single layer, and TiN/CrN multilayer thin films. (Reprinted from Sun, P. L., Su, C. Y., Liou, T. P., Hsu, C. H., and Lin, C. K., Mechanical behavior of TiN/CrN nano-multilayer thin film deposited by unbalanced magnetron sputter process, *Journal of Alloys and Compounds* 509 (6), 3197–3201, Copyright 2011, with permission from Elsevier.)

of 1100 nm. The hardness of the TiN/CrN multilayer thin film (34.8 ± 0.2 GPa) is much higher compared to the single-layer thin films. High strength of the multilayer thin film is to some extent caused by the fine grain arrangement, which is usually of the order of the thin film thickness and to some extent by dislocation-related movements, in which the critical stress needed to shift a dislocation depends on the thin film thickness and many interfaces that can block the dislocation progress. The hardness reduction at an indentation depth larger than 200 nm was also observed in TiN/TaN multilayer thin films, and it was recognized by the Si substrate effect. It is anticipated that the substrate effect becomes more significant with an increasing indentation depth. Furthermore, the indentation size effect also plays an important role. It was technically proved that the hardness reduces with an increasing indentation depth based on the existence of strain gradients in the deformation zone around the indent. However, the hardening behavior in the TiN/CrN multilayer thin film indicates that the substrate effect and size effect become insignificant in the multilayer thin films.

Figure 4.23 illustrates the hardness fluctuation as a function of the indentation depth of the as-deposited and annealed TiN/CrN multilayer thin films. It is seen that all the hardness of the four curves increases with an indentation depth near 200 nm. The as-deposited thin film exhibits a continuous hardening to 1100 nm. The annealed thin films exhibit a similar trend except there is a drop after 200 nm. Comparable observation was found in Al/Cu thin films, and it was attributed to the surface roughness effect. Yield fall phenomenon

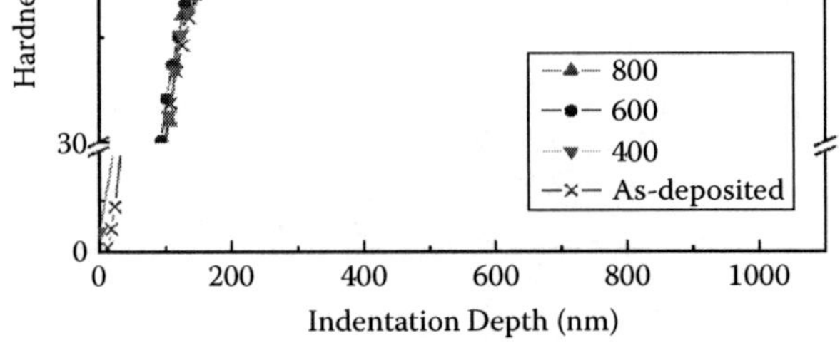

FIGURE 4.23
Nanoindentation results of TiN/CrN multilayer thin films after being annealed at different temperatures. (Reprinted from Sun, P. L., Su, C. Y., Liou, T. P., Hsu, C. H., and Lin, C. K., Mechanical behavior of TiN/CrN nano-multilayer thin film deposited by unbalanced magnetron sputter process, *Journal of Alloys and Compounds* 509 (6), 3197–3201, Copyright 2011, with permission from Elsevier.)

was observed in ultrafine-grained pure aluminum. Clarification of yield drop in LiF crystal was claimed by the shortage of movable dislocations in the specimen to complete the applied strain rate. In a crystal with quite low movable dislocations, the moving dislocations start with a high velocity and therefore a high stress is needed. When the dislocations begin to shift, dislocations increase rapidly and contribute to a lower velocity. The low velocity gives to little stress and thus yield drop occurs. It is then speculated that dislocation activity related work has caused an important effect on the hardness drop.

4.5.2 Mechanical Properties of Thin Films Using Indentation Techniques

The role of nano- and microcracks on mechanical characteristics is still not well recognized. The failure mechanism of coatings through scratch tests is very complex and depends on a lot of factors, for example the hardnesses of the coating and the substrate, the coating thickness, the scratch indenter radius, the deposition methods, the substrate material, and the scratch method. The collapse of coatings is also recognized by the deformation of both the substrate and the coating. When the coating is not thick enough, a high compressive stress field caused by the indenter during a scratch test may be transferred to the substrate; hence, the scratch response is controlled by plastic deformation of the substrate. Coatings with the same hardness may show dissimilar resistances to plastic deformation, and the deformation of a hard coating is strongly dependent upon the mixture of its hardness (H) and elastic modulus (E).

Researchers report on the mechanical properties, failure, and fracture modes in two kinds of engineering materials—that is, transparent silicon oxide thin films onto poly(ethylene terephthalate) (PET) membranes and glass-ceramic materials. The first arrangement was considered by the quasi-static indentation technique at the nanoscale and the next by the static indentation method at the microscale. Nanocomposite laminates of silicon oxide thin films onto PET were indicated to maintain higher scratch-induced stresses and were useful as protective coating material for PET membranes. Glass-ceramic materials with separated crystallites of dissimilar morphologies sustained a mixed crack propagation outline in brittle fracture mode.

Figure 4.24 illustrates the variation in hardness and modulus with penetration depth (displacement) obtained from 10 indentations in each displacement, covering the loading range 18 to 1000 μN, into $SiO_{1.8}$/PET system. Figure 4.24 illustrates an obvious rise in hardness and modulus as the indentation depth falls. Both H and E are raised in the surface region (0.47 and 5.70 GPa) and the near-surface region (i.e., 0.39, 6.65 GPa), as for higher penetration depths, corresponding in the bulk of the PET membranes, the H(E) values tend to reduce approaching the bulk H and E values (0.3 and 3.0 GPa, correspondingly).

Figure 4.25 depicts an optical micrograph of the path of an indentation-induced radial crack. The indentation was done on the glass-ceramic

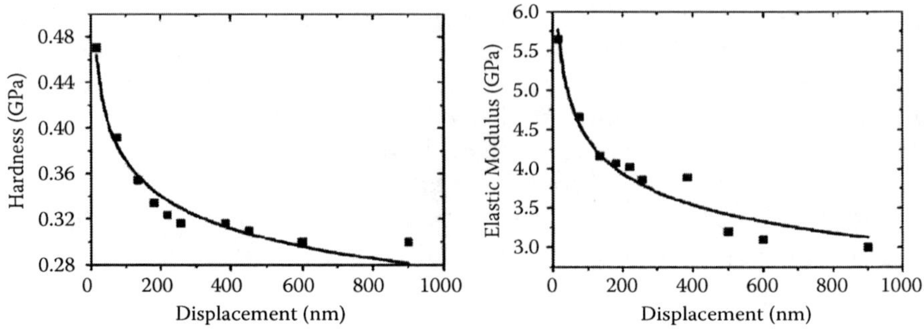

FIGURE 4.24
Hardness and elastic modulus versus displacement (solid lines are a guide for the eye). (Reprinted from Charitidis, C. A., Probing mechanical properties of thin film and ceramic materials in micro- and nano-scale using indentation techniques, *Applied Surface Science* 256 (24), 7583–7590, Copyright 2010, with permission from Elsevier.)

product. This micrograph depicts a mixed crack spread outline. In places indicated by the arrows numbered by 1, the crack is deflected by the crystallites. More exclusively, the crack spreads following the boundaries of the divided crystallites causing interfacial debonding (i.e., by separating the amorphous from the ceramic phase). In places showed by the arrows numbered by 2, the crack spreads in a direct line as it cuts through the microcrystallites, with no being deflected by the crystallite/amorphous matrix interfaces. As a result, the fracture mode is both transgranular and intergranular for this product.

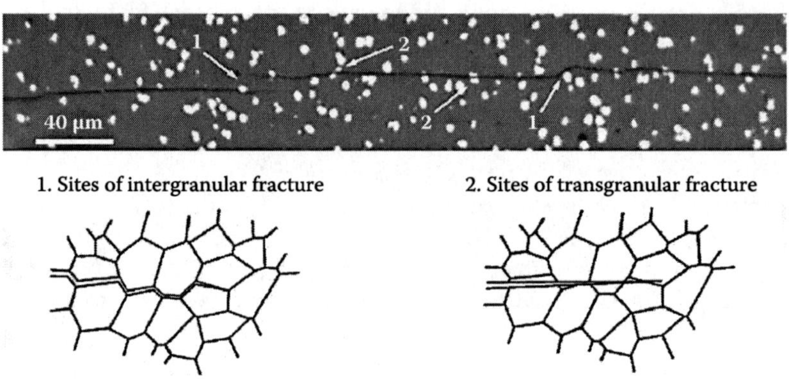

FIGURE 4.25
Optical micrograph shows a single microcrack in a glass-ceramic material. The microcrack propagation is both inter- and transgranular. (Reprinted from Charitidis, C. A., Probing mechanical properties of thin film and ceramic materials in micro- and nanoscale using indentation techniques, *Applied Surface Science* 256 (24), 7583–7590, Copyright 2010, with permission from Elsevier.)

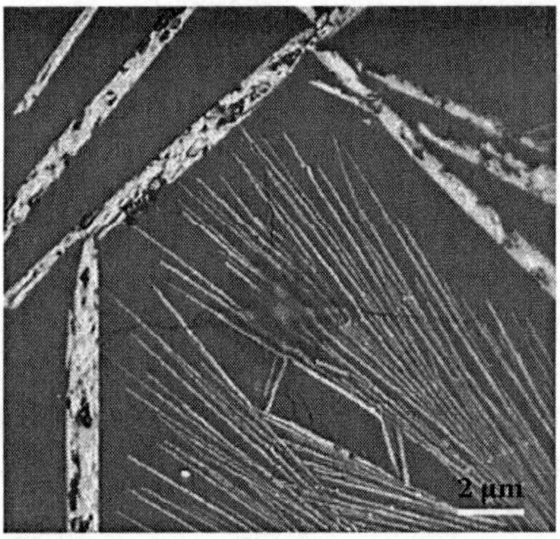

FIGURE 4.26
When the microcrack propagates perpendicularly with respect to the needle-like crystallites, the cracking occurs in a transgranular manner. When the microcrack propagates at 45° with respect to the needle-like crystallites, the cracking occurs in a mixed manner (both trans- and intergranularly). The crack has the shape of a broken line. (Reprinted from Charitidis, C. A., Probing mechanical properties of thin film and ceramic materials in micro- and nano-scale using indentation techniques, *Applied Surface Science* 256 (24), 7583–7590, Copyright 2010, with permission from Elsevier.)

In another product, the morphology is different, with accidentally oriented elongated crystallites of different widths dispersed in the remaining glass matrix. Whether these crystallites will act as a fence to crack propagation depends on their width and relative orientation with respect to the direction of the crack propagation. In Figure 4.26 all radial cracks, excluding the lower one, follow mixed-style propagation. The cracks chase for a short distance the crystalline/glass matrix interface (intergranular mode) and then they cut through it (transgranular mode). This interchange between the two propagation modes is most obviously studied in the upper crack, where the crack path has a step-like look. The lower crack finds a thicker elongated crystallite and is annihilated. The upper and left side cracks were finally annihilated from thicker elongated crystallites.

4.5.3 Nanostructured Carbon Nitride Thin Films

Carbon nitride films have emerged to be an ideal material due to their mechanical and tribological properties with a strong biocompatibility that can lead to functional uses for engineering and biomedical plans. Reports revealed that the strange combination of the properties of carbon nitride is due to the hybridization between nitrogen and carbon existing in the material. The percentage of nitrogen incorporation in the material can even alter the electrical response from conductive to highly resistive. The sputtering method is considered to be the most successful for the fabrication of carbon nitride films among different deposition methods like chemical vapor deposition (CVD), ion beam deposition, laser ablation, and so forth. Even though a great number of studies have been made, enough correlation between the bonding arrangement of nitrogen and carbon with its mechanical properties has not yet been fully explored.

Researchers studied nanostructured carbon nitride thin films that were fabricated at different radio frequency (RF) powers and a steady gas ratio of (argon:nitrogen) by RF magnetron sputtering. The atomic percentage of nitrogen:carbon (N/C) content and impedance of the films increased with a rise in RF power. The hardness of the deposited films increased from 3.12 GPa to 13.12 GPa. The increase in sp3 hybridized C–N sites and reduction of grain size with an increase in RF power is responsible for such a difference of achieved mechanical properties.

The hardness of the as-deposited films was measured by nanoindentation. Variation of hardness with RF power is shown in Figure 4.27. The plot illustrates an increase of hardness with increasing RF power. The greatest hardness of the film was revealed to be 13.12 GPa for 225 W RF power, and the minimum was 3.12 GPa at 150 W of RF power. The increase in hardness is due to the increase in N/C content inside the coating, chemical bonding between carbon and nitrogen, and surface roughness. The sp3 hybridized C–N has tetrahedral 3D geometry, which has a constant and compact arrangement. Because the increase in RF power illustrates an increase in the amount of C–N sites and surface smoothness, hardness increases.

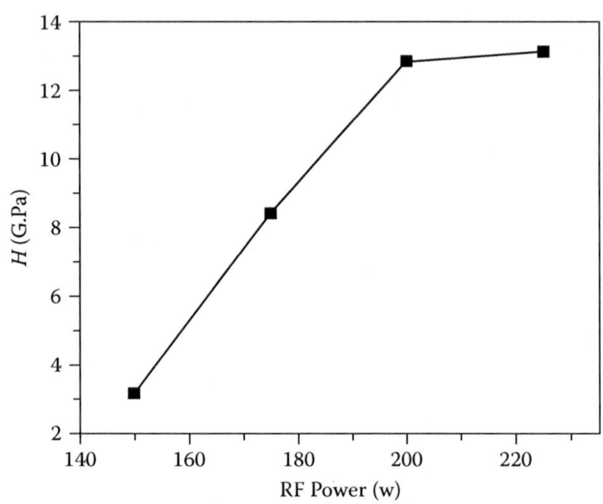

FIGURE 4.27
Variation of hardness for the films deposited at different RF power. (Reprinted from Banerjee, I., Kumari, N., Singh, A. K., Kumar, M., Laha, P., Panda, A. B., Pabi, S. K., Barhai, P. K., and Mahapatra, S. K., Influence of RF power on the electrical and mechanical properties of nanostructured carbon nitride thin films deposited by RF magnetron sputtering, *Thin Solid Films* 518 (24), 7240–7244, Copyright 2010, with permission from Elsevier.)

Nitrogen has five electrons in its external shell (1s2, 2s2, 2p3) and so is capable of bonding in sp, sp2, and sp3 hybridizations with carbon (1s2, 2s2, 2p2). Nitrogen also has lone pairs (where one bond is doubly filled by two electrons) that induce the polarization effect in carbon nitride. The bondings produced are C–N, C=N, and C≡N that are linear, trigonal, and tetrahedral, having a one-dimensional, two-dimensional, and three-dimensional arrangement, respectively. The C–N is sp3 hybridized having a linear geometry that is more compact and firm than the other two (C≡N, C=N). The increase in sp3 hybridized species raises the hardness of the materials.

4.5.4 Au–TiO_2 Nanocomposite Thin Films

In the current decade, there is increasing interest in nano metal-ceramic composite thin films due to their fascinating properties in optical non-linearity, specific heat, magnetism, photocatalysis, and their mechanical performance. So, the preparation techniques and characterization of these films have been major topics in the field of composite materials.[74–77] Nano metal-ceramic composite thin films can be fabricated by several methods such as RF sputtering, ion implantation, and the sol-gel process. The sol-gel process, as an ideal method, has gained a lot of attention. The sol-gel process can result in molecular-scale homogeneity of the initial solution, is of low cost, is fairly simple in controlling the deposition factors, and can include a variety of metal dopants into various ceramics matrixes. In addition, the sol-gel thin films show outstanding antiwear and friction reduction performances under low loads. Because of the self-lubricating and nonlinear optical characteristics of nano Au particles, Au-ceramic composite thin films fabricated by sol-gel processes are among the most attractive composite materials.

Researchers prepared nano Au-TiO_2 composite thin films on Si (100) and glass substrates with a simple sol-gel process followed by sintering. The Au particles, of diameter 14 to 22 nm depending on the sintering temperatures applied, were found to be well dispersed in the TiO_2 matrix, with a little amount of the fugitive particles from the film. The surfaces of the films were homogeneous, dense, and crack-free. Hardness and elastic modulus of the films were calculated by using the nanoindentation method. Friction and wear characteristics were inspected by a one-way reciprocating tribo-meter. It was found that the maximum hardness and elastic modulus values were found for the films prepared with a 500°C sintering temperature. The films showed superior antiwear and friction reduction performances in sliding against an AISI 52100 steel ball. They found that the friction coefficient and wear life reduced with increasing sliding speed and load. The failure mechanism of the Au-TiO_2 films was recognized to be light scuffing and abrasion. Those films can be potentially used as ultrathin lubricating layers.

TABLE 4.6

Hardness and Elastic Modulus of 5 mol% Au-TiO_2 Films on Glass at Various Sintering Temperatures

Sintering Temperature (°C)	Depth (nm)		Maximum Load (mN)	Hardness (GPa)	Elastic Modulus (GPa)
	Maximum	Plastic			
300	109.0	74.6	1.12	3.61	37.4
400	100.6	79.3	1.68	4.97	86.3
500	99.7	75.0	1.58	6.48	90.2
600	102.6	77.0	2.04	6.45	89.9

Source: Reprinted from Liu, W. M. et al. Characterization and mechanical/tribological properties of nano Au-TiO_2 composite thin films prepared by a sol-gel process, *Wear* 254 (10), 994–1000, Copyright 2003, with permission from Elsevier.

Table 4.6 shows the hardness and elastic modulus values of the films with 5 mol% Au measured by nanoindentation tests. The hardness and elastic modulus raised as the sintering temperature increased from 300°C to 500°C. At 500°C these values reached the highest value. The oxide films were porous. Sintering at a certain high temperature decreased the dimension of the pores, which in turn enhanced the mechanical properties of the films. Higher temperatures could make an adverse result on the micromechanical properties. This could be because of the growth of TiO_2 crystallites.

Figure 4.28a illustrates the friction coefficient and wear life of Au-TiO_2 composite thin films on Si (100) with different Au amounts. It is obvious that the antiwear and friction reduction performances of these films were

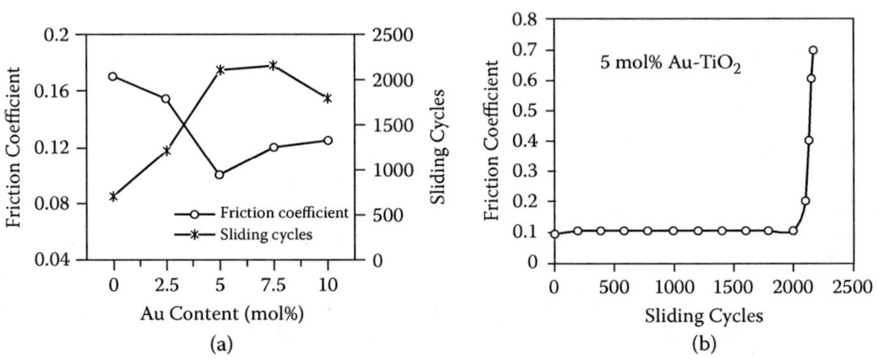

FIGURE 4.28

(a) Friction coefficient and wear life of Au-TiO_2 films with different Au contents and (b) a typical evolution of the friction coefficient versus sliding cycles for 5 mol% Au-TiO_2 film. Conditions: substrate Si (100), sintering temperature 500°C, sliding speed 90 mm/min, load 1 N. (Reprinted from Liu, W. M., Chen, Y. X., Kou, G. T., Xu, T., and Sun, D. C., Characterization and mechanical/tribological properties of nano Au-TiO_2 composite thin films prepared by a sol-gel process, *Wear* 254 (10), 994–1000, Copyright 2003, with permission from Elsevier.)

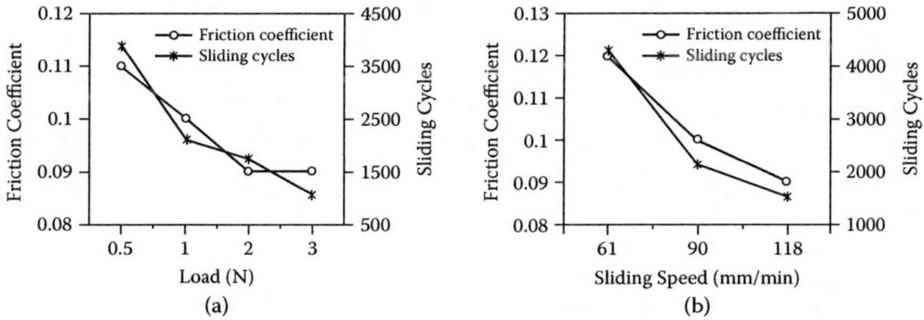

FIGURE 4.29
Friction coefficient and wear life of 5 mol% Au-TiO_2 films under different conditions (substrate Si (100), sintering temperature 500°C). (a) At 90 mm/min and (b) at 1 N. (Reprinted from Liu, W. M., Chen, Y. X., Kou, G. T., Xu, T., and Sun, D. C., Characterization and mechanical/tribological properties of nano Au-TiO_2 composite thin films prepared by a sol-gel process, *Wear* 254 (10), 994–1000, Copyright 2003, with permission from Elsevier.)

enhanced with the small contents of Au doping compared with the pure TiO_2 films. The low shear strength of Au makes its transfer easy to the surface of the steel ball through rubbing, which changed the friction into one between the Au transfer film and the composite films. As a result, the films showed improved tribological properties than pure TiO_2 films. At 5 mol% Au, the friction coefficient was as low as 0.10 and the wear life went over 2000 sliding cycles, while the pure TiO_2 film registers a quite higher friction coefficient (0.17) and a comparatively shorter wear life (only 700 sliding cycles). The typical progress of the friction coefficient versus sliding cycles for 5 mol% Au-TiO_2 film is shown in Figure 4.28b. It can be seen that the friction coefficient was kept stable with very little variation until the films failed. When the Au content became more than 5 mol%, the tribological properties got worse because the composite thin films became less homogeneous and compact. Figure 4.29 shows the tribological performances of the films with 5 mol% under different work conditions. It can be seen that the friction coefficient and wear life both reduced with increasing load and sliding speed. The reduction of the friction coefficient showed that the transfer of Au to the steel ball was simpler under higher load and sliding speed. Also the wear life of the films was reduced under high load and speed owing to high shearing force, as anticipated. These results also suggested that the performances of sol-gel films were sensitive to the sliding speed and applied load.

4.5.5 Effect of Internal Stress on the Mechanical Property of Thin CN_x Films

Almost all films are in a state of stress. The total stress is composed of a thermal stress and a fundamental stress. Depending on the deposition procedure and the deposition factors, these stresses can be either compressive or tensile. Information about the origins and characteristics of internal stress in thin films such as polycrystalline silicon films, diamond, and amorphous carbon nitride superhard films is vital to advances in micro-electromechanical systems (MEMS) and super-high-density magnetic recording systems. Inappropriate internal stress in thin films may cause MEMS device and hard disk failure by instability, curling, or fracture. Also, internal stress affects how MEMS devices act (e.g., low-pressure CVD fabricated polycrystalline silicon films are used in gate electrodes in complementary metal-oxide-semiconductor (CMOS) technology, in thin film transistor (TFT) liquid crystal displayer, and in micromachining), and a stress gradient may cause deflection of released structures.

Determination of hardness and modulus of films at the nano and bigger scales in the nanoindentation test has been extensively investigated. Researchers found that the internal stress has an effect on microtribological properties of CN_x film. They fabricated thin hard CN_x with various internal stresses by an ion beam-assisted deposition technique. A nanoindentation method was employed to estimate the micro-/nanomechanical properties of CN_x film. The resultant load-displacement data were analyzed, and it was found that the internal stress has an effect on the elastic modulus and hardness of CN_x film consistent with the amount and type of internal stress. They used a finite element method to simulate the nanoindentation process and to inspect the reliance of hardness and modulus on internal stress.

According to fabrication condition and material selection, the source of the fundamental stress can be classified as follows: (1) differences in the expansion coefficients of film and substrate; (2) incorporation of atoms (e.g., residual gases) or chemical reactions; (3) differences in the lattice spacing of monocrystalline substrates and the film during epitaxial growth; (4) variation of the interatomic spacing with the crystal size; (5) recrystallization process; (6) microscopic voids and special arrangement of dislocations; and (7) phase transformation. Thermal effects present a significant contribution to film stress. Figure 4.30 illustrates elastic-plastic behavior of films during the nanoindentation process.

Figure 4.31a illustrates the experimental results of hardness of films as a function of contact depth. It was noticed that the hardness of films with a compressive internal stress of –2 GPa (Figure 4.31a, sample 1) reduced about 1 GPa when it was annealed in a vacuum (Figure 4.31a, sample 2). It was also noticed that the hardness of films with tensile internal stress (Figure 4.31a, sample 3) alters a bit when annealed in a vacuum (Figure 4.31a, sample 4). The falling of hardness with increasing contact depth is because of substrate effects. The hardness of films with an internal stress of –0.2 GPa (samples 5 and 6) changes also and was not illustrated in the figure. Figure 4.31b illustrates the elastic modulus of films as a function of contact depth. It was also noticed that there was a

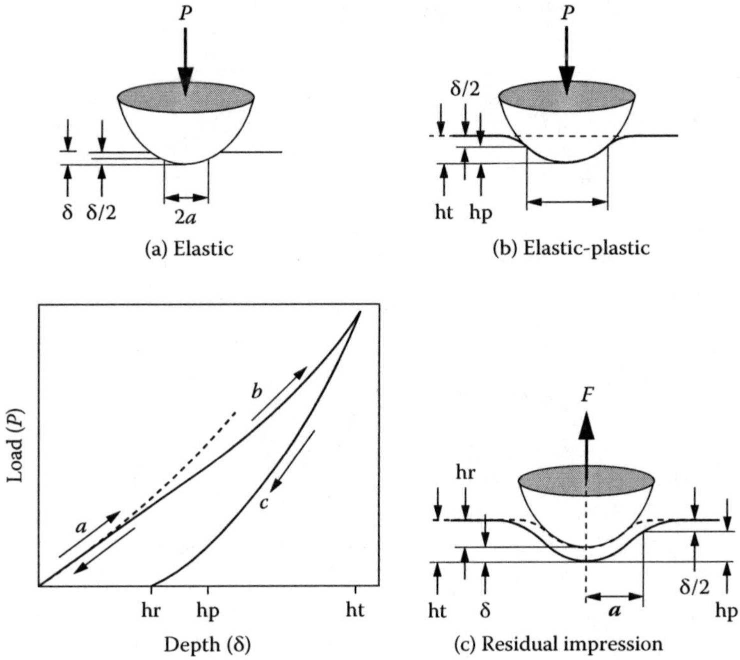

FIGURE 4.30
Elastic-plastic behavior of films during nanoindentation: (a) elastic regime, (b) plastic deformation above a critical load on loading, (c) elastic behavior on unloading. (Reprinted from Bai, M., Kato, K., Umehara, N., and Miyake, Y., Nanoindentation and FEM study of the effect of internal stress on micro/nano mechanical property of thin CN_x films, *Thin Solid Films* 377–378, 138–147, Copyright 2000, with permission from Elsevier.)

considerable decrease of elastic modulus for both films with compressive internal stress (Figure 4.31b, sample 1) and the film with tensile internal stress (sample 3) after the film was annealed in a vacuum (Figure 4.31b, samples 2 and 4).

The nodal stresses due to the presence of internal stresses in films and the substrate were created by simulation of film/substrate bending due to interface mismatch. The nanoindentation dynamic simulations were performed under various applied forces and various internal stresses in films. Figure 4.32 illustrates the applied load as a function of vertical displacement of indenter for a film under different internal stresses. The loading and unloading curve moves to the left when the film was exposed to compressive internal stress. The higher the internal stress, the more the curve moves. Though, for the films exposed to tensile internal stress, the curves move to the right. The higher the tensile internal stress is, the more the curve moves.

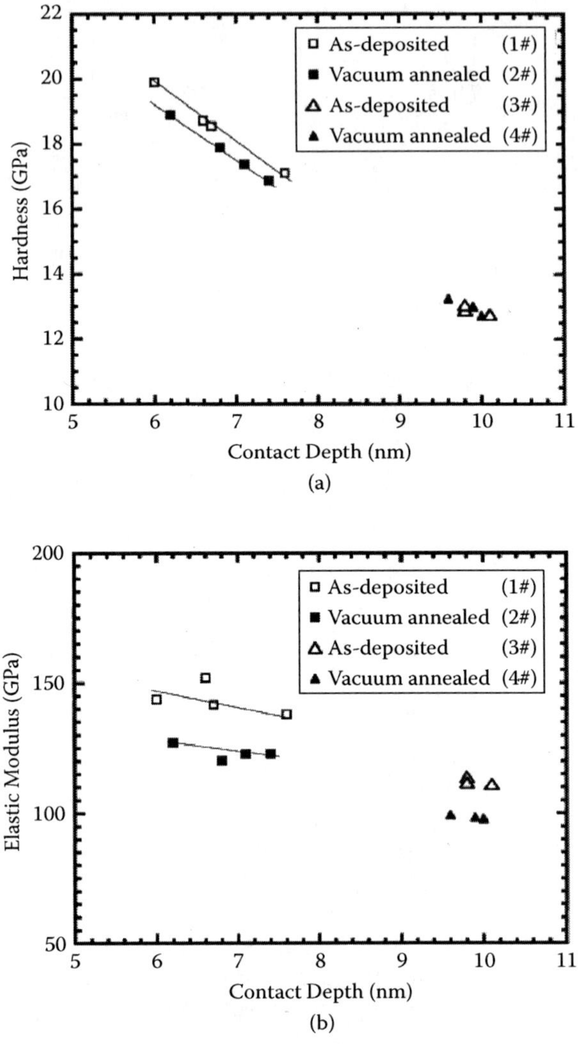

FIGURE 4.31
Measured (a) hardness and (b) modulus of films as a function of contact depth for samples 1 through 4. Empty symbol: as-deposited, filled symbol: vacuum annealed. (Reprinted from Bai, M., Kato, K., Umehara, N., and Miyake, Y., Nanoindentation and FEM study of the effect of internal stress on micro/nano mechanical property of thin CN_x films, *Thin Solid Films* 377–378, 138–147, Copyright 2000, with permission from Elsevier.)

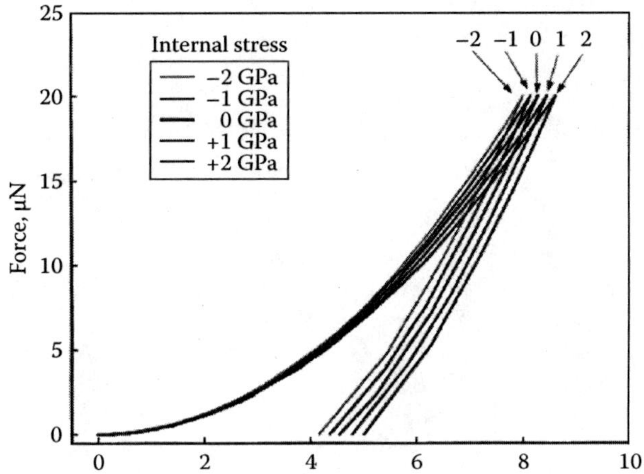

FIGURE 4.32
Simulated loading and unloading curves for the films with internal stresses of −2, −1, 0, 1, and 2 GPa, respectively. (Reprinted from Bai, M., Kato, K., Umehara, N., and Miyake, Y., Nanoindentation and FEM study of the effect of internal stress on micro/nano mechanical property of thin CN_x films, *Thin Solid Films* 377–378, 138–147, Copyright 2000, with permission from Elsevier.)

References

1. Aliofkhazraei, M., and Sabour Rouhaghdam, A., Wear and coating removal mechanism of alumina/titania nanocomposite layer fabricated by plasma electrolysis, *Surface and Coatings Technology* 205 (Suppl. 2), S57–S62, 2011.
2. Aliofkhazraei, M., and Sabour Rouhaghdam, A., Fabrication of TiC/WC ultra hard nanocomposite layers by plasma electrolysis and study of its characteristics, *Surface and Coatings Technology* 205 (Suppl. 1), S51–S56, 2010.
3. Aliofkhazraei, M., Yousefi, M., Ahangarani, S., and Rouhaghdam, A. S., Synthesis and properties of ceramic-based nanocomposite layer of aluminum carbide embedded with oriented carbon nanotubes, *Ceramics International* 37 (7), 2151–2157, 2011.
4. Cho, J., Joshi, M. S., and Sun, C. T., Effect of inclusion size on mechanical properties of polymeric composites with micro and nano particles, *Composites Science and Technology* 66 (13), 1941–1952, 2006.
5. Fong, H., Sarikaya, M., White, S. N., and Snead, M. L., Nano-mechanical properties profiles across dentin-enamel junction of human incisor teeth, *Materials Science and Engineering C* 7 (2), 119–128, 2000.
6. Gao, L., Wang, H. Z., Hong, J. S., Miyamoto, H., Miyamoto, K., Nishikawa, Y., and Torre, S. D. D. L., Mechanical properties and microstructure of nano-SiC-Al_2O_3 composites densified by spark plasma sintering, *Journal of the European Ceramic Society* 19 (5), 609–613, 1999.

7. Gojny, F. H., Wichmann, M. H. G., Fiedler, B., Bauhofer, W., and Schulte, K., Influence of nano-modification on the mechanical and electrical properties of conventional fibre-reinforced composites, *Composites Part A: Applied Science and Manufacturing* 36 (11), 1525–1535, 2005.
8. He, G., Eckert, J., Löser, W., and Hagiwara, M., Composition dependence of the microstructure and the mechanical properties of nano/ultrafine-structured Ti-Cu-Ni-Sn-Nb alloys, *Acta Materialia* 52 (10), 3035–3046, 2004.
9. Hong, Z., Zhang, P., He, C., Qiu, X., Liu, A., Chen, L., Chen, X., and Jing, X., Nano-composite of poly(L-lactide) and surface grafted hydroxyapatite: Mechanical properties and biocompatibility, *Biomaterials* 26 (32), 6296–6304, 2005.
10. Kumar, R., Prakash, K. H., Cheang, P., and Khor, K. A., Microstructure and mechanical properties of spark plasma sintered zirconia-hydroxyapatite nano-composite powders, *Acta Materialia* 53 (8), 2327–2335, 2005.
11. Mukai, T., Kawazoe, M., and Higashi, K., Dynamic mechanical properties of a near-nano aluminum alloy processed by equal-channel-angular-extrusion, *Nanostructured Materials* 10 (5), 755–765, 1998.
12. Park, J. H., and Jana, S. C., The relationship between nano- and micro-structures and mechanical properties in PMMA-epoxy-nanoclay composites, *Polymer* 44 (7), 2091–2100, 2003.
13. Tien, Y. I., and Wei, K. H., The effect of nano-sized silicate layers from montmorillonite on glass transition, dynamic mechanical, and thermal degradation properties of segmented polyurethane, *Journal of Applied Polymer Science* 86 (7), 1741–1748, 2002.
14. Tsuji, N., Ueji, R., Minamino, Y., and Saito, Y., A new and simple process to obtain nano-structured bulk low-carbon steel with superior mechanical property, *Scripta Materialia* 46 (4), 305–310, 2002.
15. Kraft, O., and Volkert, C. A., Mechanical testing of thin films and small structures, *Advanced Engineering Materials* 3 (3), 99–110, 2001.
16. Ma, K. J., Bloyce, A., and Bell, T., Examination of mechanical properties and failure mechanisms of TiN and Ti-TiN multilayer coatings, *Surface and Coatings Technology* 76–77 (1 –3 pt 1), 297–302, 1995.
17. Ding, J., Meng, Y., and Wen, S., Mechanical properties and fracture toughness of multilayer hard coatings using nanoindentation, *Thin Solid Films* 371 (1), 178–182, 2000.
18. Okumiya, M., and Griepentrog, M., Mechanical properties and tribological behavior of TiN-CrAlN and CrN-CrAlN multilayer coatings, *Surface and Coatings Technology* 112 (1–3), 123–128, 1999.
19. Papo, M. J., Catledge, S. A., Vohra, Y. K., and Machado, C., Mechanical wear behavior of nanocrystalline and multilayer diamond coatings on temporomandibular joint implants, *Journal of Materials Science: Materials in Medicine* 15 (7), 773–777, 2004.
20. Weber, F. R., Fontaine, F., Scheib, M., and Bock, W., Cathodic arc evaporation of (Ti,Al)N coatings and (Ti,Al)N/TiN multilayer-coatings-correlation between lifetime of coated cutting tools, structural and mechanical film properties, *Surface and Coatings Technology* 177–178, 227–232, 2004.
21. Gubisch, M., Liu, Y., Spiess, L., Romanus, H., Krischok, S., Ecke, G., Schaefer, J. A., and Knedlik, C., Nanoscale multilayer WC/C coatings developed for nanopositioning: Part I. Microstructures and mechanical properties, *Thin Solid Films* 488 (1–2), 132–139, 2005.

22. Wei, C., Fin Lin, J., Jiang, T. H., and Ai, C. F., Tribological characteristics of titanium nitride and titanium carbonitride multilayer films. Part I. The effect of coating sequence on material and mechanical properties, *Thin Solid Films* 381 (1), 94–103, 2001.
23. Chen, L., Wang, S. Q., Du, Y., and Li, J., Microstructure and mechanical properties of gradient Ti(C, N) and TiN/Ti(C, N) multilayer PVD coatings, *Materials Science and Engineering A* 478 (1–2), 336–339, 2008.
24. Lee, C., Wei, X., Li, Q., Carpick, R., Kysar, J. W., and Hone, J., Elastic and frictional properties of graphene, *Physica Status Solidi (B) Basic Research* 246 (11–12), 2562–2567, 2009.
25. Holleck, H., Lahres, M., and Woll, P., Multilayer coatings—Influence of fabrication parameters on constitution and properties, *Surface and Coatings Technology* 41 (2), 179–190, 1990.
26. Ji, J., Tan, Q., Fan, D. Z., Sun, F. Y., Barbosa, M. A., and Shen, J., Fabrication of alternating polycation and albumin multilayer coating onto stainless steel by electrostatic layer-by-layer adsorption, *Colloids and Surfaces B: Biointerfaces* 34 (3), 185–190, 2004.
27. Liu, X., Dai, B., Zhou, L., and Sun, J., Polymeric complexes as building blocks for rapid fabrication of layer-by-layer assembled multilayer films and their application as superhydrophobic coatings, *Journal of Materials Chemistry* 19 (4), 497–504, 2009.
28. Chiba, K., and Kaminishi, S., Fabrication and optical properties of low-emissivity coatings of AlSiN and AgCuNd-alloy multilayer films on glass, *Japanese Journal of Applied Physics* 47 (1), 240–243, 2008.
29. Ji, J., Tan, Q., Fan, D. Z., Sun, Y. F., Barbosa, M. A., and Shen, J., Fabrication of alternating polycation and albumin multilayer coating by electrostatic layer-by-layer adsorption, *Journal of Materials Science* 39 (1), 349–351, 2004.
30. Okudera, H., and Nonami, T., Fabrication of silica-anatase multilayer coating on a K-Ca-Zn-Si glass substrate, *Thin Solid Films* 441 (1–2), 50–55, 2003.
31. Ueda, Y., Hataya, N., and Zaman, H., Magnetoresistance effect of Co/Cu multilayer film produced by electrodeposition method, *Journal of Magnetism and Magnetic Materials* 156 (1–3), 350–352, 1996.
32. Péter, L., Kupay, Z., Cziráki, Á, Pádár, J., Tóth, J., and Bakonyi, I., Additive effects in multilayer electrodeposition: Properties of Co-Cu/Cu multilayers deposited with NaCl additive, *Journal of Physical Chemistry B* 105 (44), 10867–10873, 2001.
33. Lachenwitzer, A., and Magnussen, O. M., Electrochemical quartz crystal microbalance study on the kinetics of nickel monolayer and multilayer electrodeposition on (111)-oriented gold films, *Journal of Physical Chemistry B* 104 (31), 7424–7430, 2000.
34. Chaure, N. B., Samantilleke, A. P., Burton, R. P., Young, J., and Dharmadasa, I. M., Electrodeposition of p^+, p, i, n and n^+-type copper indium gallium diselenide for development of multilayer thin film solar cells, *Thin Solid Films* 472 (1–2), 212–216, 2005.
35. Fei, J. Y., and Wilcox, G. D., Electrodeposition of zinc-nickel compositionally modulated multilayer coatings and their corrosion behaviours, *Surface and Coatings Technology* 200 (11), 3533–3539, 2006.
36. Yamada, A., Houga, T., and Ueda, Y., Magnetism and magnetoresistance of Co/Cu multilayer films produced by pulse control electrodeposition method, *Journal of Magnetism and Magnetic Materials* 239 (1–3), 272–275, 2002.

37. Onoda, M., Shimizu, K., Tsuchiya, T., and Watanabe, T., Preparation of amorphous/crystalloid soft magnetic multilayer Ni-Co-B alloy films by electrodeposition, *Journal of Magnetism and Magnetic Materials* 126 (1–3), 595–598, 1993.
38. Crousier, J., Hanane, Z., and Crousier, J. P., Electrodeposition of NiP amorphous alloys: A multilayer structure, *Thin Solid Films* 248 (1), 51–56, 1994.
39. Jayasena, B., and Subbiah, S., A novel mechanical cleavage method for synthesizing few-layer graphenes, *Nanoscale Research Letters* 6 (1), 1–7, 2011.
40. Kim, K. S., Zhao, Y., Jang, H., Lee, S. Y., Kim, J. M., Kim, K. S., Ahn, J. H., Kim, P., Choi, J. Y., and Hong, B. H., Large-scale pattern growth of graphene films for stretchable transparent electrodes, *Nature* 457 (7230), 706–710, 2009.
41. Mirkarimi, P. B., McCarty, K. F., and Medlin, D. L., Review of advances in cubic boron nitride film synthesis, *Materials Science and Engineering R: Reports* 21 (2), 47–100, 1997.
42. Safadi, B., Andrews, R., and Grulke, E. A., Multiwalled carbon nanotube polymer composites: Synthesis and characterization of thin films, *Journal of Applied Polymer Science* 84 (14), 2660–2669, 2002.
43. Martinu, L., and Poitras, D., Plasma deposition of optical films and coatings: A review, *Journal of Vacuum Science and Technology A: Vacuum, Surfaces and Films* 18 (6), 2619–2645, 2000.
44. Shchukin, D. G., and Sukhorukov, G. B., Nanoparticle synthesis in engineered organic nanoscale reactors, *Advanced Materials* 16 (8), 671–682, 2004.
45. Yu, Y. Y., Chen, C. Y., and Chen, W. C., Synthesis and characterization of organic-inorganic hybrid thin films from poly(acrylic) and monodispersed colloidal silica, *Polymer* 44 (3), 593–601, 2002.
46. Moore, J. J., and Feng, H. J., Combustion synthesis of advanced materials: Part II. Classification, applications and modelling, *Progress in Materials Science* 39 (4–5), 275–316, 1995.
47. Pujari, V. B., Mane, S. H., Karande, V. S., Dargad, J. S., and Deshmukh, L. P., Mercury-cadmium-selenide thin film layers: Structural, microscopic and spectral response studies, *Materials Chemistry and Physics* 83 (1), 10–15, 2004.
48. Sutrave, D. S., Shahane, G. S., Patil, V. B., and Deshmukh, L. P., Microcrystallographic and optical studies on $Cd_{1-x}Zn_xSe$ thin films, *Materials Chemistry and Physics* 65 (3), 298–305, 2000.
49. Pellicer, E., Varea, A., Pané, S., Nelson, B. J., Menéndez, E., Estrader, M., Suriñach, S., Baró, M. D., Nogués, J., and Sort, J., Nanocrystalline electroplated Cu-Ni: Metallic thin films with enhanced mechanical properties and tunable magnetic behavior, *Advanced Functional Materials* 20 (6), 983–991, 2010.
50. Foecke, T., and Lashmore, D. S., Mechanical behavior of compositionally modulated alloys, *Scripta Metallurgica et Materiala* 27 (6), 651–656, 1992.
51. Proville, L., and Bakó, B., Dislocation depinning from ordered nanophases in a model FCC crystal: From cutting mechanism to Orowan looping, *Acta Materialia* 58 (17), 5565–5571, 2010.
52. Bandyopadhyay, S. N., A note on Orowan mechanism below the root of an yielding crack like notch in a strain-hardening metal, *Engineering Fracture Mechanics* 16 (6), 889–893, 1982.
53. Bandyopadhyay, S. N., Orowan mechanism for abrupt localized flow and fracture initiation in metals ahead of an yielding notch, *Engineering Fracture Mechanics* 16 (6), 871–887, 1982.

54. Li, Z., Hou, C., Huang, M., and Ouyang, C., Strengthening mechanism in micro-polycrystals with penetrable grain boundaries by discrete dislocation dynamics simulation and Hall-Petch effect, *Computational Materials Science* 46 (4), 1124–1134, 2009.
55. Pozdnyakov, V. A., Mechanisms of plastic deformation and the anomalies of the Hall-Petch dependence in metallic nanocrystalline materials, *Fizika Metallov i Metallovedenie* 96 (1), 114–128, 2003.
56. Wolf, D., Yamakov, V., Phillpot, S. R., and Mukherjee, A. K., Deformation mechanism and inverse Hall-Petch behavior in nanocrystalline materials, *Zeitschrift fuer Metallkunde/Materials Research and Advanced Techniques* 94 (10), 1091–1097, 2003.
57. Pozdnyakov, V. A., Mechanisms of plastic deformation and the anomalies of the Hall-Petch dependence in metallic nanocrystalline materials, *Physics of Metals and Metallography* 96 (1), 105–119, 2003.
58. Takeuchi, S., The mechanism of the inverse Hall-Petch relation of nanocrystals, *Scripta Materialia* 44 (8–9), 1483–1487, 2001.
59. McKeown, J., Misra, A., Kung, H., Hoagland, R. G., and Nastasi, M., Microstructures and strength of nanoscale Cu-Ag multilayers, *Scripta Materialia* 46 (8), 593–598, 2002.
60. Deyneka-Dupriez, N., Herr, U., Fecht, H. J., Zhang, X. W., Yin, H., Boyen, H. G., and Ziemann, P., Mechanical and tribological properties of epitaxial cubic boron nitride thin films grown on diamond, *Advanced Engineering Materials* 10 (5), 482–487, 2008.
61. Baral, D., Ketterson, J. B., and Hilliard, J. E., Mechanical properties of composition modulated Cu-Ni foils, *Journal of Applied Physics* 57 (4), 1076–1083, 1985.
62. Subramanian, C., and Strafford, K. N., Review of multicomponent and multilayer coatings for tribological applications, *Wear* 165 (1), 85–95, 1993.
63. Rincón, C., Zambrano, G., Carvajal, A., Prieto, P., Galindo, H., Martínez, E., Lousa, A., and Esteve, J., Tungsten carbide/diamond-like carbon multilayer coating on steel for tribological applications, *Surface and Coatings Technology* 148 (2–3), 277–283, 2001.
64. Mollart, T. P., Haupt, J., Gilmore, R., and Gissler, W., Tribological behaviour of homogeneous Ti-B-N, Ti-B-N-C and TiN/h-BN/TiB_2 multilayer coatings, *Surface and Coatings Technology* 86–87 (Part 1), 231–236, 1996.
65. Simmonds, M. C., Savan, A., Van Swygenhoven, H., Pflüger, E., and Mikhailov, S., Structural, morphological, chemical and tribological investigations of sputter deposited MoS_x/metal multilayer coatings, *Surface and Coatings Technology* 108–109, 340–344, 1998.
66. Mikhailov, S., Savan, A., Pflüger, E., Knoblauch, L., Hauert, R., Simmonds, M., and Van Swygenhoven, H., Morphology and tribological properties of metal (oxide)-MoS_2 nanostructured multilayer coatings, *Surface and Coatings Technology* 105 (1–2), 175–183, 1998.
67. Chen, Y. H., Polonsky, I. A., Chung, Y. W., and Keer, L. M., Tribological properties and rolling-contact-fatigue lives of TiN/SiN_x multilayer coatings, *Surface and Coatings Technology* 154 (2–3), 152–161, 2002.

68. Luo, Q., Hovsepian, P. E., Lewis, D. B., Münz, W. D., Kok, Y. N., Cockrem, J., Bolton, M., and Farinotti, A., Tribological properties of unbalanced magnetron sputtered nano-scale multilayer coatings TiAlN/VN and TiAlCrYN deposited on plasma nitrided steels, *Surface and Coatings Technology* 193 (1–3 Spec. Iss.), 39–45, 2005.
69. Kok, Y. N., Hovsepian, P. E., Luo, Q., Lewis, D. B., Wen, J. G., and Petrov, I., Influence of the bias voltage on the structure and the tribological performance of nanoscale multilayer C/Cr PVD coatings, *Thin Solid Films* 475 (1–2 Spec. Iss.), 219–226, 2005.
70. Hovsepian, P. E., Lewis, D. B., Luo, Q., Münz, W. D., Mayrhofer, P. H., Mitterer, C., Zhou, Z., and Rainforth, W. M., TiAlN based nanoscale multilayer coatings designed to adapt their tribological properties at elevated temperatures, *Thin Solid Films* 485 (1–2), 160–168, 2005.
71. Sun, P. L., Su, C. Y., Liou, T. P., Hsu, C. H., and Lin, C. K., Mechanical behavior of TiN/CrN nano-multilayer thin film deposited by unbalanced magnetron sputter process, *Journal of Alloys and Compounds* 509 (6), 3197–3201, 2011.
72. Charitidis, C. A., Probing mechanical properties of thin film and ceramic materials in micro- and nano-scale using indentation techniques, *Applied Surface Science* 256 (24), 7583–7590, 2010.
73. Banerjee, I., Kumari, N., Singh, A. K., Kumar, M., Laha, P., Panda, A. B., Pabi, S. K., Barhai, P. K., and Mahapatra, S. K., Influence of RF power on the electrical and mechanical properties of nano-structured carbon nitride thin films deposited by RF magnetron sputtering, *Thin Solid Films* 518 (24), 7240–7244, 2010.
74. Mirzamohammadi, S., Aliov, M. K., Aghdam, A. S. R., Velashjerdi, M., and Naimi-Jamal, M. R., Tribological properties of tertiary Al_2O_3/CNT/ nanodiamond pulsed electrodeposited Ni-W nanocomposite, *Materials Science and Technology* 27 (2), 546–550, 2011.
75. Mirzamohammadi, S., Aliov, M. K., Sabur, A. R., and Hassanzadeh-Tabrizi, A., Study of wear resistance and nanostructure of tertiary Al_2O_3/Y_2O_3/CNT pulsed electrodeposited Ni-based nanocomposite, *Materials Science* 46 (1), 76–86, 2010.
76. Mirzamohammadi, S., Kiarasi, R., Aliov, M. K., Sabur, A. R., and Hassanzadeh-Tabrizi, A., Study of corrosion resistance and nanostructure for tertiary Al_2O_3/Y_2O_3/CNT pulsed electrodeposited Ni based nanocomposite, *Transactions of the Institute of Metal Finishing* 88 (2), 93–99, 2010.
77. Aliov, M. K., and Sabur, A. R., Formation of a novel hard binary SiO_2/quantum dot nanocomposite with predictable electrical conductivity, *Modern Physics Letters B* 24 (1), 89–96, 2010.
78. Liu, W. M., Chen, Y. X., Kou, G. T., Xu, T., and Sun, D. C., Characterization and mechanical/tribological properties of nano Au-TiO_2 composite thin films prepared by a sol-gel process, *Wear* 254 (10), 994–1000, 2003.
79. Bai, M., Kato, K., Umehara, N., and Miyake, Y., Nanoindentation and FEM study of the effect of internal stress on micro/nano mechanical property of thin CN_x films, *Thin Solid Films* 377–378, 138–147, 2000.

5

Chemical/Electrochemical Fabrication/ Properties of Two-Dimensional Nanostructures

5.1 Introduction

The chemical/electrochemical method is one of the oldest methods to produce nanomaterials. In this method the primary materials would be solved in an ordinary solvent first, and then the deposition agent would be added. Nonaqueous solvents are usable in this method so that they would have many advantages. For instance, more parameters would exist to improve the conditions of interaction (stable deposition agent in a nonaqueous environment, reduction potentials, etc.). The deposition agent can be a complex, reducing, or oxidation factor. Then it is possible to stabilize nanoparticles in various ways after their formation, for instance by space methods, electrostatic methods, or a combination of both. One of the most important advantages of this method is that there is no need to use professional equipment, and it can be used by equipment and tools that exist in a chemistry or physics laboratory easily.[1–5]

5.2 History

One of the simplest ways to fabricate nanoparticles is to deposit low-soluble materials from their aqua solvents that would change to oxide then due to thermal decomposition. The codeposition reaction includes nucleation, growth, and coagulation (making clots) procedure. Due to problems in studying and separating each process, the exact mechanism of the codeposition method is not completely recognized. Reactions of deposition are the oldest techniques of procuring nanomaterials. Primary materials are solved in an ordinary solvent first in this reaction, and then the deposition agent would be added.

The deposition agent can be a complex, a reducing, or an oxidizing agent. Adding these materials to the solution would make insoluble cores. Following the progress of reaction, Oswald ripening would be done. By continuing the reaction, deposited cores would be ready for more reactions during processing and prepared drying processes.[6–11]

Mineral reactants' reduction does not make deposition all the time. For instance, $HAuCl_4$ (II) would be reduced by sodium citrate and would make a red colloid solution. When $HAuCl_4$ is heated by distilled water and would be titered with sodium citrate, the solution would be changed from light yellow to red. Particle size and the color of the colloid solution would be controlled by various concentrations of sodium citrate. Sodium citrate is a reducing agent, and citrate ions would help to stabilize Au colloids. By decreasing sodium citrate concentration, the stability of particles would be decreased and therefore Au particles would be larger. If the surface energy of particles is controllable, interaction among them would be reduced, and more stable nanoparticles would be made that prevent saturation.

The main advantage of deposition is the high quality of produced nanomaterials. Although production of nanomaterials with the concerned size can be difficult, existing kinetic factors can be used for growth of particles. Adding complexion agents and surfactants can be effective in controlling the size of particles.[12–15]

For instance, in construction of Au nanoparticles, surfactants and nonaqueous materials can be used. Some researchers could find Au nanoparticles by using tetractyl ammonia bromide surfactant and toluene. These materials would make a micelle in which Au ions are reduced by sodium borohydride. Surfactants would help the stability of the surface of Au particles and let them be spread in the toluene phase. Side products would be removed through washing by sulfuric acid.

Generally, key properties in deposition reactions are as follows:

1. Products of these reactions have low solubility in supersaturation conditions.
2. Nucleation is the key step that would be done in a supersaturation condition. Many small particles (primary cores) would be made in this step.
3. Secondary processes are necessary, like Oswald ripening and making a clot that affects size, morphology, and properties of the products.
4. The supersaturation condition is usually being made during a chemical reaction and affects particle size, geometrical shape, and distribution of particle size.

5.3 Direct Writing of Metal Nanostructures

Continued development in the fast-growing field of nanoplasmonics will need the improvement of new techniques for the manufacture of metal nanostructures. Optical lithography offers a constantly increasing toolbox. Two-photon procedures, as demonstrated by some researchers, enable the manufacture of gold nanostructures encapsulated in dielectric material in an easy, direct procedure and offer the view of three-dimensional production. At higher resolution, scanning probe methods allow nanoparticle placement by localized oxidation, and near-field sintering of nanoparticulate films enables direct writing of nanowires. Direct laser "printing" of single gold nanoparticles offers a noteworthy ability to control manufacture of model structures for fundamental studies, particle by particle. Optical techniques continue to provide powerful support for research into metamaterials.

The amazing optical characteristics of nanostructured metals are attracting strong interest. Gold and silver nanocrystals occupy large improvements in the signals of common optical reporters (e.g., fluorescent labels, Raman dyes), guiding in a new era of ultrasensitive biomolecular analysis. Carefully tailored morphologies (hemispheres, rods, crosses, holes, bowties, crescents, and others) allow the design of structures to achieve optical phenomena. Metamaterials interact in new and unprecedented ways with electromagnetic radiation, promising to render science fiction into reality in the form of cloaking devices, plasmonic waveguides, and other new technologies.

In a lot of these fast-developing areas, the improvement of fabrication tools remains a main focus of consideration, and often, the lack of appropriate tools is a barrier to the improvement of new technologies. With no doubt, electron-beam lithography continues to set the bar for a final performance. An illustration of its capability for fundamental studies was provided by some researchers. They described the manufacture of gold nanoparticle pairs, in which the particles were formed (90 to 110 nm rods or 100 nm diameter rings) and placed in different preparations on surfaces to achieve varying extinction spectra. However, e-beam techniques rely upon a series of processing and utilize expensive instrumentation that is not readily easy to get to many workers. In the semiconductor tool industry, e-beam methods remain a means of fabricating photolithography barriers but not a manufacturing tool; in the study of metamaterials, likewise, e-beam techniques are an accurate tool for the manufacture of model materials but undoubtedly suffer limitations.

The simplest, cheapest way to manufacture metal nanostructures, and one that has been employed by many researchers in the field of plasmonics, is colloidal lithography. A film of colloidal particles is shaped on a substrate and used as a shadow cover: metal is evaporated onto the particle film,

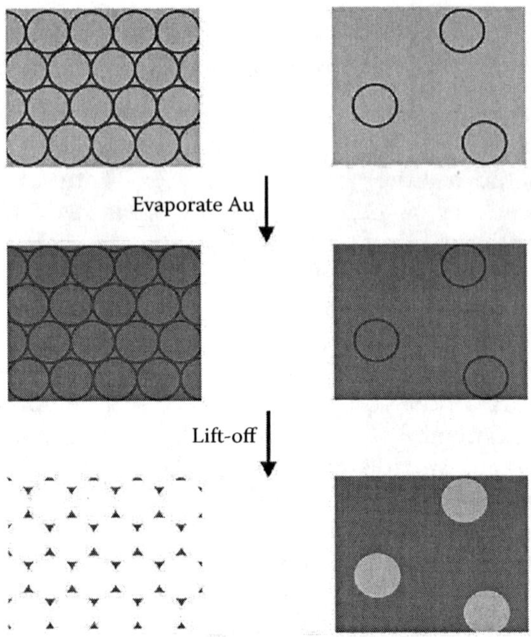

FIGURE 5.1
Fabrication of gold nanoprisms (left) and nanohole arrays (right) using colloidal lithography. (Reprinted from Leggett, G. J., Direct writing of metal nanostructures: Lithographic tools for nanoplasmonics research, *ACS Nano* 5 (3), 1575–1579, Copyright 2011, with permission from the American Chemical Society.)

which classically exhibits hexagonal close packing under an optimal situation. The particles cover the substrate efficiently during metal deposition, but the metal atoms can penetrate the particle film at the interstices, where pyramidal structures are shaped in a regular, periodic array (Figure 5.1). The colloidal particle cover can then be removed in an easy lift-off step. Some control of the periodicity is achievable by controlling the particle diameter. An alternate but analogous way is to deposit colloidal particles more lightly on the substrate; metal deposition then leads to a permanent metal film punctuated by holes where nanoparticles masked the substrate.

Regardless of the beautiful simplicity of such approaches, they present access to a limited range of architectures, and there has thus been an enormous deal of attention in the expansion of methods that would allow much greater control to be exercised over the morphologies of the metal nanostructures. A wish-list for manufacturing tools for nanoplasmonics would include high-resolution, low-cost, high-throughput, ease of process, and maybe, also, the capacity for three-dimensional manufacture.

Two-photon processes are beautiful for materials processing. In the high passing intensities achieved in femtosecond pulses delivered by a

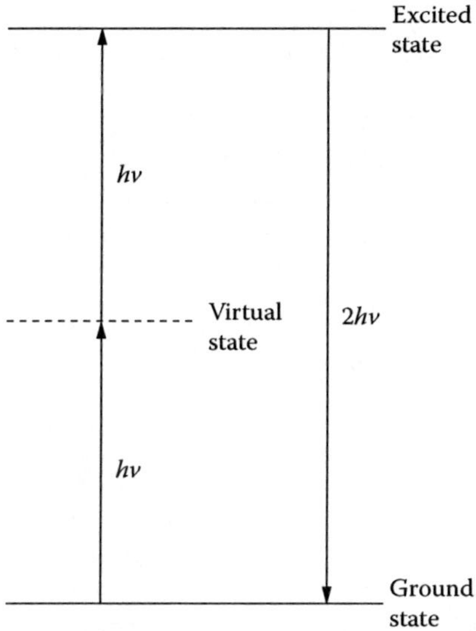

FIGURE 5.2
A two-photon excitation via a virtual quantum state. (Reprinted from Leggett, G. J., Direct writing of metal nanostructures: Lithographic tools for nanoplasmonics research, *ACS Nano* 5 (3), 1575–1579, Copyright 2011, with permission from the American Chemical Society.)

Ti-sapphire laser, nonlinear optical responses are energized. In a two-photon transition, an arrangement is excited from a ground state to an excited state via a virtual quantum state; for incident photons of frequency ν, the absorbed energy is equivalent to $2h\nu$, where h is Planck's constant (Figure 5.2). In photolithography, two-photon processes could be used to get enhanced resolution. The strength of two-photon absorption fluctuates quadratically with the power of the incident illumination; hence, there is an effective sharpening of the excitation beam. By cautious selection of the excitation source, photons could be used that have energies that are too little to perform modification of the photoresist but, when combined in two-photon absorption, deliver adequate energy. Because the cross section for two-photon absorption is minute, high transient powers are necessary (high denoted powers would degrade the resist through heating), and therefore we have the use of femtosecond-pulsed laser arrangements.

Optical methods go on surprising us. Lithographic methods based upon the application of light are varied, powerful, flexible, and usually need much less complex instrumentation than electron beam lithography. Optics will certainly continue not only to be the base of the exciting new science created by metamaterials but also to offer the means to meet a lot of the manufacturing requirements of the nanoplasmonics society.

5.4 Theory and Thermodynamic Method of Codeposition

Most chemical reactions of this method take place in a way that the products have low solubility to achieve the supersaturation condition rapidly.

Various chemical reactions are used in this method. For instance, a simple addition reaction is used to make an electrolyte like A_xB_y:

$$XA^{y+}(aq) + yB^{x-}(aq) \rightarrow A_xB_s(s) \tag{5.1}$$

The equilibrium relation of the reaction is as follows, and K_{sp} is the solvent constant of the product:

$$K_{sp} = (a_A)^x(a_B)^y \tag{5.2}$$

a_A and a_B are the activity of existing cations and anions of the solution. K_{sp} and subsequently solubility are very low for many hydroxides, carbonates, oxalates, and chalcogenides in aqueous solutions. Tables of K_{sp} of various materials are shown in reference books.*

In simple exchange/addition reactions, deposition can be produced through other methods like chemical reaction, reduction by light, oxidation, and hydrolysis. It is possible to make the deposition by changing certain parameters (like concentration and temperature).

While the product of the reaction includes one or two elements (like a metal, dual oxide, etc.), the deposition reaction would be so simple. However, most of the time the system is dual or quadrate, and it is difficult to perceive processes in this condition because various types should deposit with each other. Mostly there is no guarantee for making nanometric products or single sputtering during deposition of a complex. Nucleation and growth processes would control the size and geometrical shape of the particles. When the deposition reaction starts, a large amount of small crystals would be made first (nucleation step). However, these deposits are unstable and so tend to ratify and clot (growth process) to make larger particles that are more stable thermodynamically.[17–27]

5.4.1 Nucleation

A key point of each deposition process is the supersaturation degree (S) that is found as follows:

$$S = \frac{a_A.a_B}{K_{sp}} \quad \text{or} \quad S = \frac{C}{C_{eq}} \tag{5.3}$$

* References such as D. L. Reger, S. R. Goode and D. W. Ball, *Chemistry: Principles and Practice*, Belmont, CA: Brooks/Cole, Cengage Learning, c. 2010.

while C and C_{eq} show the concentration of soluble matter in saturation and equilibrium forms. The difference between them shows the force required for deposition ($\Delta C = C - C_{eq}$).

Nucleation would start in a supersaturation solution with equilibrium critical radius (R^*) with the following equation:

$$R^* = \frac{\alpha}{\Delta C} \tag{5.4}$$

α would be found through the following equation:

$$\alpha = \left[\frac{2\sigma_s L}{KT \ln S}\right] \upsilon C_\infty \tag{5.5}$$

$\sigma_s L$ is the stress of the surface phase between liquid and solid. υ is the atomic volume of the soluble matter. K is the Boltzman constant, and T is the temperature. S is the supersaturation parameter that is found in Equation (5.6).

If $R > R^*$, the growth process would continue. While $R < R^*$ the formed particles would solve. Activation energy of the particle-making process would be supplied through the following equation:

$$\Delta G^\circ = \frac{4\pi\sigma_{SL} R^{*2}}{3} = \frac{16\pi\sigma_{SL}^3 \upsilon^2}{3K^2T^2 \ln^2 S} \tag{5.6}$$

Therefore, in constant conditions, homogenous nucleation speed (R_N) would be as follows:

$$R_N = \left(\frac{dN}{dt}\right)\frac{1}{V} = A \exp\left[\frac{-(\Delta G^*)}{KT}\right] \tag{5.7}$$

N shows the number of cores formed in a time unit and volume unit. V and A are constants that are generally between 10^{25} and 10^{56} $s^{-1}m^{-3}$.

Equation (5.8) would be found out of compounding Equations (5.6) and (5.7):

$$R_N = A \exp\left[\frac{-16\pi\sigma_{SL}^3 \upsilon^2}{3K^3T^3 \ln^2 S}\right] \tag{5.8}$$

This equation shows that R_N is a function of S. R_N is very low until the supersaturation critical condition (S^*) is reached.

In 1896, Friedrich Wilhelm Ostwald stated that the crystallization process would be done during a procedure called Ostwald. He reiterated that exchanging unstable or semistable conditions to stable ones is generally being done during several steps. Primary products are mostly unstable thermodynamically, and medium products have a free energy close to that of primary material. For instance, if citric acid is crystallized in a temperature higher than 34°C, a crystal without water that is stable thermodynamically would be formed, while a monohydrate crystal would be stable at room temperature. In similar crystallization systems like ferrous sulfate and sulfate hydrogen sodium, similar phenomena are observed. Crystals of the primary product have a structure the same as that of the solvent. In the example of citric acid, the structure of a monohydrate is similar to that of the hydrous one. While the solution temperature is increased, solubility of the crystal and structure of the solvent would be changed; consequently, nonaqueous product would be more stable. Exchange between two kinetic and thermodynamic products in critical transition temperature would happen. In lower temperature the product would tend to a hydrous structure (kinetic product), and in higher temperatures a nonaqueous structure (thermodynamic product) would be preferred. The other important point is that free energy distribution among surfaces, for crystals larger than 1 μm, is so low in the semistable condition. Ostwald ripening would be performed near equilibrium conditions. During this ripening larger clusters would be made. Therefore, despite the large number of small crystals in the primary system, they would disappear gradually and deposit on larger crystals. In polycrystals, larger crystals with lower surface free energy would make smaller crystals; therefore, particle growth would be decreased progressively, and larger crystals with smaller size distribution would be formed.

Ostwald's research was the basis of crystallization theories for several decades. In 1950 an American chemist, Victor Kuhn LaMer, developed concepts of colloid sols crystallization. LaMer theorized that crystallization is an influx-buoyancy process. He explained his model in three steps. These steps are shown in Figure 5.3. In primary steps of the reaction, molar concentration would be increased until critical concentration. This concentration would be named *supercritical concentration* and it can be said that

$$\Delta\mu = k_B \ln\left(\frac{C}{C_s}\right) \tag{5.9}$$

K_B is Boltzmen's constant, and T is temperature. A supercritical condition would be explained by the relation below:

$$\frac{\Delta\mu}{k_B T} = \frac{\ln C}{C_s} \tag{5.10}$$

At this point, materials start to make a clot, and molecular clusters are in balance with molecule islands. At the second step, clusters are more stable

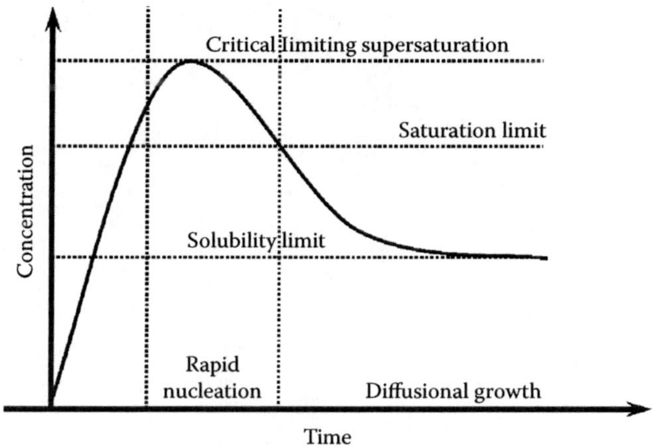

FIGURE 5.3
Nucleation process, material concentration would be increased until supercritical concentration. At this point, the nucleation process would happen and then concentration would be decreased.

to lead to nucleation processes. Meanwhile, the first stable nucleuses would be formed. Consequently, molar concentration of the materials would be decreased partly and lead to the supersaturation condition. Nucleation can be done in two ways: homogenous and heterogeneous nucleation. At this step, a material's concentration would be decreased to under supercritical concentration immediately, and nucleation would be stopped and no longer continued. Then cores would be stable under Ostwald and diffusive processes.

Crystal growth would continue until the molar concentration reached the solubility limit and finally the formed crystals would be in balance with the solved ones. Many research was performed on this model, and many advantages are fulfilled based on a simple LaMer's model.[28–32]

5.4.2 Growth

In the growth process, deposited particles agglomerate and make larger particles. Gradients of concentration and temperature are among factors determining the growth speed of particles. Balance relation of particles as a monomer with the surface of spherical crystals is as follows:

$$\frac{dr}{dr} = D\Omega\left(\frac{1}{\delta} + \frac{1}{r}\right)(C_b - C_i) \tag{5.11}$$

r is the crystal radius, t is time, D is monomer permeability, Ω is molar volume, and δ_i is the thickness of the layer that changes following the change of C_b (concentration of solved massive material) to C_i (concentration of solved material around the crystal surface).

The relation between monomer concentration and crystal size is being estimated by the Gibbs–Thomson's relation:

$$C_e(r) \approx C_\infty \left[\frac{1 + 2\Omega\gamma}{R_G T R} \right] \tag{5.12}$$

γ is the stress among dimensions, R_G is the universal gas constant, T is the temperature, and C_∞ is the equation constant. Finally, the relation between the growth speed (G) and relativity of supersaturation would be as follows (shown as a low-pressure equation):

$$G = \frac{dl}{dt} = K_G S^g \tag{5.13}$$

K_G is the growth speed constant, and g is the level of the growth process. Figure 5.4 is a schematic figure of the deposition process.

5.4.3 Ostwald Ripening

The first mathematical equations to explain this process were found by Lifshitz, Slyozv, and Wagner. Their model is called the LSW theory. A great deal of research has been done on this theory. The most prominent defect of

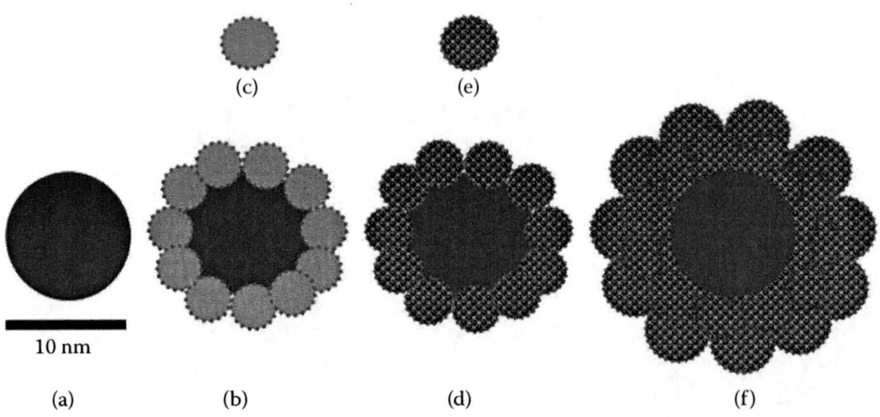

FIGURE 5.4
Schematic nucleation and growth mechanism of EL (Ni-P)/radar absorbing material (RAM) nanocomposite powder. (a) RAM particle, (b) deposition of single (Ni-P) layer onto RAM particle consists of (c) amorphous (Ni-P globule) matrix with Ni nucleation sites (in lighter color) onto it, (d) annealed single layer of Ni-P matrixes consist of (e) precipitation of Ni and Ni_3P nanocrystals onto the Ni-P globule, and (f) optimum thickness of annealed EL Ni-P layers. (Reprinted from Sharma, R., Agarwala, R. C., and Agarwala, V., Development of electroless (Ni-P)/$BaNi_{0.4}Ti_{0.4}Fe_{11.2}O_{19}$ nanocomposite powder for enhanced microwave absorption, *Journal of Alloys and Compounds* 467 (1–2), 357–365, Copyright 2009, with permission from Elsevier.)

LSW theory is that the mutual effects among cores and steady changes of solved material with time are not estimated sufficiently.

The new research justifies deficiencies of the LSW theory to some extent to make a balance between the results of the theory and practice. In other words, the most complete theory for this process is still based on the main LSW theory. This theory is still appropriate due to observation of some important points on the deposition of nanoparticles.

The main LSW theory is briefly as follows:

1. Average radius of deposited particles for the controlled process of permeation is a function of time (t):

$$\bar{r}(t) = \sqrt[3]{Kt} \tag{5.14}$$

   K is the clot speed that is equal to ($4\alpha D/9$), and D is the permeation flow of solved material among the particles' border. Particle size is proportionate to the second power of time.

2. Density (N) is as follows during the controlled process of permeation:

$$N(t) = \frac{0.22Q_0}{\bar{R}(t)^a} = \frac{0.22}{2D\alpha t} \tag{5.15}$$

   Q_0 is equal to the total innate supersaturation. The number of particles solved with t^{-1} would be decreased during the clotting process.

3. Size distribution of particles is as follows:

$$\rho(t) = \frac{R}{R(t)}, \; f(R,t) = \left[\frac{N(t)}{R(t)}\right] p_0(\rho(t)) \tag{5.16}$$

   $p(\rho)_0$ is a function of time that its unit is the number of particles. Considering Equations (5.1) through (5.16), there are useful views regarding deposition of nanoparticles. In the fabrication of nanoparticles, the nucleation process is relatively rapid and the growth process is relatively slow. Making particles with low size distribution requires simultaneous formation of all cores and no formation of smaller particles in the next step.[34–45]

5.4.4 Final Growth and Formation of Stable Nanoparticles

Considering the thermodynamic referred to above, the proportion of surface to volume of the particles would reach its maximum level. Consequently, if the particles are released in the solution without any stabilizer they would practically be clotted. However, in most research, nanoparticles would be separated in the form of a colloid solution or a powder-like production. It should be noticed that although clotting can be done in each step of the fabrication of nanomaterials, it is more important during deposition.

The stability of nanomaterials would be generally done through the methods below:

1. Space barrier: Among particles that would be formed through bonds among surfactants, polymers, or other organic materials (that are generally called capping ligands) with surface of nanoparticles.

   Lu et al.[46] procured cobalt nanoparticles by various capping ligands. They found that although using pure stearic acid would make nanoparticles with a smaller size distribution, because of a weak bond with nanoparticles the particles cannot have a good self-assembly during evaporation of the solvent. This can be solved by mixing stearic acid with oleic acid or the transisomer of elaidic acid. Arrangement of these ligands around Au nanoparticles is shown in Figure 5.5. Figure 5.6 illustrates transmission electron microscopy (TEM) images of a regular two-dimensional (2D) array of spherical Co nanoparticles synthesized with oleic acid added as a capping agent. The molar ratio of oleic acid to dicobalt octacarbonyl was 0.3. It also shows a selected-area electron diffraction (SAED) pattern taken from an assembly of cobalt nanoparticles, with all rings indexed to diffractions from ε-Co.

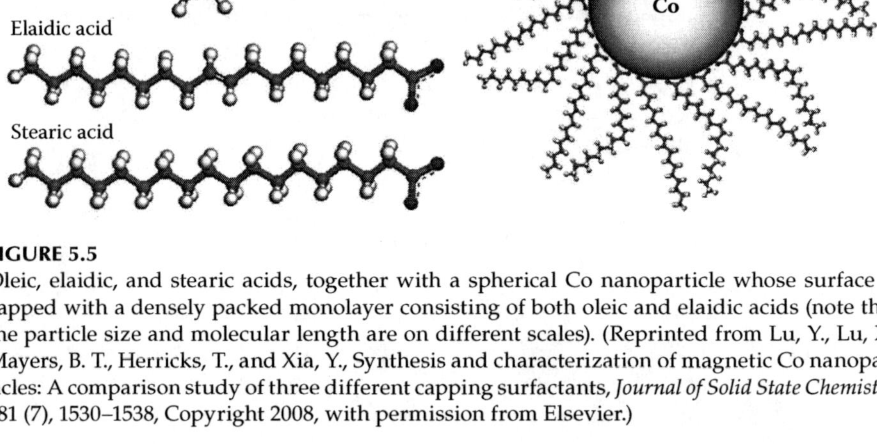

FIGURE 5.5
Oleic, elaidic, and stearic acids, together with a spherical Co nanoparticle whose surface is capped with a densely packed monolayer consisting of both oleic and elaidic acids (note that the particle size and molecular length are on different scales). (Reprinted from Lu, Y., Lu, X., Mayers, B. T., Herricks, T., and Xia, Y., Synthesis and characterization of magnetic Co nanoparticles: A comparison study of three different capping surfactants, *Journal of Solid State Chemistry* 181 (7), 1530–1538, Copyright 2008, with permission from Elsevier.)

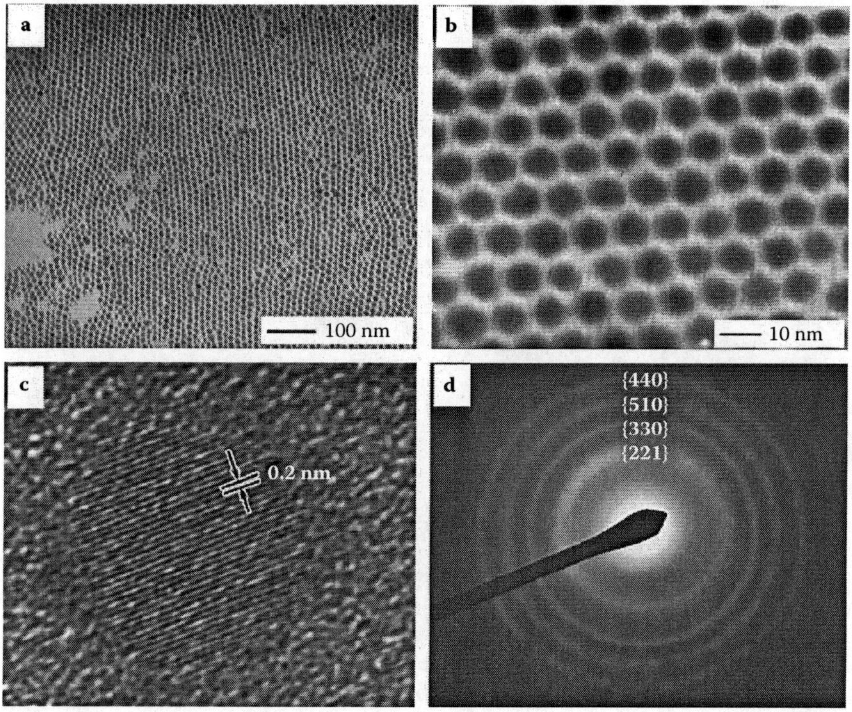

FIGURE 5.6
(a, b) Transmission electron micrograph (TEM) images (at two different magnifications) of a regular two-dimensional array of spherical Co nanoparticles synthesized with oleic acid added as a capping agent. The molar ratio of oleic acid to dicobalt octacarbonyl was 0.3 (i.e., oleic acid/cobalt = 0.6). (c) High-resolution TEM image of a single Co nanoparticle, with the lattice spacing matching the distance between (221) planes of ε-Co. (d) Selected-area electron diffraction (SAED) pattern taken from an assembly of Co nanoparticles, with all rings indexed to diffractions from ε-Co. (Reprinted from Lu, Y., Lu, X., Mayers, B. T., Herricks, T., and Xia, Y., Synthesis and characterization of magnetic Co nanoparticles: A comparison study of three different capping surfactants, *Journal of Solid State Chemistry* 181 (7), 1530–1538, Copyright 2008, with permission from Elsevier.)

2. Electrostatic barrier: This is due to the surface charge of particles and is generally H^+ or OH^- (although it can be other particles). Scanning tunneling microscopy (STM) and TEM studies have been conducted on Pd clusters that are covered by tetra botyl bromide ammonium that has made monolayers on the surface of particles at least in certain capping ligands.

   Au colloid particles are surrounded by a dual electric layer of chlorine anions and citrate and cations absorbed by them. This dual layer might make colonic propulsion among particles. This network is shown in Figure 5.5. Space stability is definitely more general. It can be due to chemical stability of nanoparticles in high pHs.

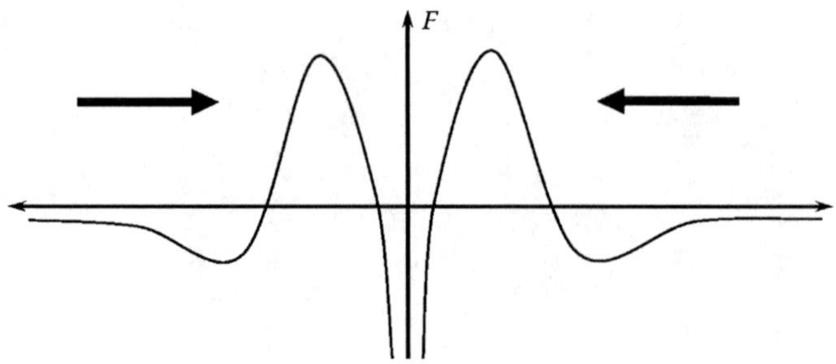

FIGURE 5.7
Electrostatic stability of colloid nanoparticles.

3. Electrospace stability: It would be made by electrostatic stability with space stability. The key point here is the absorption of large molecules like polymers or surfactants on the surface of particles. These active protecting layers on the surface would coordinate to the surface of the particle and would be solved in the environment. This is the main advantage of using halides of tetraoctylammonium as protectors. Halide anions would bond to the metal surface through their own negative charge, while long chains of alkyl would cover the metal core like an umbrella. When the concentration of a stabilizer in the solution is high, the particles would have a particular configuration, and this decreases disorder and increases free energy instead (Figure 5.7).

Surface ligands have two main duties: avoiding clots of the particles and their distribution in concerning solvents. For each of the above applications, ligands should have a particular space chemistry. In the former, radicals that bond with donor atoms should tend toward the surface of particles in a way that covers the surface of particles as much as possible. In the second application that dissolves nanoparticles more in nonpolar solvents, ligand radicals should take nanoparticles into a solvent's molecules as much as possible (Figure 5.8).

Figure 5.9 also shows two external configurations of hydrophobic ligands on the surface of a nanoparticle. While the bond is strong and there are no other strong ligands in the solution, ligand configuration would be changed from (b) to (a) by increasing the amount of (N-L) ligand. This change in ligand configuration would be recognizable following the change in surface curve through change of particle size and length of hydrocarbon chain.

Some researchers have stabilized Au nanoparticles by using macrocyclic polyammonium cations. The scheme of a used cation is shown

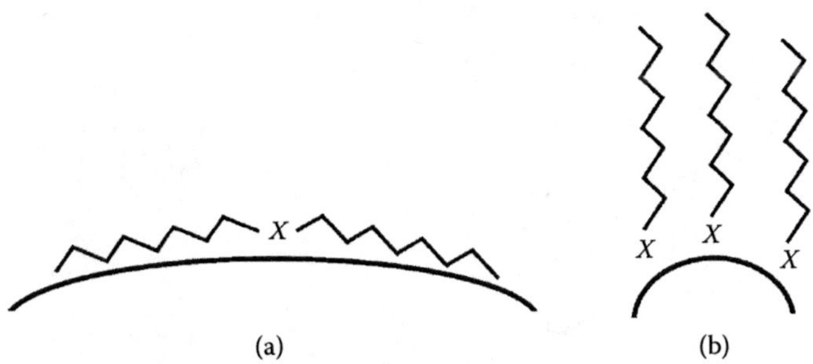

FIGURE 5.8
Two kinds of ligands configuration around nanoparticles in (a) and (b).

in Figure 5.10. Figure 5.11 also shows the electrostatic barrier among Au nanoparticles by these cations. It should be mentioned that due to the large amount, this cation can stabilize Au nanoparticles spatially and electrostatically simultaneously.

4. Stability by solvent: solvents like tetrahydrofuran (THF), THF/NaOH, and propylene carbonate can act as colloid stabilizers. Long-chain alcohols also can be used as colloid stabilizers successfully for metallic nanoparticles.

Size and morphology of nanoparticles can be controllable to some extent. This can be done through controlling certain conditions like pH and cation concentration and deposition agents (Figure 5.12). The most effective parameter on size and its distribution is the cation's concentration. When the metal's concentration increases, particles will be larger. Their isomorphism would be decreased as well.

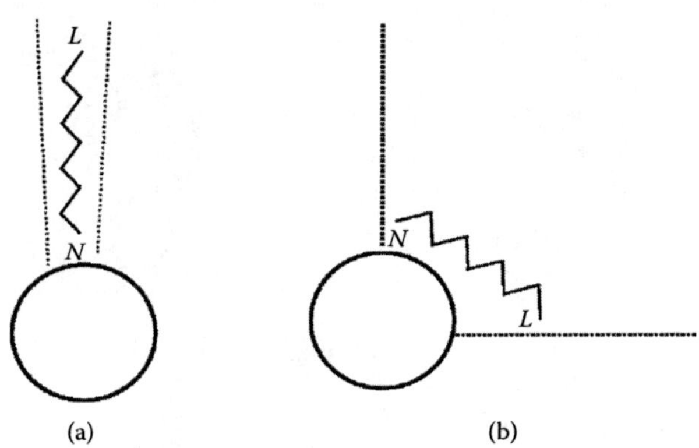

FIGURE 5.9
Two models for configuration of hydrophobic hydrocarbon chains in (a) and (b). (*N* is the donor, and *L* is ligands hydrocarbon.)

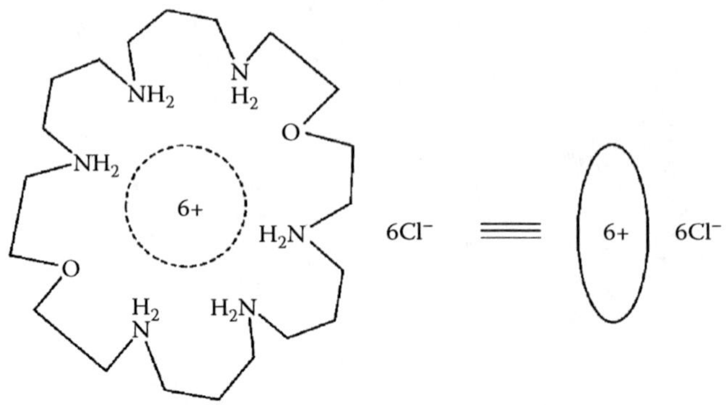

FIGURE 5.10
Macromolecule [28]ane-$(NH^{+}_{2})_6O_2$.$6Cl^-$ or 28-MCPAC. (Reprinted from Misra, T. K., and Liu, C. Y., Synthesis of 28-membered macrocyclic polyammonium cations functionalized gold nanoparticles and their potential for sensing nucleotides, *Journal of Colloid and Interface Science* 326 (2), 411–419, Copyright 2008, with permission from Elsevier.)

5.5 Phase Transition of Two-Dimensional Nanostructure by Electrochemical Potential

Phenomena related to chirality have held a deep attraction for scientists since Pasteur physically divided left-handed and right-handed sodium ammonium tartrate crystals. It is one of the most beautiful areas in chemistry due to the important usage in heterogeneous and asymmetric catalysis, enantiomeric selectivity, and chemical and pharmaceutical industries. During recent

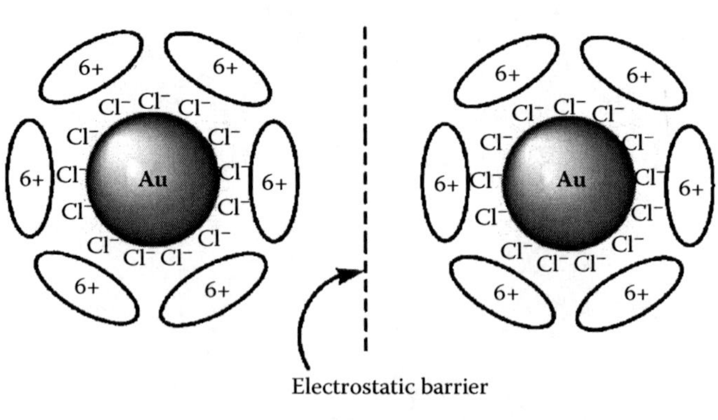

FIGURE 5.11
Au nanoparticles stabilized by 28-MCPAC. (Reprinted from Misra, T. K., and Liu, C. Y., Synthesis of 28-membered macrocyclic polyammonium cations functionalized gold nanoparticles and their potential for sensing nucleotides, *Journal of Colloid and Interface Science* 326 (2), 411–419, Copyright 2008, with permission from Elsevier.)

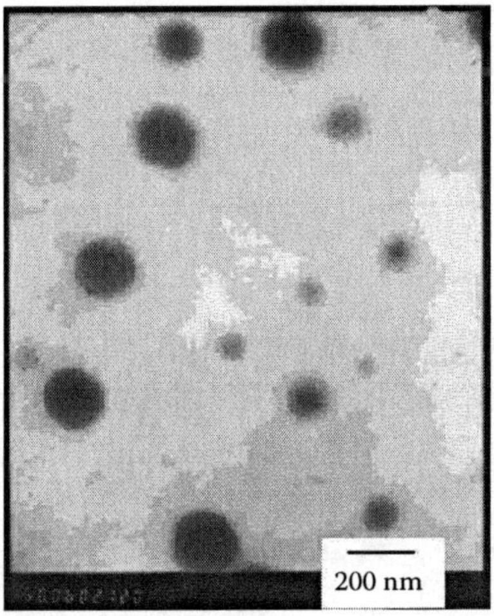

FIGURE 5.12
Transmission electron micrograph (TEM) of nanocopper-sulfonated polybutadiene composite particles in THF. (Reprinted from Nandi, A., Dutta Gupta, M., and Banthia, A. K., Sulfonated polybutadiene random ionomer as stabiliser for colloidal copper nanoparticles, *Colloids and Surfaces A: Physicochemical and Engineering Aspects* 197 (1–3), 119–124, Copyright 2002, with permission from Elsevier.)

decades, the study of the expression of the chirality on solid surfaces is receiving more and more consideration, particularly with the progress of scanning tunneling microscopy (STM), which is capable of straightly visualizing the chirality at the atomic/submolecular resolution at different interfaces. On solid surfaces, chirality can be approached by adsorption of chiral molecules or be possible during chiral ordering. For instance, the racemic mixtures were found to unexpectedly separate into chiral domains on Cu(111), Au(111), and highly oriented pyrolytic graphite (HOPG). The supramolecular chiral channels were shaped by the chiral tartaric acids on a Cu(110) surface. In addition, for simple geometrical reasons, three-dimensional (3D) achiral molecules may turn to two-dimensional (2D) chiral species upon adsorption due to the reduced symmetry at the surface. This case is called *adsorption-induced chirality*.

Researchers studied the molecular chirality and phase transformation of p-phenylenedi(α-cyanoacrylicacid) di-n-ethyl ester (p-CPAEt) collected on Au(111) in the electric double-layer region in 0.1 M $HClO_4$ by the electrochemical scanning tunneling microscopy (ECSTM) method. Three types of chiral supramolecular nanostructures were resolved at dissimilar charged interfaces. Based on the high-resolution STM images, it was uncertainly said that three types of chiral supramolecular nanostructures were formed by two-dimensional adorption-induced chiral p-CPAEt species jointly with a lateral hydrogen-bonding interaction. Interestingly, ECSTM images allow in situ observation of the phase transformation process of these chiral adlayers driven by the electrochemical potential. The full dynamic results demonstrated that the chiral two-dimensional adlayers could be reversibly tuned purely by the used electrode potential.

A newly prepared p-CPAEt film on Au(111) was straightly transferred into the electrochemical STM cell in contact with 0.1 M $HClO_4$ electrolyte solution at the open circuit potential (OCP) around 0.71 V. As illustrated in Figure 5.13a, the long-range ordered stripe patterns were seen on the large

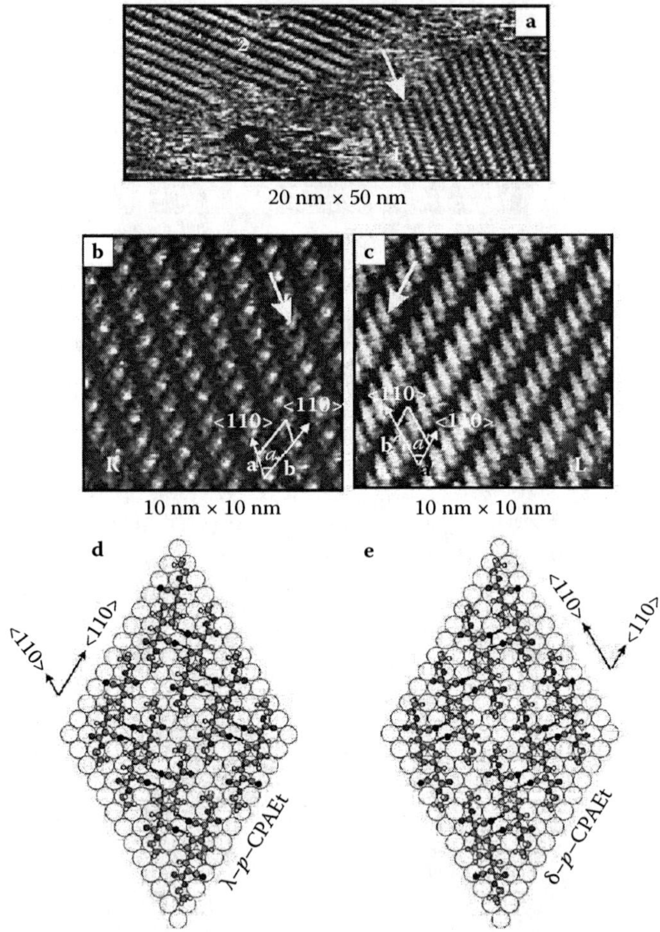

FIGURE 5.13
(a) Large-scale scanning tunneling microscopy (STM) image of the stripe pattern on Au(111) in 0.1 M $HClO_4$ (20 nm × 50 nm). (b, c) High-resolution STM images (10 nm × 10 nm) of R and L domains. $E_s = 0.71$ V. (d, e) Tentatively suggested structural models for λ-phase and δ-phase of stripe pattern. (Reprinted from Su, G. J., Li, Z. H., and Aguilar-Sanchez, R., Phase transition of two-dimensional chiral supramolecular nanostructure tuned by electrochemical potential, *Analytical Chemistry* 81 (21), 8741–8748, Copyright 2009, with permission from the American Chemical Society.)

terrace of the Au(111) surface with parallel bright ridges and dark troughs. Two domains, 1 and 2, were divided by a fuzzy domain boundary.

High-resolution STM images for domains 1 and 2 are illustrated in Figures 5.13b and 5.13c, correspondingly. It was observed that each p-CPAEt molecule adsorbed on the Au(111) surface by linear pattern. The central bright rods related to the conjugated backbone of p-CPAEt molecules. The central brightest part was recognized to a phenyl ring. Its width is 0.57 nm and similar to the diameter of a phenyl ring, suggesting a planar surface orientation. Based on these results, it could be concluded that p-CPAEt molecules might take the planar direction via the interaction between the conjugated π arrangement and substrate in the stripe model.

Based on the molecular arrangement and chirality as well as the intermolecular interaction such as the hydrogen bonding, structural models were proposed in Figures 5.13d and 5.13e for R and L chiral assemblies. The chiral motifs of L and R domains were built correspondingly by the adsorption-induced chiral λ-p-CPAEt and δ-p-CPAEt, which is more logical than vice versa. The benzene rings were strongly fixed in the three hollow sites of Au(111) surfaces. The van der Waals interaction exists between the molecular rows.

Scanning the electrode potential in the region $0.2 < E \leq 0.65$ V, a rather unusual STM contrast reveals the electrode surface accompanied by the dissolution of the stripe phases. After the dynamic equilibration, the steady-state STM images are measured (Figure 5.14). As shown in Figure 5.14a, four different domains (1, 1′, 2, 3) could be distinguished, showing as structurally inhomogeneous. Domain 1, and its translational domain 1′, and domain 2 are indicated as network I, whereas domain 3 is indicated as network II. It seems that reconstructed surface pieces preferentially support network structure I, but on an unreconstructed surface pieces of network structure II are more welcomed.

Figure 5.15 illustrates the full dynamic changes of three kinds of chiral supramolecular nanostructures tuned by the used electric field. The experiment began by a newly prepared p-CPAEt film, which is transferred into the electrochemical STM cell at 0.55 V within the thermodynamically steady region of the network model. The illustrated spot is chosen by an area with coexistence of the deposited stripe structure, from the sample ethanol solution, and the evolved network structure (Figure 5.15a). It means that a number of patches of the stripe model were previously tuned into the network structures after the Au(111) electrode contacted the electrolyte within a thermodynamically stable potential region of network patterns. Scanning the electrode potential positively at 10 mV s^{-1}, but within the thermodynamic stability range of the network structure, it has been found that the stripe nanostructure was slowly decaying and network nanostructures were made bigger. At $E = 0.65$ V (Figure 5.15e), the stripe model was totally transferred into the network structure, but at the same time, it is also noted that the network structures began to dissolve at 0.65 V from the upper right corner. This proposed that network nanostructures are not constant on less negatively charged electrodes.

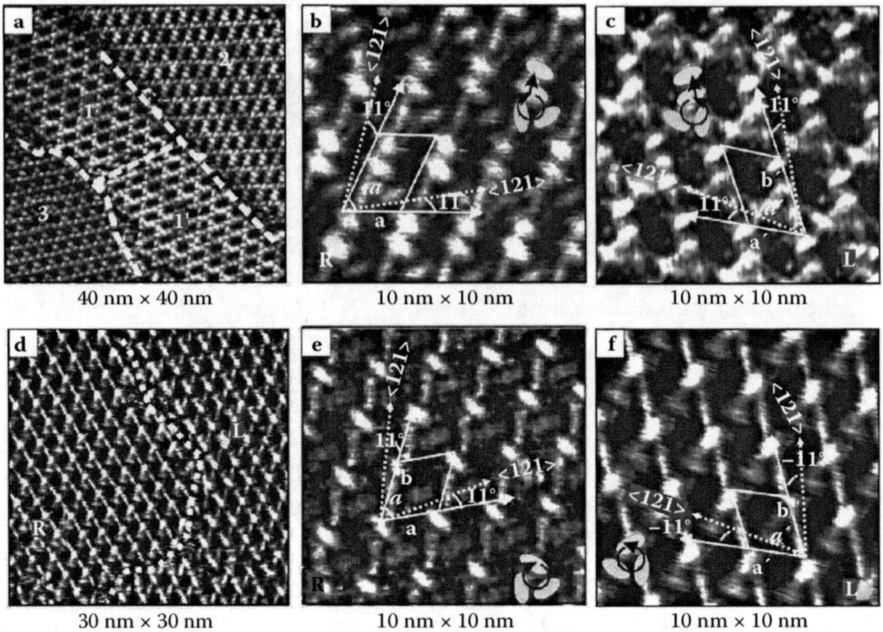

FIGURE 5.14
(a) Large-scale scanning tunneling microscopy (STM) image (40 nm × 40 nm) of p-CPAEt film on Au(111) in 0.1 M $HClO_4$ showing the coexistence of networks I and II. (b,c) High-resolution STM images (10 nm × 10 nm) of R and L domains for network I. (d) Large-scale STM image (25 nm × 25 nm) of network II. (e,f) Close-up STM images (10 nm × 10 nm) of R and L domains for network II. For all the images from a to f, Es is equal to 0.55 V. (Reprinted from Su, G. J., Li, Z. H., and Aguilar-Sanchez, R., Phase transition of two-dimensional chiral supramolecular nanostructure tuned by electrochemical potential, *Analytical Chemistry* 81 (21), 8741–8748, Copyright 2009, with permission from the American Chemical Society.)

5.6 Procurement of Nanomaterials through Deposition

5.6.1 Procurement of Metals from Aqueous Solutions

Compounding metals from their aqueous or nonaqueous solutions need chemical reductive processes of cation of that metal. There are many reducing agents that among them H_2 gas, dissolved ABH_4 (A = alkali metal), monohydrate hydrazine ($N_2H_4.H_2O$), and dihydrochloride hydrazine are the most popular. To perform the usual reduction process of the cation of intermediary metal, its corresponding oxidation process should exist.

$$nMt + ne^- \rightarrow M^\circ \tag{5.17}$$

$$X^m - ne^- \rightarrow X^{m-n} \tag{5.18}$$

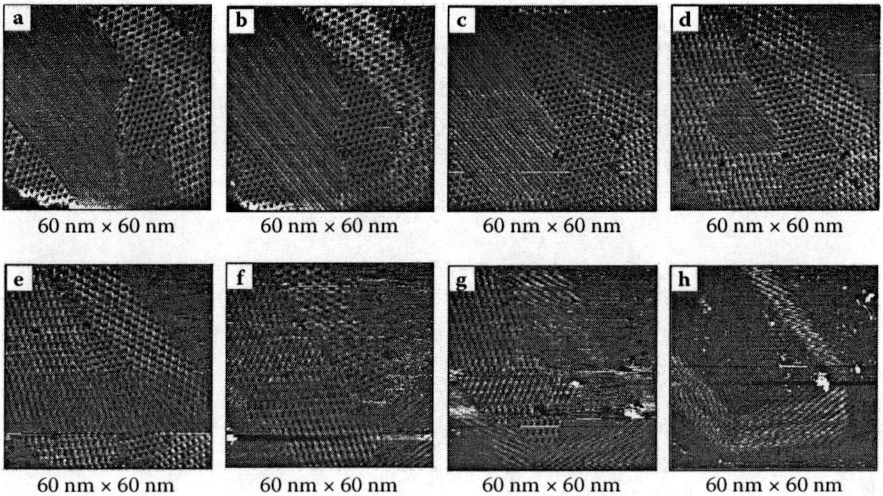

FIGURE 5.15
Potential-dependent series of scanning tunneling microscopy (STM) images from Es = 0.55 V to Es = 0.75 V: (a) t = 0 s, Es = 0.55 V; (b) t = 253 s, Es = 0.6 V; (c) t = 896 s, Es = 0.65 V; (d) t =1003 s, Es = 0.65 V; (e) t = 1172 s, Es = 0.65 V; (f) t = 1503 s, 0.67 V; (g) t = 1678 s, Es = 0.7 V; (h) t = 2478 s, Es = 0.75 V. (Reprinted from Su, G. J., Li, Z. H., and Aguilar-Sanchez, R., Phase transition of two-dimensional chiral supramolecular nanostructure tuned by electrochemical potential, *Analytical Chemistry* 81 (21), 8741–8748, Copyright 2009, with permission from the American Chemical Society.)

In other words, there is an electron transfer. Consequently, change of free energy ΔG should be proportional. An oxidation-reduction reaction depends on an electrochemical half-reaction of the concerned element.

The amount of E^0 in all reactions depends on an H_2 half-reaction that is 0 V in standard temperature and pressure (STP):

$$2H^+ + 2e^- \rightarrow H_2, E^0 = 0.00V \tag{5.19}$$

Electrochemical half-reaction and E^0 of borohydride ion is as follows:

$$B(OH)_3 + 7H^+ + 8e^- \rightarrow BH_4^- + 3H_2O \tag{5.20}$$

Although we should be cautious in using borohydrides, it would be used in aqueous systems to reduce cations and to convert them to metallic boride. Recently, borohydrides are used to deposit metallic nanoparticles.

Hydrate hydrazine would be dissolved easily in water, because N_2H_4 is basic. Most ions in $N_2H_5^+$ are active:

$$N_2H_4.H_2O \rightarrow N_2H_5^+ + OH^- \tag{5.21}$$

And it would be as follows for hydrazine di-hydro chloride:

$$N_2H_4.2HCl \rightarrow N_2H_5^+ + H^+ + 2Cl^- \tag{5.22}$$

The reduction potential of hydrazonium ion $N_2H_5^+$ is as follows:

$$N_2 + 5H^+ + 4e^- \rightarrow N_2H_5^+,\ E^0 = -0.23\ \text{V} \tag{5.23}$$

As theory, for each metal that its E^0 is more positive than −0.481 V or −0.23 V, can be reduced by an appropriate amount of reducing agent and control of pH at room temperature. Therefore, it might include many elements such as a large number of first series intermediary metals ions like Fe^{2+}, Fe^{3+}, Co^{2+}, Ni^{2+}, and Cu^{2+} and second and third series of intermediary metals and some nonmetals as well. Table 5.1 shows certain nanoparticles that are procured in this way.

But the reduction of some metallic ions with $-0.481\ \text{V} < E^0$ is not applicable practically or it would be hard. This is usually because of instability of metallic cations in aqueous environments or in certain cases the cation of intermediary metals like Rh^{3+} would make stable complexes with hydrazine. There are many limits in performing the reduction interaction of metallic ions. It should be realized that pH and the relative potential of oxidation-reduction are the best guides to predict deposition of metals and answer the question of whether it is possible to procure metallic particles in this way or not?

Reduction of Au cations to metal form is so easy. Au cations in the form of $AuCl_4^-$ are easily reduced by gas H_2 to Au metal:

$$2AuCl_4^-(aq) + 3H_2(g) \rightarrow 2Au(s) + 6H^+(aq) + 8Cl^-(aq) \tag{5.24}$$

TABLE 5.1

Deposition Metallic Nanoparticles from Their Aqueous Solution

Average Diameter of Particles (nm)	Stabilizer	Reducing Agent	Primary Material	Metal
20~	None	$N_2H_4.H_2O$	$Co(OAc)_2$	Co
10–36	CTAB	$N_2H_4.H_2O$ + NaOH	$NiCl_2$	Ni
(10–20) × (200–300) rods	None	$N_2H_4.H_2O$ + NaOH	$Ni(OAc)_2$	Ni
35~	SDS	$N_2H_4.H_2O$	$CuSO_4$	Cu
15–26	Daxad 19	Ascorbic acid	$AgNO_3$	Ag
3–5	TADDD	$NaBH_4$	$AgNO_3$	Ag
<1,5	TDPC	Potassium bitartrate	H_2PtCl_6	Pt
—	S3MP	Trisodium citrate	$HAuCl_4$	Au

Source: Szot, K. et al. Sol-gel processed ionic liquid-hydrophilic carbon nanoparticles multilayer film electrode prepared by layer-by-layer method. *Journal of Electroanalytical Chemistry* 623 (2), 170–176, 2008.

$AuCl^{-}_4$ has a standard potential of +1.002 that would be easily reduced with weaker reducing agents like carboxylates or alcohols. Some researchers procured nanoparticles of Au, Pt, Pd, and Ag by reducing with potassium bitartrate. By using an appropriate stabilizer, stable colloids of said nanoparticles are procurable.

In many cases, organic covering agents that are generally used to prevent clot are also being used as reducing agents. Turkevich is one of the recognized processes in this context that is used to construct Au colloid nanoparticles. They explained how to construct Au colloid nanoparticles in their papers in 1951 by an ebullient and thin solution of $HAuCl_4$ and sodium citrate. This method is still used, and it is the base of constructing Au self-arraying monolayers.

Generally, if the sulfuric stabilizing factors are used, borohydrides or a similar reducing agent would be used to reduce $AuCl_4^-$ ions in aqueous solutions because complexes formed between $AuCl_4^-$ and tiols are more stable than other covering agents and consequently would not be reduced by citrate or weak reducing agents. However, if citrate and tiols are used simultaneously, Au nanoparticles would be reached. The size of reached Au particles is 2 to 10 nm. To reduce metals with high negative amounts of reduction potentials, strong reducing agents like most of the amines, hydroxy carboxylate acids, or alcohols are required. Reduction of ions such as those in Table 5.2 would be done by borohydride salts (Equation 5.18).

Silver also can procure silver nanoparticles with the size of 3.3 nm by borohydride from an Ag^+ aqueous solution that includes bis(11-tri methyl ammonium decan oleil amino ethyl) disulfide diboromide (TADDD). Extra borohydride would reduce disulfide to a tiol that would be used as a capping ligand. Particles would be spread in an acidic solution easily.[50–52]

5.6.2 Deposition of Metals by Reduction of Nonaqueous Solutions

Au nano particles in aqueous solutions would be stable by capping ligands like citrate. Brust et al. procured Au nanoparticles that are stabilized by the tiols' alcan in nonpolar solutions. These particles show a relatively good capability to spread in solutions after separating in the form of dried powder.

In the Brust method, a compound of a two-phase reaction is used to reduce Au particles that are similar to Faraday's technique. Tetrachloroaurate ions

TABLE 5.2

Certain Metallic Ions That Would Be Reduced by Borohydride Salts (by Their Chemical Potentials)

E^0 (V)	Metal
–0.44 V	Fe^{2+}
–0.28	Co^{2+}
–0.25 V	Ni^{2+}

($AuCl^-_4$) would be transferred to an organic phase by compounding an aqueous solution with a solution of tetraoctylammonium bromide (TOAB) dissolved in toluene. (TOAB is known as a phase transfer catalyst.) After the addition of dodecanthiol to the organic phase, $NaBH_4$ aqueous solution would be added to the compound gradually. Colloid Au (1 to 3 nm) would be formed in an organic phase and would be separated from it through evaporating in a vacuum or gradual deposition by methanol. Researchers found that dried powder can be converted to stable colloid suspension in each nonpolar solvent or the one with low polarity like toluene, pentane, and chloroform (but hydrate and alcohol cannot be used). These researchers have shown by using infrared (IR) and x-ray photoelectron spectrometry (XPS) that the found powder still has some thiols after being washed, and Au particles are like Au^0. Thiol bands also prefer RSH to RS^-.

Brust's research is the basis for many researches on colloid of stabilized nanoparticles by thiol composites. In many researches, new composites with thiol or those of amine, phosphine silate, or with disulfide are used as capping legands. Steova improved Brust's method and could procure Au nanoparticles by thiols in a gram scale. Some researchers procured Au nanoparticles by borohydride as a reducing agent and then stabilized it by mercaptosuccinic acid (MSA) as a capping ligand. They showed that if capped nanoparticles could deposit as sodium salt, they could be spread in a hydrate as well. The size of Au particles is controllable by change of [MSA]/[Au] proportion, and such particles with sizes of 1 to 3 nm are reachable. This method also can be optimized to construct Ag metallic nanoparticles with sizes of 1.4 to 5.7 nm.[53–57]

Some researchers have found a method for reducing Au ions in nonaqueous solutions. Formamid ($HCONH_2$) is used as a reducing agent in this method. The reaction mechanism has two steps, and it did not perform by the presence of oxygen. Nanoparticles with an average size around 30 nm are stabilized by poly vinyl pyrrolidon (PVP), and the size distribution of particles is very narrow. It is not completely clear, but it is possible to control the size of Au nanoparticles with the proportion of [PVP]/[Au].

They found silver nanoparticles through reducing $AgNO_3$ or $AgClO_4$ by N, N-dimethylformamide (DMF) in a similar reaction and used 3-(amino propil) tri-metoxy silane as a stabilizer. DMF is probably oxidized to carboxylic acid:

$$HCONMe_2 + 2Ag^+ + H_2O \rightarrow 2Ag^\circ + Me_2NCOOH + 2H^+ \tag{5.25}$$

The size of silver nanoparticles is variable from 6 to 20 nm, and the size of particles is controllable by temperature and molar proportion of [DMF]/[Ag]. Stabilizers with a silane base would be hydrolyzable and dense in high temperatures, especially under acidic conditions. A thin layer of SiO_2 would deposit on certain Ag particles that would avoid clots. Figure 5.16 illustrates the morphologies of the product when the reaction solution of DMF with $AgNO_3$ has been used.

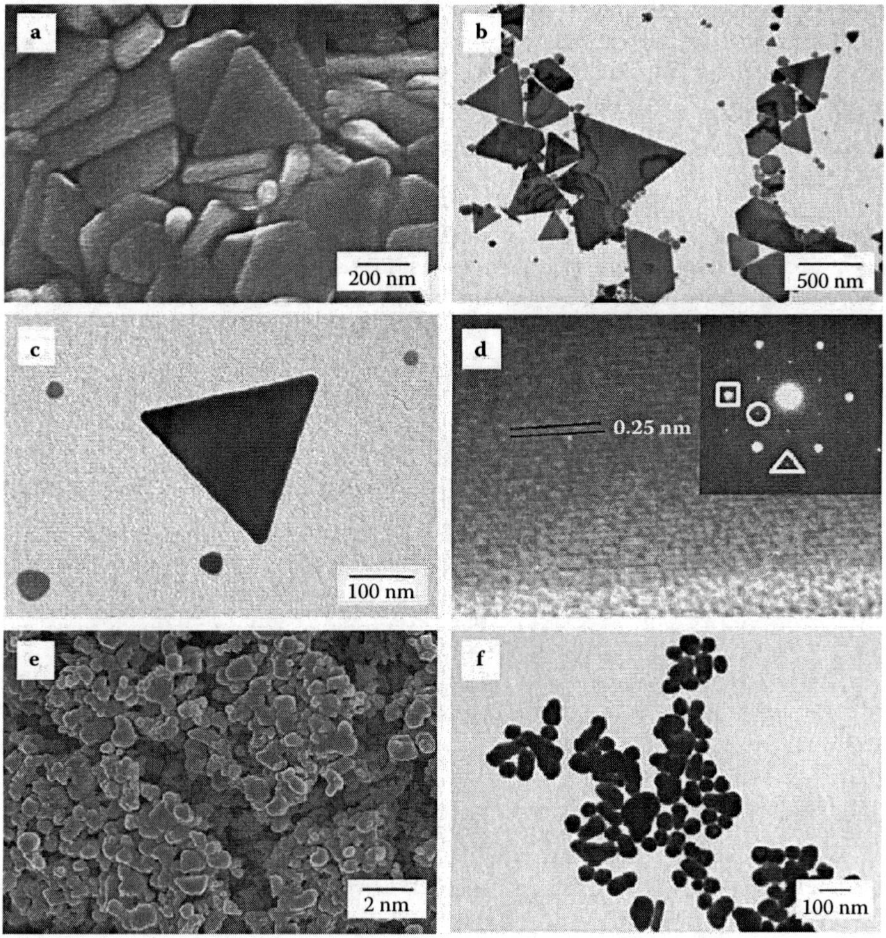

FIGURE 5.16
Scanning electron microscopy (SEM) (a) and transmission electron microscopy (TEM) (b) images of the product when MR = 1.5. The reaction solution of 40 mL of DMF with $AgNO_3$ (50 mM) and poly vinyl pyrrolidon (PVP) (75 mM) was kept in an oven at 200°C for 4 h. (c) TEM image of an individual triangular plate. (d) High-resolution TEM (HR-TEM) of the edge of part (c). The inset shows the electron diffraction pattern taken relatively. (e) Silver nanoparticles obtained under the same conditions without PVP added. (f) Silver nanoparticles obtained under the same conditions when 1 mL of water was added to the solution before it was transferred to the autoclave. (Reprinted from Lu, Q., Lee, K. J., Lee, K. B., Kim, H. T., Lee, J., Myung, N. V., and Choa, Y. H., Investigation of shape controlled silver nanoplates by a solvothermal process, *Journal of Colloid and Interface Science* 342 (1), 8–17, Copyright 2010, with permission from Elsevier.)

The capability of such alcohols like ethanol to be strong reducing agents for oxidation of cations is clearly recognized. However, particles tend to make clots in the environment of the reaction even in the presence of stabilizers, which is one of the problems of this method. Researchers use polyalcohols like ethylene, glycol, or 1,2 diol propane. Using these complexes increases the efficiency of products. They have been used in many samples as reducing agents and stabilizers. For instance, some researchers have procured mono-spray 4 nm Ru nanoparticles by reducing $RuCl_3$ with various poly Ls. These poly Ls are used as reducing agents, and 2 dokan thiol is used as a stabilizer.

Through a similar technique and by using hydrous hydrazine in an ethylene glycol solution, Ni^{2+} is converted to Ni^0 (Equation 5.19). The size of found particles is in the range of 9 nm. Equation (5.21) shows that hydrazinium ions cannot convert Ni^{2+} to Ni^0 appropriately. Therefore, extra NaOH would be added to the solution. Finally,

$$2Ni^{2+}+N_2H_5^+ +5OH^- \rightarrow 2Ni^\circ + N_2 + 2H_2O \tag{5.26}$$

Increasing pH at a temperature of 60°C would increase the reducing capacity of $N_2H^+_5$.

When equation $M^{n+} + ne \rightarrow M^\circ$ has negative E^0, aqueous solvents are not usable because instead of metallic ions, water molecules would be reduced (according to Equation 5.25). In these cases more stable solvents should be used:

$$2H_2O + 2e \rightarrow H_2 + 2OH^-, E^0 = -0.8277 \text{ V} \tag{5.27}$$

Thermodynamically, the strongest reducing agents exist in a dissolved electron solution (e_s^-) that is replaced directly by an metal anion (A^-). In these cases, the alkali metal would be dissolved in solvents without protons like dimethyl ether or tetrahydrofuran (THF) and complexing agents such as alkali metals like 15-crown-5:

$$2A^\circ + (15-crown-5) \rightarrow A^+(15-crown-5)A^- \tag{5.28}$$

An active reducing agent procured in Equation (5.26) is generally called an *alkalide*. The following reaction would be performed in the presence of an extra complexing agent:

$$A^\circ + 2(15-crown-5) \rightarrow A^+(15-crown-5)_2 e_s^- \tag{5.29}$$

These products are called an electride. The above reactions would go toward right due to strong complexing agents easily. Although electride and alkalide solutions are strong reducing agents, they are unstable thermodynamically and would be decomposed in air. Consequently, their reaction should be performed in low temperatures and under an ineffective atmosphere.

Some researchers reported the construction of Au, Pt, Cu, Te, Ni, Fe, Zn, Ga, Si, Mo, W, In, Sn, Sb, and Ti metallic nanoparticles and Au-Cu, Au-Zn, Cu-Te, and Zn-Te alloys from alkalide and electride solutions. These reactions are being generally performed in –50°C, and the size of found particles is 3 to 15 nm. Products found during a wash with methanol tend to be oxidized. Some researchers used a Li metal solution in naphthalene to construct a composite of the Mg-Co nanoparticle.

Composites like tetra-alkyl borohydride, $ABEt_3H$ (A= Li, Na, K) and dissolved magnesium Mg^* are used as strong reducing agents. Some researchers showed that these reducing agents can procure many metals from their salts in organic solvents like toluene, dioctyl ether, or THF:

$$xMX_y + yABEt_aH \rightarrow xM + yAX + yBEt_a + \frac{xy}{2H_2} \tag{5.30}$$

$M^{y+} = Cr^{3+}, Mn^{2+}, Fe^{2+}, Fe^{3+}, Co^{2+}, Ni^{2+}, Cu^{2+}, Zn^{2+}, Ru^{3+}, Rh^{3+}, Pd^{2+}, Ag^{+}, Cd^{2+}, In^{3+}, Sn^{2+}, Re^{3+}, Os^{3+}, Ir^{3+}, Pt^{2+}, Au^{+}$

A is the symbol of lithium or sodium, and *X* shows a metal ion. By optimizing the conditions it is possible to reduce Mo^{3+} and W^{4+} to their metals as well. It is possible to procure colloid nanoparticles with sizes of 1 to 5 nm by using Equation (5.28), as their crystallization depends on the reaction temperature ($T < 70$°C). Basically this reaction depends on the type of reducing agent and the solvent. Exchanging $LiBEt_3H$ with $NaBEt_3H$ and THF with toluene would make a metallic colloid carbide (MC_2) with a size of 1 to 5 nm.

Some researchers have used similar methods to construct Ni, Co nanocrystals, or their alloys. Chloride solution or metals' acetate is mostly used in this method. Two types of capping ligands are also used so that one of them has strong bonds (carboxylic acid) and the other one weaker bonds (tetra-alkyl phosphine and phosphine oxide) with nanoparticles. Reactions would be done at a relatively high temperature. Adding a solution of a long-chain reducing complex like 1,2 dyol or $LiBEt_3H$ to a metallic salt solution by covering factors that are heated at boiling point (dioctyl or diphenyl ether) might increase the speed of nucleation and therefore make smaller nanocrystals. Change in proportion of two covering factors and chain length would affect the size of the crystals. The presence of the carboxylic acid (oleic acid) that is placed on the surface of nanocrystals in a covalent manner might make nanocrystals monospray and stable. Figures 5.17 and 5.18 are examples of the images for zero- and two-dimensional nanostructures of nickel amorphous alloys.

The advantages of using organic solvents are briefly as follows:

1. Organic reducing agents like citrate ion, alcohols, and so forth besides the previous reducing agents can be used.
2. These materials can be used as solvents, reducing factors, and stabilizers for nanomaterials.

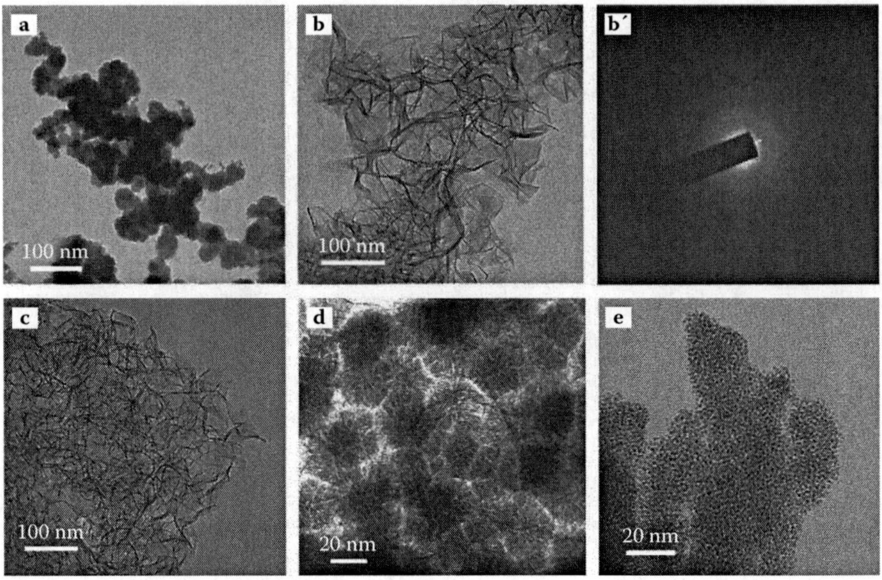

FIGURE 5.17
Transmission electron microscopy (TEM) images of zero- (0D) and two-dimensional (2D) nanostructures of Ni amorphous alloys: (a) 0D aggressive nanoparticles prepared in aqueous solution without template; (b) 2D ultrathin nanofilms fabricated by hard collodion membrane; (b′) selected-area electron diffraction pattern (SADP) of 2D ultrathin nanofilms of Ni; (c) 2D ultrathin nanofilms directed by the synergistic effect of hard collodion membrane and soft reverse microemulsion; (d) 0D nanospheres obtained by the synergistic effect of hard collodion membrane and soft reverse microemulsion with ethylenediamine as ligand; and (e) 0D nanospheres leaded by hard collodion membrane template with ethylenediamine as ligand. (Reprinted from Wen, M., Wang, Y., Wu, Q., Jin, Y., and Cheng, M., Controlled fabrication of 0 & 2D NiCu amorphous nanoalloys by the cooperation of hard-soft interfacial templates, *Journal of Colloid and Interface Science* 342 (2), 229–235, Copyright 2010, with permission from Elsevier.)

3. While the potentials of metal reduction are negative, aqueous solvents are not usable because aqueous molecules would be reduced instead of metal.
4. Reaction, nucleation, and particle growth speed are better controlled due to better control of power and activity of reducing factors.
5. By using this method more diffused nanomaterials with fewer percentage of clots are reachable because most organic molecules are very massive and might prevent agglomeration of the particles.
6. Due to a higher volatility of organic solvents, it would be easier to omit impurities and solvent from the solution (to form complete pure nanomaterials).

One of the disadvantages of using organic solvents is that many metal salts are just solved in water or are just polar solvents; therefore, their salt would

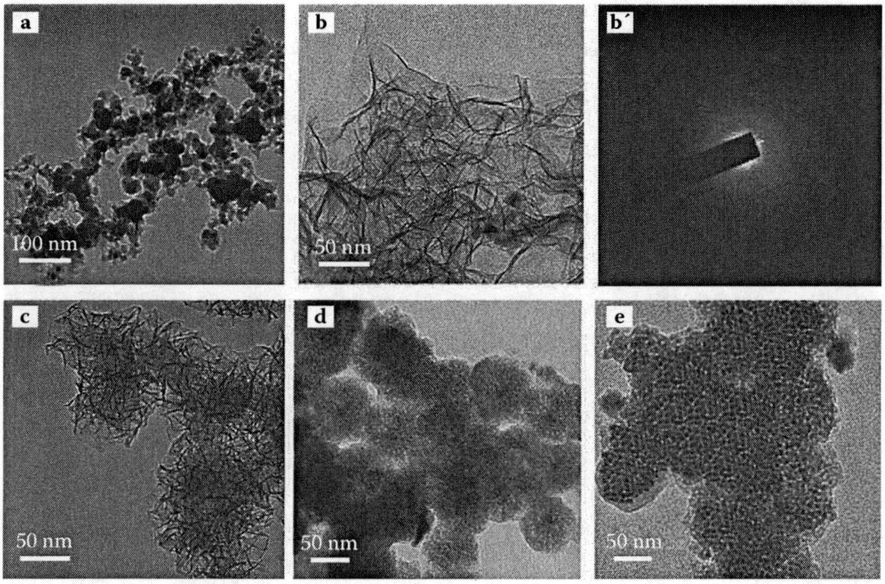

FIGURE 5.18
Transmission electron microscopy (TEM) images of zero- (0D) and two-dimensional (2D) nanostructure of NiCu amorphous alloys: (a) 0D aggressive nanoparticles prepared in aqueous solution without template; (b) 2D ultrathin nanofilms fabricated by hard collodion membrane; (b′) SADP of 2D ultrathin nanofilms of NiCu; (c) 2D ultrathin nanofilms directed by the synergistic effect of hard collodion membrane and soft reverse microemulsion; (d) 0D nanospheres obtained by the synergistic effect of hard collodion membrane and soft reverse microemulsion with ethylenediamine as ligand; and (e) 0D nanospheres leaded by hard collodion membrane template with ethylenediamine as ligand. (Reprinted from Wen, M., Wang, Y., Wu, Q., Jin, Y., and Cheng, M., Controlled fabrication of 0 & 2D NiCu amorphous nanoalloys by the cooperation of hard-soft interfacial templates, *Journal of Colloid and Interface Science* 342 (2), 229–235, Copyright 2010, with permission from Elsevier.)

not be solved in organic solvents. Table 5.3 includes a summary of the construction of metal nanoparticles that are procured by deposition method.

5.6.3 Deposition of Oxides from Aqueous Solutions

The deposition of oxides from aqueous and nonaqueous solutions is not as understandable as is deposition of metals. Reactions of procuring oxides would be divided into two categories: oxide construction directly and indirectly during which more processes should be performed on them (like drying, calcification, etc.). In both types, capping ligands or stabilizers should be used to procure monospray oxide nanoparticles to prevent the formation of clots.

While calcification or annealing is required, formation of clots is usually seen. Nanoparticles without coating also are observable, but there is little

TABLE 5.3

Some Metallic Nanoparticles or Metallic Alloys That Are Procured by Deposition from Nonaqueous Solutions

Metal	Primary Material	Solvent	Reducing Factor	Stabilizer	Average Size of Particles (nm)
Fe	$Fe(OEt)_2$	THF	$NaBEt_3H$	THF	10–100
Fe	$Fe(acac)_3$	THF	Mg^*	THF	~8
$Fe_{20}Ni_{80}$	$Fe(OAc)_2$ $Ni(OAc)_2$	EG	EG	EG	6(A)
Co	$Co(OH)_2$	THF	$NaBEt_3H$	THF	10–100
Co	$CoCl_2$	THF	Mg^*	THF	~12
$Co_{20}Ni_{80}$	$Co(OAc)_2$ $Ni(OAc)_2$	EG	EG	EG	18–22(A)
Ni	$Ni(acac)_2$	HAD	$NaBH_4$	HDA	3, 7(C)
Ni	$NiCl_2$	THF	Mg^*	THF	~94
Ni	$Ni(OAc)_2$	EG	EG	EG	25(C)
Ru	$RuCl_3$	1, 2-PD	1, 2-PD	Na(OAc) and DT	1–6(C)
Ag	$AgClO_4$	DMF	DMF	3-APTMS	7–20(C)
Ag	$AgNO_3$	Methanol	$NaBH_4$	MSA	1–6(C)
Au	$AuCl_3$	THF	$K^+(15C5)_2K$	THF	6–11(C)
Au	$HAuCl_4$	Formamide	Formamide	PVP	30(C)

Note: *A* means agglomerated, and *C* means colloidal/monodispersed.

chance for the formation of monodispersed nanoparticles. Products of codeposition reactions that are usually formed in room temperature or close to it are amorphous. In samples that are being taken as deposited from carbonate compound or metal hydroxide, it is difficult to take a deposit from a homogenous solid phase from oxide compounds, carbonates, or metal hydroxide.

Some researchers performed their first synthesis of controlled magnetic nanoparticles by an alkali deposition technique in 1981. In this method, Fe_3O_4 nanoparticles would be reached from deposition of $FeCl_3$ and $FeCl_2$ solutions in pH = 8.2. These particles are irregular spheres with a diameter equal to 10 nm and a particle distribution of more than 50%. By spreading particles in the solvent, a stable colloid suspension would be made. Consequently, distribution of particle size can reach less than 1%. The solution would be a titer with an electrolyte solution or a solvent that removes stability of the colloid, and larger nanoparticles would finally deposit. Suspended particles in the solution are spread. The temperature of the reduction reaction is the same as the temperature that would remove stability of the solution. These particles are being collected by centrifuge or filtration, and the process would be performed again. This can decrease the distribution of particles' size up to less than 1% that generally make a very spread colloid solution.

They then improved this method to produce an alkali deposition of the ferrite compound (M = Co, Mn, MnZn) MFe_2O_4. Procuring these ferrites has lots of difficulties due to various solubilities of metal hydroxides. For instance, in procuring (MnZn), FeI_3 in pH = 2.6 would deposit easily while $Mn(OH)_2$ would deposit in pHs more than 9.4. Zn^{2+} is an amphoter and $Zn(OH)_2$ deposits in pH = 7.6. But in pH = 9 it would be resolved and $Zn(OH)_4^{-2}$ would be formed. Then pH should be set as 8.6 to make a metal amorphous deposition. Fe mostly tends to oxidation and forming α-Fe_2O_3 that usually makes an impure phase, and its magnetic properties would be reduced as a consequence. Fe^{3+} is an oxide in pH > 9. In the titration of metal with a base, pH is increased a little more than the time the base is titered with metal. Titration increases problems gradually due to heterogeneity of the reaction. pH would be gradually increased. First, Fe^{3+} deposits and the reaction would be completed gradually, and then Zn would deposit and finally Mn would deposit. The final deposit is a central core that is oxide with γ-Fe_2O_3 and scales made up of ferrite with Fe deficiency.

This disorder is often due to a difference in solubility and can be useful in manufacturing. Deposition reactions can be performed orderly in a way that the second material deposits on the previous material. For instance, CdS and CdSe lattices are relatively similar. In deposition from the solution, CdS would nucleate due to less solubility, and then CdSe would grow on that.

Some researchers procured CdS-CdSe and CdSe-CdS core-shell structures in 1996. They used cadmium perchlorate and hydrogen sulfate or hydrogen selenide as reactive and produced these nanostructures through manufacturing microemulsion conditions. This is the first example to procure core-shell structures by the difference of their solubility.

Many nanoparticles of metal oxides are procured by calcification of codeposition products of their hydroxide. Some researchers produced the spinal structure of $Ni_{0.5}Zn_{0.5}Fe_2O_4$ by deposition of Fe, Ni, Zn nitrates with NaOH and its calcification in 300°C and higher. The limit for particle size is between 9 and 90 nm, which depends on calcification temperature. Nanoparticles of $MgFe_2O_4$ and $Sm_{1-x}Sr_xFeO_{3.6}$ are procured in ways similar to this.

Codeposition of metallic cations in the form of carbonates, bicarbonates, or oxalates and then their calcification and damage are the most ordinary methods of procuring nanoparticles of metallic oxide. Calcification is often performed beside forming clots or at high temperatures, making clots besides sintering. If calcification of prematerials is being done at low temperatures (<400°C), we will have the most surface area and minimal clots. For instance, $Ce_{0.8}Y_{0.2}O_{1.9}$ is being procured by adding oxalic acid to aqueous solution of $Ce(NO_3)_3$ and $Y(NO_3)_3$ and then calcifying the formed deposition. Symmetrical spherical particles with a size of 10 nm will be formed by calcification in 500°C with little clots. Calcification in 1000°C will make irregular particles 100 nm with more clots.

In a few cases, deposit can be taken from the crystal oxide of metals from their aqueous solution without any need for calcification that is mostly possible for dual oxides. For instance, titanium oxide with a rutile phase is capable of depositing from an aqueous solution of $TiCl_3$ by NH_4OH. Particles are between 50 and 60 nm, which will be stabilized by poly-methyl methacrylate.

Direct codeposition of most complexes of complicated triplet oxides is relatively impossible, especially when thermodynamic stable structures like a spinal are concerned. In these cases, deposition reactions are usually done in extra temperatures (50 to 100°C). In this case, hydroxide intermediaries would be converted to oxides in the reaction container. In single-step construction methods, there is no need for the calcification process (e.g., procurement of Fe_3O_4 nanoparticles). These particles would be produced through simple codeposition of $Fe^{2+} + 2Fe^{3+}$ with NaOH in temperatures higher than 70°C, similarly 5 to 25 nm particles of $MnFe_2O_4$ would be produced by codeposition of Mn^{2+} and Fe^{2+} in temperatures higher than 100°C.[60–72]

Procured oxides through the codeposition method from their aqueous solutions are summarized in Table 5.4.

Effective factors on the size of particles are

1. Interaction temperature
2. Concentration of reactants
3. Speed of adding materials to each other

5.6.4 Deposition of Oxides from Nonaqueous Solutions

Deposition reactions would be performed in nonaqueous solutions in certain cases. This method is useful when nonsimilar metals cannot deposit simultaneously in aqueous solutions due to various conditions required for deposition as in various pHs, like $LiCoO_2$ that is being used in chargeable Li batteries. It is not possible to take a deposit of LiOH from aqueous solutions, but their solubility in alcohols is less than that.

TABLE 5.4

Summaries of Reactions of Deposition of Oxides from Their Aqueous Solutions

Metal	Primary Material	Deposition Agent	Stabilizer	Particles Size (nm)
VO_2(B)	NH_4VO_3	$N_2H_4.H_2O$	None	35
Cr_2O_3	$K_2Cr_2O_7$	$N_2H_4.H_2O$	None	30
γ-Mn_2O_3	$KMnO_4$	$N_2H_4.H_2O$	None	8
$MnFe_2O_4$	$MnCl_2$, $FeCl_3$	NaOH	None	5–25
Fe_3O_4	$FeCl_2$, $FeCl_3$	NH_4OH	H^+	8–50
NiO	$NiCl_2$	NH_4OH	CTAB	22–28
ZnO	$ZnCl_2$	NH_4OH	CTAB	40–60
SnO_2	$SnCl_4$	NH_4OH	CTAB	11–18
Sb_2O_3	$SbCl_3$	NaOH	PVA	10–80

Some researchers procured $LiCoO_2$ nanoparticles by codeposition of LiOH + $Co(OH)_2$. They added an ethanol solution of $LiNO_3$ and $Co(NO_3)_2$ gradually to an ethanol solution of 3M KOH. Then the calcified hydroxide compound in 400 to 700°C adjacent to air and size of found $LiCoO_2$ nanoparticles was around 12 to 41 nm.

Solvents can be used as the factor for deposition in certain cases. Some researchers procured γ-Fe_2O_3 5 to 20 nm nanoparticles by mixing molten stearic acid and hydrated $Fe(NO_3)_3$ in 125°C and calcifying found a deposit in 200°C adjacent to air. This product would change phases from γ phase (maghemite) to α phase (hematite).

In these cases, aqueous solvents are not used for deposition to prevent early deposition of oxide or hydroxide of metals. Early deposition is more for electropositive metals with high capacity, like Ti^{4+}, Zr^{4+}, and so forth. This problem is solvable to some extent by using chloride or alchoxide of metals as a prematerial in nonaqueous solvents. Hydrolysis reactions can be controlled by limiting growth of particles and preventing clots. For instance, $BaTiO_3$ ferroelectric particles would be produced by deposition of metal alchoxide prematerials. This material consists of BaTi $(O_2C(CH_2)_6CH_3)[OCH(CH_3)_2]_5$ and is suitable for the hydrolysis process. The deposit of this article can be taken by H_2O_2 in diphenyl ether and in the presence of oleic acid as a stabilizer. Adding H_2O_2 on the hydrolysis reaction is effective. The reaction would be performed in 100°C by forming a Ba-O-Ti bond; however, growth of particles would be stopped due to the presence of an oleic acid stabilizer. Formed 6 to 12 nm nanoparticles are monodispersed and crystallize and do not need to be calcified. This reaction is similar to sol-gel reactions.[60,72–84]

Some researchers used metallic colloid prematerial to produce ferrite colloids. Metallic colloid and diethylene glycol are used simultaneously to produce this prematerial.

Diethylene glycol is used here as a solvent and stabilizer. Although glycol is normally used for assurance of the growth process instead of oleic acid, it is exchanged during the growth process. Researchers procured a series of ferrites M Fe_2O_4 (M = Mn, Fe, Co, Ni, Zn) with a 3 to 7 nm size (Figure 5.19).

5.6.5 Nanoimprint Lithography for Fabrication of Plasmonic Arrays

Localized surface plasmon resonance (LSPR), collective electron density oscillations revealed in noble metal nanostructures, has been studied widely over the past decade due to its potential utility as the backbone for a number of photonic methods able to control light at nanometric dimensions well below the diffraction limit. Research in this field has been produced by the tremendous growth in fabrication techniques able to produce a huge variety of nanoparticle (NP) systems and nanostructured films. Numerous theoretical and experimental studies have established that LSPR is sensitive to the shape, size, interparticle distance, dielectric environment, and material

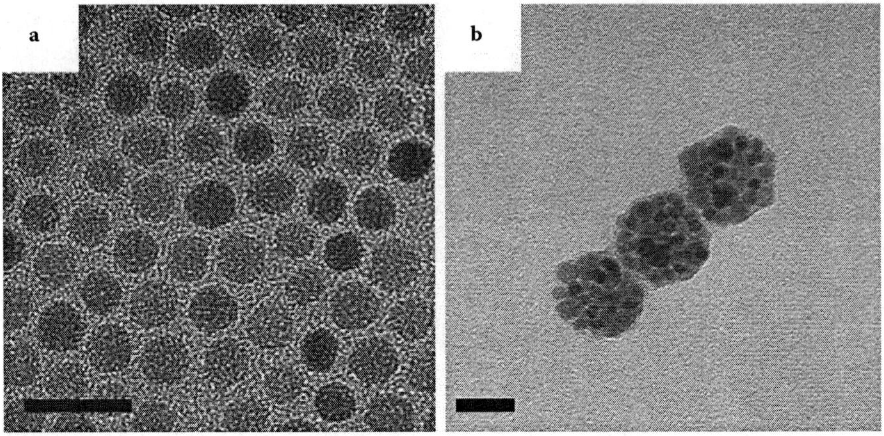

FIGURE 5.19
Transmission electron microscopy (TEM) images of (a) $MnFe_2O_4$ nanocrystals (MNCs) and (b) fluorescent magnetic nanoprobes (FMNPs) (Scale bar: 20 nm). (Reprinted from Lim, E. K., Yang, J., Dinney, C. P. N., Suh, J. S., Huh, Y. M., and Haam, S., Self-assembled fluorescent magnetic nanoprobes for multimode-biomedical imaging, *Biomaterials* 31 (35), 9310–9319, Copyright 2010, with permission from Elsevier.)

composition of the constituent NPs. In addition, optical dichroism observed from well-aligned nanoparticle arrays has confirmed the polarization dependence of their LSPR response. One of the most promising applications of nanoparticle systems is their use as real-time chemical and biological sensors that originate from the aforementioned LSPR dependence on their dielectric environment. Such systems have been confirmed using a variety of NP implementations including single-particle, one-dimensional, and two-dimensional array configurations on clear substrates as well as solution-based techniques.

An abundance of nanofabrication methods have been used to produce the desired nanostructures utilized in LSPR studies. A few examples of these methods include electron beam lithography, templates, nanosphere lithography (NSL), and colloidal solution-based nanoparticle synthesis of which the latter two have been employed quite widely. While NSL and solution-based methods have been useful for basic studies of the influence of NP characteristics on LSPR, there are still important limitations of both methods that limit their applicability to commercialized LSPR-based usages.

Chemical production methods have the benefit of creating a wide array of exotic nanostructures based on the alteration of the reaction parameters such as time, relative concentration of reactants, and temperature. However, the monodispersity and reproducibility of preferred structures can be hard to achieve using this method. More importantly, this method lacks the organization of relative NP positioning and orientation in addition to the requirement of novel surface chemistries for the reduction of

NP agglomeration and useful substrate attachment. The inability to exactly control the sample-to-sample LSPR response of immobilized NP systems is a severe limitation of this fabrication approach.

Nanosphere lithography is an alternative fabrication technique introduced by some researchers to create periodic particle arrays (PPAs) straight on a variety of substrates. This method utilizes a closed-packed nanosphere mask that permits direct deposition of noble metal NPs onto a substrate through the interstitial regions of the mask. NSL has been applied in a single-layer and double-layer approach with wide characterization and employment of triangular nanoparticles resulting from the single-layer technique. The precise control of PPA attributes governed by this method makes it a hopeful candidate as a fabrication technique relevant to commercialized LSPR applications. Limitations of NSL consist of issues with surface coverage and the geometric constraints imposed by the nanosphere mask on the PPA lattice structure and NP shape characteristics that decrease the degrees of freedom available for the designed LSPR response of the PPA system.

With the aim of address and supplement limitations faced by a number of current NP fabrication techniques, use of nanoimprint lithography (NIL) and two-dimensional nanostructure array (nanoblock) molds derived from one-dimensional gratings to produce noble metal nanoparticles (NPAs) has been introduced. Some researchers believe this approach possesses a number of attributes that will not only improve the basic study of NP systems, but may also play a key role in the manufacture of marketable LSPR technologies. First, NIL is a mold-based, high-throughput, and low-cost process capable of patterning large areas with sub-10 nm resolution. Second, the benefits of alternative nanofabrication techniques capable of creating unique LSPR nanostructures and possessing less attractive throughput and cost characteristics can be maximized through their employ for creation of NIL molds that allow the continuous reproduction of these structures. This feature of NIL to "preserve" nanostructures in a mold permits the optimization of NPAs for particular LSPR applications through empiricism. Moreover, the use of one-dimensional grating structures as the foundation to create NPAs enables resulting two-dimensional patterns of varying complexity to be geometrically understood and modeled for design reasons.

Some researchers explore a NIL manufacture approach capable of creating homogeneously oriented and homogenous noble metal NPAs. Extensions of this production technique demonstrating the flexibility of this method to produce NPAs possessing a variety of unique structural and LSPR characteristics are demonstrated. The significance of this technique is further elucidated by exploring the consequences of these attractive findings within the context of other works. Researchers believe this to be the first report of such a NIL-based fabrication approach for the production of plasmonic NPAs. As shown in Figure 5.20, the general approach is used to create large-area

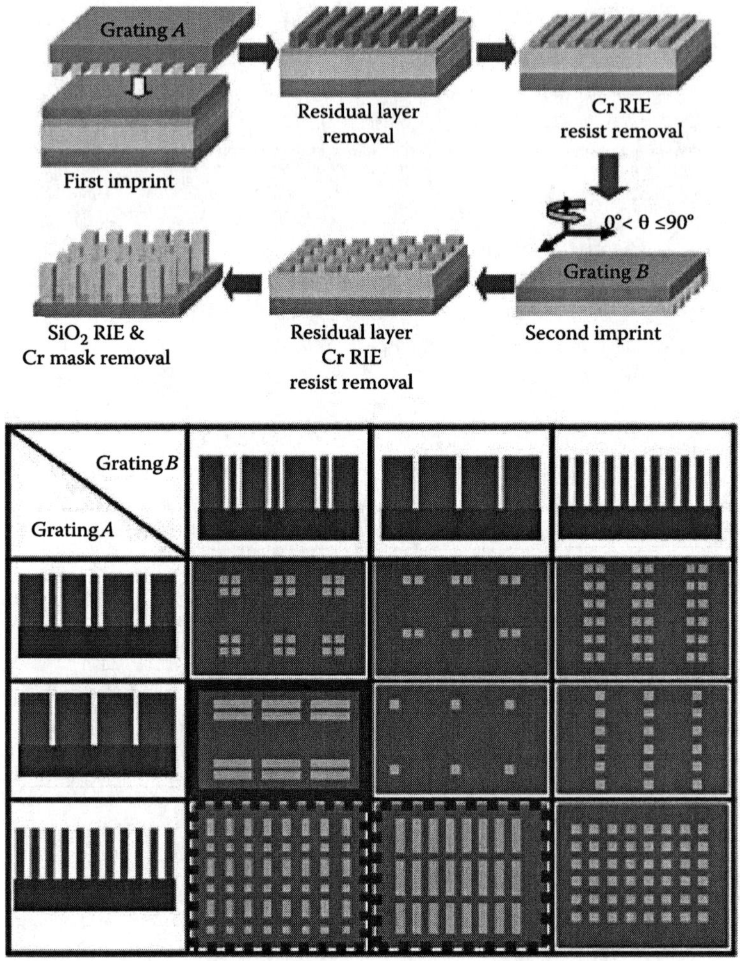

FIGURE 5.20
(Top) The general process used to fabricate "nanoblock" molds possessing different lattice and particle geometries. (Bottom) Matrix illustrating possible nanoparticle array configurations from A90B nanoblock molds produced using a collection of one-dimensional grating molds with different profiles. The array bordered by a solid line is produced using the inverse profile of grating A. Similarly, the arrays bordered by broken lines are achieved using the inverse of grating B. (From Lucas, B. D., Kim, J. S., Chin, C., and Guo, L. J., Nanoimprint lithography based approach for the fabrication of large-area, uniformly oriented plasmonic arrays, *Advanced Materials* 20 (6), 1129–1134, Copyright 2008, Wiley. Reprinted with permission.)

molds that are then used to fabricate NPAs. First, an appropriate substrate is coated with a mask layer and spin-coated with a nanoimprint resist. The resist is patterned using a one-dimensional grating mold (A), and the pattern is transferred to the underlying mask layer (e.g., Cr) through an appropriate etching procedure. After removing the formerly patterned resist, a new layer of resist is applied for a second imprint with a one-dimensional grating mold (B) rotated relative to the prior imprint. The pattern is once more transferred to the underlying mask layer by an etch process resulting in a two-dimensionally patterned mask layer that is then used to create a two-dimensional array of "nanoblocks" in the substrate material through an appropriate anisotropic etch procedure. Large-area metallic NPAs are produced by these nanoblock molds through conventional NIL processing steps consisting of imprinting, residual polymer deletion, metallization, and lift-off.

This method is enormously powerful in creating a range of structures by simply using one-dimensional gratings with variations in their lateral characteristics (i.e., duty cycle or period) and relative angular orientation for succeeding imprints. Benefiting this method are the numerous realized methods used to create, modify, and optimize one-dimensional grating structures. For example, if a similar grating mold is used and oriented orthogonally for successive imprints (A90A), a square lattice of rectangular or cylindrical (due to the reduced anisotropy of the etching procedure) nanoblocks is realized.

Different methods have demonstrated the ability to align rod-like nanostructures by incorporating them into a polymer matrix to produce nanocomposite films possessing polarization-dependent optical properties. Some researchers used stretching of a polymer-NP nanocomposite film to create a "pearl-necklace" alignment of the constituent Ag NPs. As a result of this induced alignment, the nanocomposite film achieved very different colors when viewed using linearly polarized light. For the square nanoblocks illustrated in the insert of Figure 5.21, even though the in-plane aspect ratio of the structures are approximately unity, there is still a clear distinction between the LSPR response of normally incident light polarized along the short axis and the long axis. For light linearly polarized beside the short axis, the resonance happens at the highest energy (transverse). On the other hand, light polarized along the long axis of the structure yields a resonance at the lowest energy (longitudinal). The ability to differentiate the two modes with a collection of nanostructures possessing very small shape anisotropy is a direct result of the orientational control afforded by the NIL production method as opposed to the ensemble average response of chemically synthesized nanostructures. This shape anisotropy can be attributed to slight dissimilarities in etching characteristics throughout mold manufacture using orthogonal imprints with the similar one-dimensional grating mold.

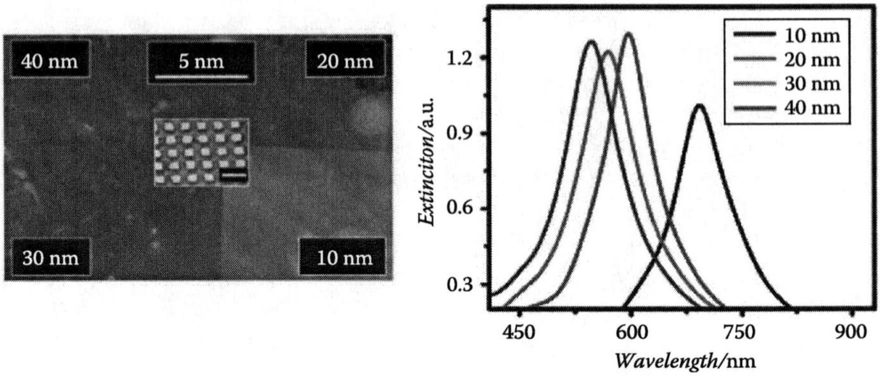

FIGURE 5.21
(Left) Ag sample possessing four distinct areas of structures that differ in metal thickness as indicated by the labels. Scanning electron micrograph (SEM) of the NPA is shown as an inset (scale bar: 350 nm). (Right) Measured extinction spectra corresponding to the four distinct areas shown in the photograph. (From Lucas, B. D., Kim, J. S., Chin, C., and Guo, L. J., Nanoimprint lithography based approach for the fabrication of large-area, uniformly oriented plasmonic arrays, *Advanced Materials* 20 (6), 1129–1134, Copyright 2008, Wiley. Reprinted with permission.)

5.7 Conclusion

The chemical/electrochemical method is one of the oldest methods used in the fabrication of nanomaterials. This method of construction is one of the wet chemical methods. In this method, first, primary material would be solved in an ordinary solvent and then the deposition agents would be added. Nonaqueous solvents are also usable in this method that can have many advantages, including more parameters for optimizing conditions of reaction (deposition agent stable in nonaqueous environment, reduction potentials, and primary material, etc.). The deposition agent can be a complex, reducing agent or an oxidation factor. After formation of nanoparticles, it is possible to make them stable in various ways, for instance, space methods, electrostatic, or a combination of both. One of the most important advantages of this method is that there is no need for modern equipment, and it can easily be used with equipment and tools that can already be found in the chemistry or physics laboratory. Another advantage of this method is its simple mechanism that can classify reactions performed in this method in the chemistry of solutions. Although this method is time consuming, complete monodispersed nanomaterials are able to be procured. The final result of this method is a simple one that requires accuracy, time, and ingenuity of the researcher to be able to find isomorphic, monodispersed nanoparticles with narrow size distribution.

References

1. Bozzini, B., D'Urzo, L., and Mele, C., Electrochemical fabrication of nano- and micrometric Cu particles: In situ investigation by electroreflectance and optical second harmonic generation, *Transactions of the Institute of Metal Finishing* 86 (5), 267–274, 2008.
2. Han, W. K., Choi, J. W., Hwang, G. H., Hong, S. J., Lee, J. S., and Kang, S. G., Fabrication of Cu nano particles by direct electrochemical reduction from CuO nano particles, *Applied Surface Science* 252 (8), 2832–2838, 2006.
3. Liu, L., Zhao, Y., Jia, N., Zhou, Q., Zhao, C., Yan, M., and Jiang, Z., Electrochemical fabrication and electronic behavior of polypyrrole nano-fiber array devices, *Thin Solid Films* 503 (1–2), 241–245, 2006.
4. Lohmüller, T., Müller, U., Breisch, S., Nisch, W., Rudorf, R., Schuhmann, W., Neugebauer, S., Kaczor, M., Linke, S., Lechner, S., Spatz, J., and Stelzle, M., Nano-porous electrode systems by colloidal lithography for sensitive electrochemical detection: Fabrication technology and properties, *Journal of Micromechanics and Microengineering* 18 (11), 2008.
5. Tao, F., Guan, M., Zhou, Y., Zhang, L., Xu, Z., and Chen, J., Fabrication of nickel hydroxide microtubes with micro- and nano-scale composite structure and improving electrochemical performance, *Crystal Growth and Design* 8 (7), 2157–2162, 2008.
6. Lu, H., Sun, H., Mao, A., Yang, H., Wang, H., and Hu, X., Preparation of plate-like nano α-Al_2O_3 using nano-aluminum seeds by wet-chemical methods, *Materials Science and Engineering A* 406 (1–2), 19–23, 2005.
7. Sarangi, P. P., Naik, B., and Ghosh, N. N., Low temperature synthesis of single-phase α-Fe_2O_3 nano-powders by using simple but novel chemical methods, *Powder Technology* 192 (3), 245–249, 2009.
8. Shashikala, V., Siva Kumar, V., Padmasri, A. H., David Raju, B., Venkata Mohan, S., Nageswara Sarma, P., and Rama Rao, K. S., Advantages of nano-silver-carbon covered alumina catalyst prepared by electro-chemical method for drinking water purification, *Journal of Molecular Catalysis A: Chemical* 268 (1–2), 95–100, 2007.
9. Lu, H. X., Hu, J., Chen, C. P., Sun, H. W., Hu, X., and Yang, D. L., Characterization of Al_2O_3-Al nano-composite powder prepared by a wet chemical method, *Ceramics International* 31 (3), 481–485, 2005.
10. Wang, A., Zhang, B., Wang, X., Yao, N., Gao, Z., Ma, Y., Zhang, L., and Ma, H., Nano-structure, magnetic and optical properties of Co-doped ZnO films prepared by a wet chemical method, *Journal of Physics D: Applied Physics* 41 (21), 2008.
11. Lu, H., Sun, H., Chen, C., Zhang, R., Yang, D., and Hu, X., Coating Cu nano-sized particles on Al_2O_3 powders by a wet-chemical method and its mechanical properties after hot press sintering, *Materials Science and Engineering A* 426 (1–2), 181–186, 2006.
12. Aliofkhazraei, M., Fartash, R., and Rouhaghdam, A. S., Nanocrystalline titanium aluminide coatings for ambient and high-temperature corrosion protection of CP-Ti, *Anti-Corrosion Methods and Materials* 56 (2), 110–113, 2009.

13. Aliofkhazraei, M., Sabour Rouhaghdam, A., Heydarzadeh, A., and Elmkhah, H., Nanostructured layer formed on CP-Ti by plasma electrolysis (effect of voltage and duty cycle of cathodic/anodic direction), *Materials Chemistry and Physics* 113 (2–3), 607–612, 2009.
14. Aliofkhazraei, M., Sabour Rouhaghdam, A., and Heydarzadeh, A., Strong relation between corrosion resistance and nanostructure of compound layer of treated 316 austenitic stainless steel, *Materials Characterization* 60 (2), 83–89, 2009.
15. Aliofkhazraei, M., Mofidi, S. H. H., Sabour Rouhaghdam, A., and Mohsenian, E., Duplex surface treatment of pre-electroplating and pulsed nanocrystalline plasma electrolytic carbonitriding of mild steel, *Journal of Thermal Spray Technology* 17 (3), 323–328, 2008.
16. Leggett, G. J., Direct writing of metal nanostructures: Lithographic tools for nanoplasmonics research, *ACS Nano* 5 (3), 1575–1579, 2011.
17. Andrews, L., Zhou, M., Willson, S. P., Kushto, G. P., Snis, A., and Panas, I., Infrared spectra of cis and trans-$(NO)_2$-anions in solid argon, *Journal of Chemical Physics* 109 (1), 177–185, 1998.
18. Fan, C., and Piron, D. L., Study of anomalous nickel-cobalt electrodeposition with different electrolytes and current densities, *Electrochimica Acta* 41 (10), 1713–1719, 1996.
19. Lee, Y. K., Manceron, L., and Pápai, I., An IR matrix isolation and DFT theoretical study of the first steps of the Ti(0) ethyle reaction: Vinyl titanium hydride and titanacyclopropene, *Journal of Physical Chemistry A* 101 (50), 9650–9659, 1997.
20. Zhou, M., and Andrews, L., Infrared spectra of the $C_2O_4^+$ cation and $C_2O_4^-$ anion isolated in solid neon, *Journal of Chemical Physics* 110 (14), 6820–6826, 1999.
21. Webb, P. R., and Robertson, N., Electrolytic codeposition of Ni-γAl_2O_3 thin films, *Journal of the Electrochemical Society* 141 (3), 669–673, 1994.
22. Selvarani, G., Selvaganesh, S. V., Krishnamurthy, S., Kiruthika, G. V. M., Dhar, S., Pitchumani, S., and Shukla, A. K., A methanol-tolerant carbon-supported Pt-Au alloy cathode catalyst for direct methanol fuel cells and its evaluation by DFT, *Journal of Physical Chemistry C* 113 (17), 7461–7468, 2009.
23. Barker, D., Electroless deposition of metals, *Transactions of the Institute of Metal Finishing* 71 (pt 3), 121–124, 1993.
24. Czerwinski, F., Grain size-internal stress relationship in iron-nickel alloy electrodeposits, *Journal of the Electrochemical Society* 143 (10), 3327–3332, 1996.
25. Borowski, P., Roos, B. O., Racine, S. C., Lee, T. J., and Carter, S., The ozonide anion: A theoretical study, *The Journal of Chemical Physics* 103 (1), 266–273, 1995.
26. Bulle-Lieuwma, C. W. T., Vandenhoudt, D. E. W., Henz, J., Onda, N., and Von Känel, H., Investigation of the defect structure of thin single-crystalline $CoSi_2$ (B) films on Si(111) by transmission electron microscopy, *Journal of Applied Physics* 73 (7), 3220–3236, 1993.
27. Nakamura, Y., Suzuki, R., Umeno, M., Cho, S. P., Tanaka, N., and Ichikawa, M., Observation of the quantum-confinement effect in individual β-$FeSi_2$ nanoislands epitaxially grown on Si (111) surfaces using scanning tunneling spectroscopy, *Applied Physics Letters* 89 (12), 2006.
28. Ko, Y. S., and Kwon, Y. U., Electrochemical deposition of platinum on fluorine-doped tin oxide: The nucleation mechanisms, *Electrochimica Acta* 55 (24), 7276–7281, 2010.

29. Khelladi, M. R., Mentar, L., Azizi, A., Sahari, A., and Kahoul, A., Electrochemical nucleation and growth of copper deposition onto FTO and n-Si(1 0 0) electrodes, *Materials Chemistry and Physics* 115 (1), 385–390, 2009.
30. Wang, Z., Kan, H., Shi, Z., Gao, B., Ban, Y., and Hu, X., Electrochemical deposition and nucleation of aluminum on tungsten in aluminum chloride-sodium chloride melts, *Journal of Materials Science and Technology* 24 (6), 915–920, 2008.
31. Fleury, V., Branched fractal patterns in non-equilibrium electrochemical deposition from oscillatory nucleation and growth, *Nature* 390 (6656), 145–148, 1997.
32. Li, F. B., and Albery, W. J., Electrochemical deposition of a conducting polymer, poly(thiophene-3-acetic acid): The first observation of individual events of polymer nucleation and two-dimensional layer-by-layer growth, *Langmuir* 8 (6), 1645–1653, 1992.
33. Sharma, R., Agarwala, R. C., and Agarwala, V., Development of electroless (Ni-P)/$BaNi_{0.4}Ti_{0.4}Fe_{11.2}O_{19}$ nanocomposite powder for enhanced microwave absorption, *Journal of Alloys and Compounds* 467 (1–2), 357–365, 2009.
34. Yang, H. G., and Zeng, H. C., Preparation of hollow anatase TiO_2 nanospheres via Ostwald ripening, *Journal of Physical Chemistry B* 108 (11), 3492–3495, 2004.
35. Voorhees, P. W., The theory of Ostwald ripening, *Journal of Statistical Physics* 38 (1–2), 231–252, 1985.
36. Voorhees, P. W., Ostwald ripening of two-phase mixtures, *Annual Review of Materials Science* 22 (1), 197–215, 1992.
37. Liu, B., and Zeng, H. C., Symmetric and asymmetric Ostwald ripening in the fabrication of homogeneous core-shell semiconductors, *Small* 1 (5), 566–571, 2005.
38. Marqusee, J. A., and Ross, J., Theory of Ostwald ripening: Competitive growth and its dependence on volume fraction, *The Journal of Chemical Physics* 80 (1), 536–543, 1984.
39. Kabalnov, A. S., and Shchukin, E. D., Ostwald ripening theory: Applications to fluorocarbon emulsion stability, *Advances in Colloid and Interface Science* 38 (C), 69–97, 1992.
40. Brailsford, A. D., and Wynblatt, P., The dependence of Ostwald ripening kinetics on particle volume fraction, *Acta Metallurgica* 27 (3), 489–497, 1979.
41. Taylor, P., Ostwald ripening in emulsions, *Advances in Colloid and Interface Science* 75 (2), 107–163, 1998.
42. Bartelt, N. C., Theis, W., and Tromp, R. M., Ostwald ripening of two-dimensional islands on Si(001), *Physical Review B—Condensed Matter and Materials Physics* 54 (16), 11741–11751, 1996.
43. Voorhees, P. W., and Glicksman, M. E., Solution to the multi-particle diffusion problem with applications to Ostwald ripening—I. Theory, *Acta Metallurgica* 32 (11), 2001–2011, 1984.
44. Voorhees, P. W., and Glicksman, M. E., Solution to the multi-particle diffusion problem with applications to Ostwald ripening—II. Computer simulations, *Acta Metallurgica* 32 (11), 2013–2030, 1984.
45. Davies, C. K. L., Nash, P., and Stevens, R. N., The effect of volume fraction of precipitate on Ostwald ripening, *Acta Metallurgica* 28 (2), 179–189, 1980.
46. Lu, Y., Lu, X., Mayers, B. T., Herricks, T., and Xia, Y., Synthesis and characterization of magnetic Co nanoparticles: A comparison study of three different capping surfactants, *Journal of Solid State Chemistry* 181 (7), 1530–1538, 2008.

47. Misra, T. K., and Liu, C. Y., Synthesis of 28-membered macrocyclic polyammonium cations functionalized gold nanoparticles and their potential for sensing nucleotides, *Journal of Colloid and Interface Science* 326 (2), 411–419, 2008.
48. Nandi, A., Dutta Gupta, M., and Banthia, A. K., Sulfonated polybutadiene random ionomer as stabiliser for colloidal copper nanoparticles, *Colloids and Surfaces A: Physicochemical and Engineering Aspects* 197 (1–3), 119–124, 2002.
49. Su, G. J., Li, Z. H., and Aguilar-Sanchez, R., Phase transition of two-dimensional chiral supramolecular nanostructure tuned by electrochemical potential, *Analytical Chemistry* 81 (21), 8741–8748, 2009.
50. Xia, H., and Wang, D., Fabrication of macroscopic freestanding films of metallic nanoparticle monolayers by interfacial self-assembly, *Advanced Materials* 20 (22), 4253–4256, 2008.
51. Kim, J. W., Lim, S. K., Kim, C. K., Kim, Y. H., and Yoon, C. S., Fabrication of metallic nanoparticle mono-layer made from selective reaction of $Ni_{100-x}Fe_x$ thin films with polyamic acid during its imidization, *Colloids and Surfaces A: Physicochemical and Engineering Aspects* 284–285, 350–354, 2006.
52. Szczech, J. B., Megaridis, C. M., Zhang, J., and Gamota, D. R., Ink jet processing of metallic nanoparticle suspensions for electronic circuitry fabrication, *Microscale Thermophysical Engineering* 8 (4), 327–339, 2004.
53. Shankar, R., Groven, L., Amert, A., Whites, K. W., and Kellar, J. J., Non-aqueous synthesis of silver nanoparticles using tin acetate as a reducing agent for the conductive ink formulation in printed electronics, *Journal of Materials Chemistry* 21 (29), 10871–10877, 2011.
54. McDonald, K. J., and Choi, K. S., Photodeposition of co-based oxygen evolution catalysts on α-Fe_2O_3 photoanodes, *Chemistry of Materials* 23 (7), 1686–1693, 2011.
55. Xu, T., Lin, C., Wang, C., Brewe, D. L., Ito, Y., and Lu, J., Synthesis of supported platinum nanoparticles from Li-Pt solid solution, *Journal of the American Chemical Society* 132 (7), 2151–2153, 2010.
56. Kydd, R., Scott, J., Teoh, W. Y., Chiang, K., and Amal, R., Understanding photocatalytic metallization of preadsorbed ionic gold on titania, ceria, and Zirconia, *Langmuir* 26 (3), 2099–2106, 2010.
57. Starowicz, M., Stypuła, B., and Banas, J., Electrochemical synthesis of silver nanoparticles, *Electrochemistry Communications* 8 (2), 227–230, 2006.
58. Lu, Q., Lee, K. J., Lee, K. B., Kim, H. T., Lee, J., Myung, N. V., and Choa, Y. H., Investigation of shape controlled silver nanoplates by a solvothermal process, *Journal of Colloid and Interface Science* 342 (1), 8–17, 2010.
59. Wen, M., Wang, Y., Wu, Q., Jin, Y., and Cheng, M., Controlled fabrication of 0 & 2D NiCu amorphous nanoalloys by the cooperation of hard-soft interfacial templates, *Journal of Colloid and Interface Science* 342 (2), 229–235, 2010.
60. Thünemann, A. F., Schütt, D., Kaufner, L., Pison, U., and Möhwald, H., Maghemite nanoparticles protectively coated with poly(ethylene imine) and poly(ethylene oxide)-block-poly(glutamic acid), *Langmuir* 22 (5), 2351–2357, 2006.
61. Martinez, C. J., Hockey, B., Montgomery, C. B., and Semancik, S., Porous tin oxide nanostructured microspheres for sensor applications, *Langmuir* 21 (17), 7937–7944, 2005.
62. Nyffenegger, R. M., Craft, B., Shaaban, M., Gorer, S., Erley, G., and Penner, R. M., A hybrid electrochemical/chemical synthesis of zinc oxide nanoparticles and optically intrinsic thin films, *Chemistry of Materials* 10 (4), 1120–1129, 1998.

63. Ebina, Y., Sasaki, T., Harada, M., and Watanabe, M., Restacked perovskite nanosheets and their Pt-loaded materials as photocatalysts, *Chemistry of Materials* 14 (10), 4390–4395, 2002.
64. Bulushev, D. A., Yuranov, I., Suvorova, E. I., Buffat, P. A., and Kiwi-Minsker, L., Highly dispersed gold on activated carbon fibers for low-temperature CO oxidation, *Journal of Catalysis* 224 (1), 8–17, 2004.
65. Kemell, M., Pore, V., Ritala, M., Leskelä, M., and Lindén, M., Atomic layer deposition in nanometer-level replication of cellulosic substances and preparation of photocatalytic TiO_2/cellulose composites, *Journal of the American Chemical Society* 127 (41), 14178–14179, 2005.
66. Choi, H. C., Kundaria, S., Wang, D., Javey, A., Wang, Q., Rolandi, M., and Dai, H., Efficient formation of iron nanoparticle catalysts on silicon oxide by hydroxylamine for carbon nanotube synthesis and electronics, *Nano Letters* 3 (2), 157–161, 2003.
67. Gao, Y., Masuda, Y., Ohta, H., and Koumoto, K., Room-temperature preparation of ZrO_2 precursor thin film in an aqueous peroxozirconium-complex solution, *Chemistry of Materials* 16 (13), 2615–2622, 2004.
68. Karthikeyan, J., Berndt, C. C., Tikkanen, J., Wang, J. Y., King, A. H., and Herman, H., Nanomaterial powders and deposits prepared by flame spray processing of liquid precursors, *Nanostructured Materials* 8 (1), 61–74, 1997.
69. Milsom, E. V., Novak, J., Oyama, M., and Marken, F., Electrocatalytic oxidation of nitric oxide at TiO_2-Au nanocomposite film electrodes, *Electrochemistry Communications* 9 (3), 436–442, 2007.
70. Kim, T. H., and Sohn, B. H., Photocatalytic thin films containing TiO_2 nanoparticles by the layer-by-layer self-assembling method, *Applied Surface Science* 201 (1–4), 109–114, 2002.
71. Wang, D., Wang, Y., Li, X., Luo, Q., An, J., and Yue, J., Sunlight photocatalytic activity of polypyrrole-TiO_2 nanocomposites prepared by "in situ" method, *Catalysis Communications* 9 (6), 1162–1166, 2008.
72. Bagkar, N., Ganguly, R., Choudhury, S., Hassan, P. A., Sawant, S., and Yakhmi, J. V., Synthesis of surfactant encapsulated nickel hexacyanoferrate nanoparticles and deposition of their Langmuir-Blodgett film, *Journal of Materials Chemistry* 14 (9), 1430–1436, 2004.
73. Stathatos, E., Lianos, P., Del Monte, F., Levy, D., and Tsiourvas, D., Formation of TiO_2 nanoparticles in reverse micelles and their deposition as thin films on glass substrates, *Langmuir* 13 (16), 4295–4300, 1997.
74. Kahn, M. L., Monge, M., Collière, V., Senocq, F., Maisonnat, A., and Chaudret, B., Size- and shape-control of crystalline zinc oxide nanoparticles: A new organometallic synthetic method, *Advanced Functional Materials* 15 (3), 458–468, 2005.
75. Kobayashi, Y., Nakashima, H., Takagi, D., and Homma, Y., CVD growth of single-walled carbon nanotubes using size-controlled nanoparticle catalyst, *Thin Solid Films* 464–465, 286–289, 2004.
76. Morfa, A. J., Rowlen, K. L., Reilly Ii, T. H., Romero, M. J., and Van De Lagemaat, J., Plasmon-enhanced solar energy conversion in organic bulk heterojunction photovoltaics, *Applied Physics Letters* 92 (1), art no. 013504, 2008.
77. Rella, R., Spadavecchia, J., Manera, M. G., Capone, S., Taurino, A., Martino, M., Caricato, A. P., and Tunno, T., Acetone and ethanol solid-state gas sensors based on TiO_2 nanoparticles thin film deposited by matrix assisted pulsed laser evaporation, *Sensors and Actuators, B: Chemical* 127 (2), 426–431, 2007.

78. Scotognella, F., Puzzo, D. P., Monguzzi, A., Wiersma, D. S., Maschke, D., Tubino, R., and Ozin, G. A., Nanoparticle one-dimensional photonic-crystal dye laser, *Small* 5 (18), 2048–2052, 2009.
79. Bhattacharjee, S., Dotzauer, D. M., and Bruening, M. L., Selectivity as a function of nanoparticle size in the catalytic hydrogenation of unsaturated alcohols, *Journal of the American Chemical Society* 131 (10), 3601–3610, 2009.
80. Kooi, S. E., Baker, L. A., Sheehan, P. E., and Whitman, L. J., Dip-pen nanolithography of chemical templates on silicon oxide, *Advanced Materials* 16 (12), 1013–1016, 2004.
81. Katagiri, K., Nakamura, M., and Koumoto, K., Magnetoresponsive smart capsules formed with polyelectrolytes, lipid bilayers and magnetic nanoparticles, *ACS Applied Materials and Interfaces* 2 (3), 768–773, 2010.
82. Szot, K., Lesniewski, A., Niedziolka, J., Jonsson, M., Rizzi, C., Gaillon, L., Marken, F., Rogalski, J., and Opallo, M., Sol-gel processed ionic liquid-hydrophilic carbon nanoparticles multilayer film electrode prepared by layer-by-layer method, *Journal of Electroanalytical Chemistry* 623 (2), 170–176, 2008.
83. Zhong, Z. C., Cheng, R. H., Bosley, J., Dowben, P. A., and Sellmyer, D. J., Fabrication of chromium oxide nanoparticles by laser-induced deposition from solution, *Applied Surface Science* 181 (3–4), 196–200, 2001.
84. Lee, K. C., Lin, S. J., Lin, C. H., Tsai, C. S., and Lu, Y. J., Size effect of Ag nanoparticles on surface plasmon resonance, *Surface and Coatings Technology* 202 (22–23), 5339–5342, 2008.
85. Lim, E. K., Yang, J., Dinney, C. P. N., Suh, J. S., Huh, Y. M., and Haam, S., Self-assembled fluorescent magnetic nanoprobes for multimode-biomedical imaging, *Biomaterials* 31 (35), 9310–9319, 2010.
86. Lucas, B. D., Kim, J. S., Chin, C., and Guo, L. J., Nanoimprint lithography based approach for the fabrication of large-area, uniformly oriented plasmonic arrays, *Advanced Materials* 20 (6), 1129–1134, 2008.

6

Physical and Other Fabrication/Properties of Two-Dimensional Nanostructures

6.1 Introduction

Sculptured thin films (STFs) are films prepared through glancing angle deposition techniques. In this technique the deposition system is combined with the rotor substrate through physical evaporation. Depending on the deposition rate, rotation speed of the substrate, and selection of the rotation axis, two STFs are created: helicoidal thin films with bi-anisotropic media and sculptured nematic thin films. Such films are columnar thin films (CTFs) whose columnar direction may be changed easily during deposition. They may be considered as a unilateral anisotropic heterogeneous continuum medium in visible, infrared, and low ultraviolet (UV) wavelengths. Porosity is the characteristic of these films so that their anisotropy would be controlled through controlling porosity. Hence, such films would be used in optics, nano-optics, and magneto-optics applications.[1–10]

6.2 Concepts of Nanostructured Thin Films

Isolated masses ended in the substrate as parallel and chain-like matchwoods introduce columnar thin films that were studied for the first time in 1960 (selecting materials for the substrate and preparing such matchwoods are approximately infinite). If these isolated and parallel arrangements of matchwoods, which redirect to the upward position of the substrate are deviated from the straight line and are twisted randomly, they introduce STFs. In electromagnetic waves with larger than columnar diameter wavelengths, CTFs and STFs are considered as homogenous anisotropic continuum media and unilateral anisotropic heterogeneous continuum media, respectively. Because STFs are porous, if they are penetrated by fluids or polymers, liquid crystals, nonorganic solids, and so forth, their constitutive features would be altered heterogeneously.[11–19]

STFs can be used in the following fields:

1. Optical sensors for biologic, chemical, or nuclear fluids
2. Linear and nonlinear optical circuits that may be combined with electronic circuits
3. Nanocomposites with low permittivity and ultra-low permittivity dams in multifilm electronic chips
4. Biomedical parts such as microscreeners, viral traps, biocompatible substrates, and fatty films
5. Special catalytic reactors for asymmetric synthesis of mega-Dalton polymers, chiral and antipest drugs

STFs are subdivided into two classes morphologically:

1. Thin film helicoidal bianistropic media (TFHBM)
2. Sculptured nematic thin films (SNTFs)

The difference between two STFs is determined through spatial changes of a unit vector, which is called a director. If a STF grows in the half space of $z \geq 0$ and the substrate is in $z = 0$, the director will be a combination of the reference unit vector U_t (z) ($z = 0$) and a dyadic function $S(z)$.

The STF concept has been developed through manipulating the CTF growth as the substrate motion control. First, suppose that the substrate is fixed and motionless and CTF grows on it. Columns have been composed of 1 to 3 nm clusters; as a result, the column in the thickness of 1000 nm will have a diameter as large as 20 nm. If relative permittivity of the columns' deposition material is equal to four in the bulk state, CTF would be considered as a continuum medium in linear frequencies $<10^{15}$ Hz. In lower frequencies, CTF may be considered with permittivity dyadic depending on the frequency as

$$\underline{\underline{\varepsilon}}_{CTF}(\omega) = \varepsilon_0 \left\{ \varepsilon_a(\omega)\left[\underline{\underline{I}} - \underline{u}_\tau \underline{u}_\tau \right] + \varepsilon_b(\omega) \underline{u}_\tau \underline{u}_\tau + \varepsilon_c(\omega) \underline{u}_\tau \times \underline{\underline{I}} \right\} \tag{6.1}$$

where ω is angular frequency, ε_0 is vacuum permittivity, and $\underline{\underline{I}} = \underline{u}_x \underline{u}_x + \underline{u}_y \underline{u}_y + \underline{u}_z \underline{u}_z$ is the unit dyadic. u_z, u_y, and u_x are Cartesian unit vectors, and $\varepsilon_a(\omega), \varepsilon_b(\omega), \varepsilon_c(\omega)$ scalars introduce constitutive anisotropy. $U_t(z)$ director is parallel with the columnar direction, and the cross section of columns is presumed to be circular and any ellipsoid cross-sectional column was disregarded. Also it is presumed that the deposition material is in an anisotropic volume (amorphous Si and MgF_2). If the deposition material is in an anisotropic volume (ZnO), columns will appear as crystal forms based on their growth conditions, which have preferential orientation so that anisotropy of a CTF would be described by using one or more directions.

If the substrate rotates around the z axis with a proper angular velocity, helicoidal columns grow instead of oblique columns, and a TFHBM will be formed instead of CTF, and sculptured nematic microstructure that is SNTF is generated through a rotation around the parallel axis of the substrate plate (xy).

The permittivity dyadic for each STF is calculated as

$$\underline{\underline{\varepsilon}}_{STF}(z,\omega)=\varepsilon_0\left\{\varepsilon_a(\omega)\left[\underline{\underline{I}}-\underline{u}_t(z)\underline{u}_t(z)\right]+\varepsilon_b(\omega)\underline{u}_t(z)\underline{u}_t(z)+\varepsilon_c(\omega)\underline{u}_t(z)\times\underline{\underline{I}}\right\} \quad (6.2)$$

$$\underline{u}_t(z)=\underline{\underline{S}}(z)\cdot\underline{u}_\tau \quad (6.3)$$

where $S(z)$ is an orthogonal dyadic and a function of z. The function for a certain STF is as the following:

$$\underline{\underline{S}}(z)=\underline{\underline{S}}_j(z),\quad z_j\le z\le z_{j+1} \quad (6.4)$$

A partial dyadic $S_j(z)$ is the product of multiplication of one or more dyadics out of three following preliminary rotational dyadics:

$$\underline{\underline{S}}_{jx}(z)=(\underline{u}_y\underline{u}_y+\underline{u}_z\underline{u}_z)\cos\xi_j(z)+(\underline{u}_z\underline{u}_y-\underline{u}_y\underline{u}_z)\sin\xi_j(z)+\underline{u}_x\underline{u}_x \quad (6.5)$$

$$\underline{\underline{S}}_{jy}(z)=(\underline{u}_x\underline{u}_x+\underline{u}_z\underline{u}_z)\cos\chi_j(z)+(\underline{u}_z\underline{u}_x-\underline{u}_x\underline{u}_z)\sin\chi_j(z)+\underline{u}_y\underline{u}_y \quad (6.6)$$

$$\underline{\underline{S}}_{jz}(z)=(\underline{u}_x\underline{u}_x+\underline{u}_y\underline{u}_y)\cos\zeta_j(z)+(\underline{u}_y\underline{u}_x-\underline{u}_x\underline{u}_y)\sin\zeta_j(z)+\underline{u}_z\underline{u}_z \quad (6.7)$$

In these equations, $S_{jy}(z)$ and $S_{jx}(z)$ lead to sculptured nematic morphology separately, and $\underline{\underline{S}}_{jz}(z)$ leads to cholesteric morphology.

The jth part of a STF is as follows:

1. It is a TFHBM if

$$\underline{\underline{S}}_j(z)=\underline{\underline{S}}_{jz}(z)\cdot\underline{\underline{S}}_{jy}(z),\ \chi_j(z)=\chi_j(z_j),\ \zeta_j(z)\propto z,\ \forall z\in[z_j,z_{j+1}]$$

2. It is a SNTF, if

$$\underline{\underline{S}}_{jz}(z)=\underline{\underline{S}}_{jy}(z)$$

3. It is a CTF, if

$$\underline{\underline{S}}_j(z)=\underline{\underline{S}}_{jx}(z)\underline{\underline{S}}_{jy}(z)\underline{\underline{S}}_{jz}(z),\ \xi_j(z)=\xi_j(z_j),\ \chi_j(z)=\chi_j(z_j),\ \zeta_j(z)=\zeta_j(z_j),\ \forall z\in[z_j,z_{j+1}]$$

Linear constitutive relations of STFs are as follows:

$$\underline{D}(\underline{r},\omega)=\underset{=STF}{\varepsilon}(z,\omega)\cdot\underline{E}(\underline{r},\omega)+\underset{=STF}{k}(z,\omega)\cdot\underline{B}(\underline{r},\omega) \tag{6.8}$$

$$\underline{H}(\underline{r},\omega)=\underset{=STF}{\gamma}(z,\omega)\cdot\underline{E}(\underline{r},\omega)+\underset{=STF}{\eta}(z,\omega)\cdot\underline{B}(\underline{r},\omega) \tag{6.9}$$

where $\underset{=STF}{\varepsilon}(z,\omega)$ with $\underset{=STF}{\eta}(z,\omega)$ and $\underset{=STF}{\gamma}(z,\omega)$, $\underset{=STF}{k}(z,\omega)$ are isomorphic, and the following constraint is applied through Maxwell's equations:

$$Trac[\underset{=STF}{k}(z,\omega)-\underset{=STF}{\gamma}(z,\omega)]=0 \tag{6.10}$$

And the mentioned equations would be generalized to nonlinear reaction features. If Fourier's spatial conversions are used, electromagnetic fields will be

$$\underline{E}(\underline{r},\omega)=\underline{e}(z,k_x,k_y,\omega)e^{i(k_xx+k_yy)} \tag{6.11}$$

$$\underline{H}(\underline{r},\omega)=\underline{h}(z,k_x,k_y,\omega)e^{i(k_xx+k_yy)} \tag{6.12}$$

By placing them in Maxwell's equations, we have

$$\nabla\times\underline{E}(\underline{r},\omega)=i\omega\,\underline{B}(\underline{r},\omega) \tag{6.13}$$

$$\nabla\times\underline{H}(\underline{r},\omega)=-i\omega\,\underline{D}(\underline{r},\omega) \qquad 6.14)$$

After some mathematical operations, the following differential equitation of a 4 × 4 matrix is obtained:

$$d/dz[\underline{f}(z,k_x,k_y,\omega)]=i[\underline{\underline{p}}(z,k_x,k_y,\omega)][\underline{f}(z,k_x,k_y,\omega)] \tag{6.15}$$

where

$$[\underline{f}(z,k_x,k_y,\omega)]=co[e_x(z,k_x,k_y,\omega),e_x(z,k_x,k_y,\omega),e_x(z,k_x,k_y,\omega),e_x(z,k_x,k_y,\omega)] \tag{6.16}$$

are four columnar vectors, and $[\underline{\underline{p}}(z,k_x,k_y,\omega)]$ is a 4 × 4 matrix. Equation (6.16) would be solved digitally using standard mathematic methods. Digital solution is easy. If constitutive dyadics have a fixed value in small cuts (as thick as *h*), the following equation would be used for $[\underline{f}(z,k_x,k_y,\omega)]$:

$$[\underline{f}(z+h,k_x,k_y,\omega)] \cong e^{i[p(z+h)/2,k_x,k_y,\omega)]h}[\underline{f}(z+h,k_x,k_y,\omega)] \tag{6.17}$$

6.3 Important Physical Fabrication Methods

In addition to the extended discussion of fabrication methods in previous chapters, here some necessary aspects, limitations, and applications for different methods will be reviewed.

6.3.1 Excimer Laser

An excimer laser (or exciplex laser) is a kind of ultraviolet chemical laser that was introduced in 1970. The term *excimer* is a contracted form of "excited dimer," while the exciplex is the shortened form of "excited complex." In an excimer laser, most of the time, a combination of a noble gas, such as Argon, Krypton, or Xenon, is used with a reactant gas, such as fluorine, or chlorine. Under suitable conditions, electrical excitement of a quasi-molecule, called a dimer, can be produced. A dimer is stable only in high levels of energy; in the lower levels it emits a beam of laser with a wavelength of UV range. Due to some characteristics of this laser, discussed in the following sections, it has some unique applications and can be applied in coating systems with a pulse laser deposition (PLD), laser ablation, and the other laser-benefited methods.

Due to some drawbacks, the initial excimer lasers generally had a short life. Some of these problems are the corrosive nature of applied gases, pollution of gases with additional chemicals, or the dust created by electrical discharge and high loads on Thyratron in the power generator, which limited Thyratron switches lives from several weeks to several months. However, using different remedies, such as applying corrosion-resistant materials, new refinery and gas circulation systems, as well as solid-state high-voltage switches, the life of the laser excimer is limited only by optical properties of the UV.[20–26]

6.3.2 Molecular Beam Epitaxy (MBE)

Molecular beam epitaxy is among the methods for precipitation of the monocrystals. The method was invented in the late 1960s. In this method, crystal growth is controlled under very restricted conditions and without any contamination. The term *epitaxy* comes from the Greek words "epi" (on) and "taxis" (array). Through this method, a material's crystal develops on the same material (homopitaxy) or a different material (heteropitaxy).

Asaro–Tiller–Grinfeld (ATG) instability, formerly known as Grinfield, is an elastic instability that occurred during the MBE method. Once there is no efficient adaption between a growing crystal's network constant and substrate crystal, elastic energy will be concentrated in the growing crystal. As the thickness of free energy film increases, the film segregation will be reduced to a few "islands," resulting in a decrease of existent stresses in the planar direction.

This critical thickness depends on the rate of discrepancy, Young modulus, and surface stresses. The discrepancy factor is defined as follows:

$$D_a = |a_f - a_s| / a_s$$

where a_f and a_s are constants of the film's planar network and substrate's planar network, respectively.

Once the discrepancy factor is above 1%, planar stresses and structural defects, such as displacements, will noticeably increase in the network. This is an unfavorable event that brings up some limitations while we want to choose a material for growth of monocrystal film on every pure substrate. It should be noticed that ATG instability has some applications such as self-assembly of the quantum points.

The surface where we want to develop the crystal must have no impurity or artifact. To control the purity of the surface in MBE, auger electron spectroscopy is commonly used. However, the substrate is usually processed for surface preparation and then put in a MBE container, and finally the growth process is done on it. Sometimes, development of an oxide layer, which serves as a protector, leads to protection of the substrate. Before the development process starts, this oxide layer is extracted from the growth container.[27–35]

6.3.3 Ion Implantation

During the ion implantation process, some definite ions are literally "implanted" in the structure of a solid and affect its physical properties. The ion implantation technique is mostly used in the semiconductor industry. Implanted ions in target materials create chemical modifications as well as structural ones, which leads to destruction or damage of the target material's crystalline network. The impact of ions with high energy content can cause so many point damages in the crystal of the target material. This is partly due to the fact that the process is accompanied by a thermal annealing to perform the correction and recovery of the damages. The annealing temperature is commonly about 900 to 1100°C in the semiconductor industry. However, during the application of this method in nanotechnology, for using the ions with low energy content, the risks of structural damages will be minimized.

Physical scatter of the substance only happens when the ions involve high energy content. This leads to unfavorable sweeping of the material from the surface. Because the energy of the ions is not that much for completely sweeping the atoms, this method is not as important as the previous method in nanotechnology. Damages to the crystalline network caused by an amorphous fabrication can increase the structural defects to the point where the crystalline network of the material is completely ruined and an amorphous film developed on the surface, which is counted as a destructive effect once our aim is not the process of amorphous fabrication. The toxic gases used in the ion implication method are very poisonous, carcinogenic, corrosive, and fire inducing, particularly when semiconductors are the matter. Therefore, the safety measures and instructions for work with the machine must be exactly obeyed. However, ion implantation method machines are generally automotive and operate in isolated rooms, where there is the least human contact with them.[36–45]

6.3.4 Focused Ion Beam (FIB) in Nanotechnology

In this method, a focused ion beam (FIB) is used for sweeping and abrasion of the piece's surface. The basics of the FIB are similar to those of the scanning electronic microscope (SEM). The sample's surface bombardment by high-energy ions results in production of secondary electrons, ions, and diffusion materials, which are mainly the basic part for different applications of the FIB and will be discussed later. Applying a negative bios of 30 KV extracts the GA^+ ions from liquid metal ion source (LMIS) and accelerates them toward the sample. A pair of electrostatics lenses, a beam driver, and a rectifier of the beam into the octupole deflector concentrates ions as a focused beam and then sweeps them by a sample surface beam. After bombardment of the sample surface with an ion beam, the physical surface scatter of the materials and separation of the material from the surface occur. This is accompanied by generation of the secondary ions and electrons. These are received by a deflector plate, so an image of the surface will be developed. The beam sweep control system makes it possible to create different patterns by the machine. In addition, injection of different gases near the sample's surface during the process results in deposition and creation of the coating layer. Some applications also facilitate surface coat removal.[36,46–53]

6.3.5 FIB-SEM Dual Beam System

The FIB-SEM dual beam system is unique in that it enables etching and coat removing, pattern generation, and construction of the nanostructures using FIB as well as their simultaneous observing and analysis with SEM. In this system, a scanning electronic microscope with high resolution is combined with a focused ion beam with a Gallium ion source. Using this system it is possible to perform etching, cutting, deposition, and coat removing, as well as simultaneous changes of surface topography with secondary excited electrons by an ion beam. In addition, with a SEM section of this system, a very concentrated electronic beam allows us to take photos with very high resolution through secondary and reflected electrons for deeper parts and imaging of the transitional electrons for defining grains and texture boundary properties. It is worth saying that the SEM system lacks the etching effects of an ion beam, as well as creating capabilities for better imaging in comparison with images obtained by FIB. Also, excited X beam analysis with electronic beam beside (energy dispersive spectrometer) EDS creates unique information for elements with an atomic number higher than 12. In addition, FIB-SEM dual beam system involves great automation and ease of access.[54–60]

6.3.6 Langmuir–Blodgett Film (LBF)

An insoluble monolayer of particles floated in an interface of the gas (air) and liquid phases and frequently deposited in a solid substrate is known as Langmuir–Blodgett film (LBF).

Lipids, polymers, and some insoluble particles (in water) are capable of developing very thin monolayer films in the interface of water and air, which is called *Langmuir film*. Langmuir film can be deposited to develop multilayered complexes with very accurate order and sequence on the surface of a solid substrate, which ends up in a production called *Langmuir–Blodgett film*. It is possible to coat the mentioned layer, typically by submerging the substrate in a liquid that is in the condition of controlled temperature, surface pressure, and molecular density. Once we repeat the mentioned method, we can deposit several layers of Langmuir film with a thickness of several nanometers. The basics of the Langmuir–Blodgett filming science was founded in the early twentieth century by Irving Langmuir and Katherine Blodgett. Atoms or molecules involved in the development of Langmuir film are mainly polar ones and have a hydrophilic end as well as a hydrophobic end.

There are many factors effective on the development of Langmuir–Blodgett film. Among these, one can name

1. Nature and material of the monolayer
2. Composition of the underlying liquid phase
3. Temperature
4. Surface pressure during the deposition of the monolayer
5. Speed of deposition
6. Type and nature of the solid substrate
7. The passed time of preserving the substrate in the air or underlying phase

The most common type of these multilayers is type Y. This type can be developed whenever the monolayer starts to develop in both tail-to-head and head-to-tail directions. Once the layers just grow in one tail-to-head or head-to-tail direction, the product will be named an *X* or *Z* type. However, in some LB multilayers, there are middle force layers known as XY types.[61–68]

6.3.7 Electrospinning

In the electrospinning method, an electrical charge is used for spinning nanometer fibers from a liquid phase. The technique once registered in 1902 and then developed during 1934 to 1944 in order to use in textile industries. Then it was not used until the 1990s, when researchers found that it is possible to produce nanometer fibers from organic polymers. Since then, studies on the development of this technique are continuing with a high growth, and its application range is widening.

The different parts of a standard electrospinning system are a spinneret that includes a syringe with an injection needle connected to a high-voltage source (10 to 50 kV), syringe pump, and a fiber mat attached to the ground. The spinneret and the mat are attached to a cathode and anode, respectively.

In this system, the solution or liquid enters the syringe and is injected with a constant speed by pump pressure. This leads to the development of a drop in the tip of the needle. Through applying the voltage the drop will be bent like a Taylor Cone. If the molecular cohesion of the liquid increases enough, this created current will not be disrupted and the charged liquid jet will be developed. This jet will be dragged by repulsive electrostatic force and travels in a tortuous path. This continues until it deposits on the fiber pad. Dragging in a charged jet by bending instability, simultaneous evaporation of the solvent, and reaction of the materials within the jet with the surrounding environment may result in the production of fibers with nanometer diameters.[69–72]

6.4 Specification of Sculptured Thin Films

STFs consist of two fundamental sets, namely thin film helicoidal bianisotropic media (TFHBM) and sculptured nematic thin films (SNTFs). Each STF is described by a unit vector (director) that is altered along the z axis. Specifications of both STFs may be combined in a single-part STF through a director. If the substrate area of the source plane is $Z = 0$ and STF grows in half space ($z \geq 0$), the director vector is steep in

1. Position angle $\theta_m(z)$ rather than z axis and in $0 \leq \theta_m(z) \leq 90°$ interval
2. Polar angle $\varphi_m(z)$ rather than x axis and in $0 \leq \varphi_m(z) \leq 360°$ interval

Both angles are functions of the z axis that determines changes of position angle of the SNTF and changes of polar angles of TFHBM. Simultaneous changes of $\theta_m(z)$ and $\varphi_m(z)$ determine the director profile of a STF.

STFs are columnar thin films whose columnar directions may change during the deposition easily. A CTF is described with $\theta_m(z) = \theta_m$ and $\varphi_m(z) = 0°$ and for a CTF deposition, vapor flux reached the substrate through a certain direction and in an atomic low mobility condition. The columnar direction of CTF can be altered through changing the substrate orientation according to the flux direction. Deposition angle (X_v) and columnar angle (X_m) are depicted in Figure 6.1. Our studies have shown that for $X_m \geq X_v$ (this unequality in $X_V = 90°$ (vertical descend) becomes equal), the columnar direction is not similar to that of the descending flux.

Density of a CTF increases in line with X_v nonlinearly, and columnar direction is not sufficient to understand optical and mechanical properties. For CTFs that have been deposited diagonally, tests have shown that mechanic tension in the xy plane is anisotropic because of porosity anisotropy, and it will be unsteady along the z axis. In the most ideal condition, porosity is considered independent from the columnar direction during designing the reaction of optical and other specifications.

The physical basis for the development of CTFs is that in atomic low mobility conditions, the atoms' random collection process toward the vapor flux direction will result in clusterification in nanometric scales. Atomic low mobility indicates that atomic rearrangement after deposition near the growing film surface is lower than the cluster size standard (i.e., 1 to 3 nm). Thus, atomic adumbration will result in the competitive growth in the final microstructure of the thin film.

In vertical deposition, regional structure models have been suggested for film surface morphology qualitatively, which vary from cauliflower-like (region 1) to matchwood (M region) and to smooth (region T).

Columnar diameter (d) and columnar thickness (t) are related through power law $d \propto t^{\frac{1}{n}}$ in region 1. If three growth key factors of the substrate

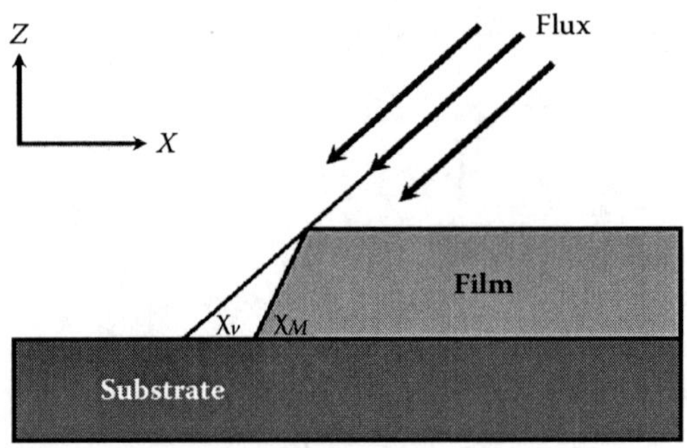

FIGURE 6.1
Deposition angle and columnar rise angle.

temperature, ion bombardment of the growing surface, and the surface chemical environment are not sufficient to alter atomic clusterification processes and adumbration processes, semicauliflower growth will change into a semifractal growth process.

With atoms' mobility increasing and propagation of columns' diameter being narrower through ion bombardment, column sizes may change so that the exponent of power law (n) increases during the growth, and the growth will become cylindrical (region M). In other words, column diameter must be fixed as the film's thickness increases (10 to 100 nm). It is called *matchwood morphology* and occurs when ion bombardment energy is more than luminal energy of sputtering (50 eV). In higher energies, columnar growth with forward sputtering is delayed, and resputtering effects will counteract the adumbration process and will lead to the growth of noncolumnar films (region T) with an homogenous structure in sizes bigger than the cluster sizes (1 to 3 nm).

The fourth key factor is the descending flux direction that affects growth and was recently identified. In oblique deposition, the clusters' nucleation density in the film and substrate interface decrease. If the angle $\theta_v = 90 - X_v$ increases, mass density decreases and the columnar form, even without ion bombardment, will become cylindrical (region M). The low nucleation density is a geometric effect that is controlled through atomic adumbration.

The columnar structure of films prepared through oblique methods is the result of adumbration effects. Islands, in the initial stages of growth, shades on some parts of substrate that are far from vapor flux, and porosity of GLAD films due to adumbration is only possible in low superficial propagation. In oblique flux conditions ($\alpha > 80°$) and substrate's lower temperatures, adumbration and atoms' limited superficial propagation form the main processes of the substrate. When atoms move from the source to substrate, nucleation sites will create irregular topography because of adumbration. If the substrate is not warmed, atoms' propagation length will be limited and porosity sites will remain empty, and when the film grows, the mentioned condition will lead to the formation of columnar structures in nanometric scales and the substrate movement will create various morphologies. Therefore, film porosity is a function of the deposition angle, which increases in line with such angles. Because adumbration length is proportionate to $\tan\alpha$ (α: the angle between descending flux and perpendicular line of substrate), column density (n_c) is proportionate to $\sqrt{\cot\alpha}$ (number of columns per area), which is larger for smaller amounts of α. Measuring $\frac{T_p}{T_s}$ (T_p and T_S are light spectrum for parallel and perpendicular polarization) helps us to analyze constitutive anisotropy in the film surface or helps us to analyze it when perpendicular to the film plane.

Figure 6.2 shows nanostructural films prepared by GLAD. In these figures, the glancing angle deposition, physical evaporation deposition in oblique descending have been combined with the rotator substrate and the average

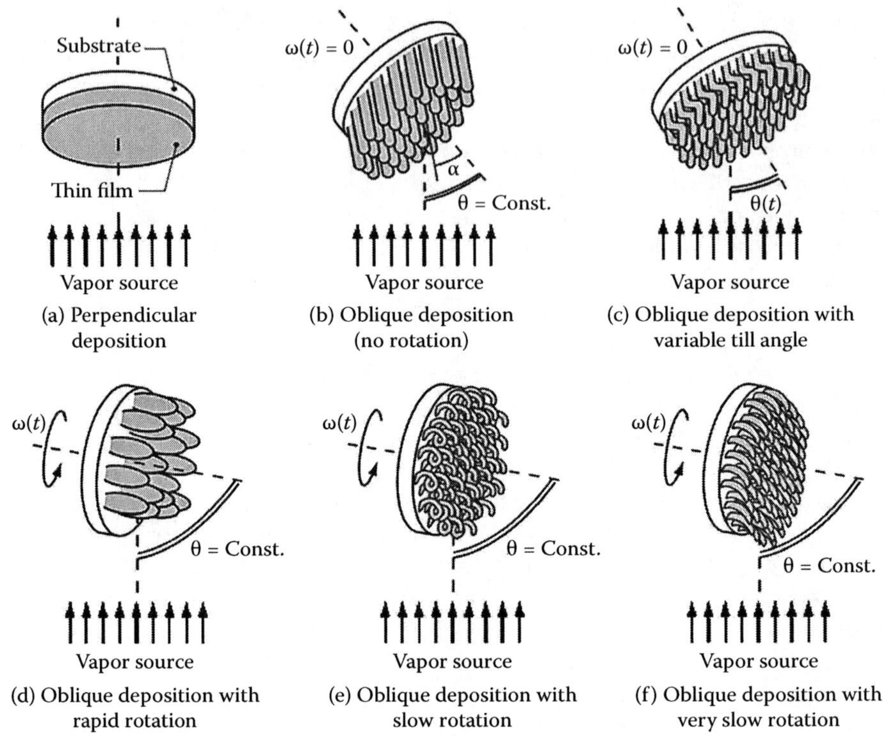

FIGURE 6.2
Glancing angle deposition geometry and resulting nanostructures: (a) deposition with normal incidence vapor flux, resulting in a nonporous thin film; (b) oblique deposition without substrate rotation (the column tilt α is always smaller than the substrate tilt angle θ) resulting in inclined columnar structures oriented toward the vapor flux direction; (c) oblique deposition with variable substrate tilt resulting in zigzag nanostructures; (d) oblique deposition with rapid rotation resulting in pillars grown normal to the substrate; (e) oblique deposition with slow azimuthal rotation resulting in an array of nanohelices; and (f) oblique deposition with very slow azimuthal rotation resulting in C-shaped nanocolumns. (Reprinted from Buzea, C., Kaminska, K., Beydaghyan, G., Brown, T., Elliott, C., Dean, C., and Robbie, K., Thickness and density evaluation for nanostructured thin films by glancing angle deposition, *Journal of Vacuum Science and Technology B: Microelectronics and Nanometer Structures* 23 (6), 2545–2552, Copyright 2005, with permission from the American Vacuum Society.)

of column diameters depend on deposition parameters that increase through increasing the flow density (electronic evaporation) and the deposition time and lower temperature of the container will lead to the growth of higher columns. Figure 6.3 illustrates scanning electron microscope (SEM) images of lateral views of nanosculptured thin films on Si(100) substrates, cleaved to reveal an interior section.

Columnar direction has a complicated dependence on some parameters such as deposition materials and bond features of atoms. X_m dependence on X_v follows the tangent law for larger amounts of X_V and the cosine law for

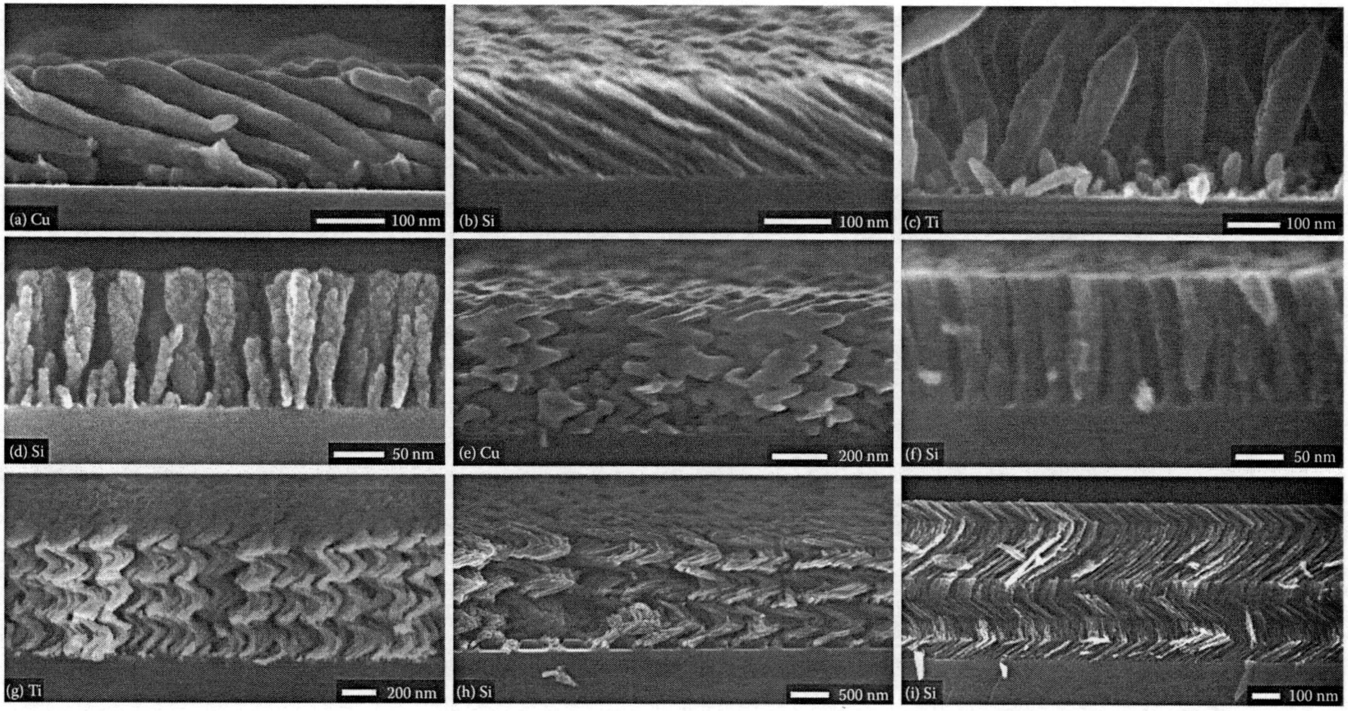

FIGURE 6.3
Scanning electron micrographs (SEMs) of lateral views of nanosculptured thin films on Si(100) substrates, cleaved to reveal an interior section: (a) copper oblique film, (b) silicon oblique, (c) titanium pillars on silicon, (d) silicon pillars on silicon, (e) bideposited copper film, (f) bideposited silicon film, (g) titanium helix, (h) silicon square helix film, and (i) silicon square helix. All depositions were performed at room temperature and at pressures below 10^{-8} Torr. (Reprinted from Buzea, C., Kaminska, K., Beydaghyan, G., Brown, T., Elliott, C., Dean, C., and Robbie, K., Thickness and density evaluation for nanostructured thin films by glancing angle deposition, *Journal of Vacuum Science and Technology B: Microelectronics and Nanometer Structures* 23 (6), 2545–2552, Copyright 2005, with permission from the American Vacuum Society.)

smaller amounts of X_v ($X_v < 40°$). This relation varies with changing materials and deposition conditions, and there is no universal relation for it. For $X_v > 0°$ (small X_v), the smallest value for X_m is 20° to 30°. Whenever the relation between X_m and X_v is determined for a certain material and deposition condition, deposing a STF with $X_m(z)$, which may be altered suddenly from a value to another inside a cone $X_m(z) > 20°$, is possible. In fact, it is the essence of the STF concept.[73–82]

6.5 Phase, Length, and Time Sandwich

Phase, length, and time (PLT) are three nanotechnological key factors of optics. They combine with each other so that optics researchers might consider that sandwich nanotechnology is composed of PLT, and this consideration is materialized through STFs' optical reaction tests.

Macroscopically, a helicoidal STF is an alternative heterogeneous dielectric continuum medium, and its constitutive relations of frequency are as follows:

$$\tilde{D}(r,\omega) = \varepsilon_0 \hat{\underline{\underline{S}}}_z\left(h,\frac{z}{\Omega}\right)\cdot \hat{\underline{\underline{S}}}_y(\chi)\cdot \tilde{\underline{\underline{\varepsilon}}}^{\circ}_{ref}(\omega)\cdot \hat{\underline{\underline{S}}}_y^{T}(\chi)\cdot \hat{\underline{\underline{S}}}_z^{T}\left(h,\frac{z}{\Omega}\right)\cdot \tilde{E}(r,\omega) \tag{6.18}$$

$$\tilde{B}(r,\omega) = \mu_0 \tilde{H}(r,\omega) \tag{6.19}$$

where ε_0 and μ_0 are vacuum permittivity and vacuum permeability, respectively; ω is angular frequency; 2Ω is constitutive duration; r is a position vector; and T is the transpose of an operator. The columnar declination angle (X) is applied in a tilt dyadic:

$$\hat{\underline{\underline{S}}}_y(\chi) = u_y u_y + (u_x u_x + u_z u_z)\cos\chi + (u_z u_x - u_x u_z)\sin\chi \tag{6.20}$$

Rotational dyadic $\hat{\underline{\underline{S}}}_z(h,\sigma)$ introduces helicoidal morphology:

$$\hat{\underline{\underline{S}}}_z(h,\sigma) = u_z u_z + (u_x u_x + u_y u_y)\cos(h\sigma) + (u_y u_x - u_x u_y)\sin(h\sigma) \tag{6.21}$$

The heterogeneity axis is chosen as parallel with the Z axis, and $h = \pm 1$ indicates the constitutive direction (left-handed or right-handed) of the film. In limit Ω, if $\Omega \to \infty$, constitutive relations of helicoidal STFs will change to equations for CTFs.

The nature of local orthromobicity of STFs is depicted in a relative permittivity dyadic of the source:

$$\underset{=ref}{\hat{\varepsilon}^{\circ}}(\omega) = \tilde{\varepsilon}_a(\omega)u_z u_z + \tilde{\varepsilon}_b(\omega)u_x u_x + \tilde{\varepsilon}_c(\omega)u_y u_y \tag{6.22}$$

$\varepsilon_\sigma(\omega)$ scalars of frequency and mixed are obtained from the following single resonance Lorentz model:

$$\tilde{\varepsilon}_\sigma(\omega) = 1 + \frac{p_\sigma}{1 + \left(\frac{1}{2\pi M_\sigma} - i\frac{\lambda_\sigma}{\lambda_0}\right)^2},\ \sigma = a, b, c \tag{6.23}$$

where $\lambda_\sigma = \frac{2\pi}{\omega\sqrt{\varepsilon_0\mu_0}}$ is the vacuum wavelength, and ρ_σ is oscillator power. Three resonance wavelengths are obtained from the following equation:

$$\lambda_{res\sigma} = \frac{\lambda_\sigma}{\sqrt{1 + \left(\frac{1}{2\pi M_\sigma}\right)^2}},\ \sigma = a, b, c \tag{6.24}$$

Widths of resonance bands are

$$(\Delta\lambda_{res})_\sigma = \frac{\lambda_\sigma}{2\pi M_\sigma},\ \sigma = a, b, c \tag{6.25}$$

In order to analyze time range, $\varepsilon_\sigma(t)$ must be measured through $\tilde{\varepsilon}_\sigma(t)$ and regarding Equation (6.23), we have

$$\varepsilon_\sigma(t) = \delta(t) + p_\sigma\omega_\sigma \sin(\omega_\sigma t)e^{-\frac{\omega_\sigma t}{2\pi M_\sigma}}u(t),\ \sigma = a, b, c \tag{6.26}$$

where $\delta(t)$ is the Dirac delta, $u(t)$ is the step function, and $\omega_\sigma = \frac{2\pi}{\lambda_\sigma\sqrt{\varepsilon_0\mu_0}}$. Thus, time-dependent constitutive relations of helicoidal STF would be rewritten as

$$D(r,t) = \varepsilon_0\left\{\underset{=z}{\hat{S}}\left(h, \frac{z}{\Omega}\right)\cdot\underset{=y}{\hat{S}}(\chi)\cdot\underset{=ref}{\varepsilon^{\circ}}(t)\cdot\underset{=y}{\hat{S}^T}(\chi)\cdot\underset{=z}{\hat{S}^T}\left(h, \frac{z}{\Omega}\right)\right\}\times E(r,t) \tag{6.27}$$

$$B(r,t) = \mu_0 H(r,t) \tag{6.28}$$

where in limit Ω ($\Omega \to \infty$), the right side of Equation (6.27) will change into the CTF equation.

If $z < z_1$ and $z > z_1$ are occupied with a helicoidal STF, consider an optical signal that starts from $z = 0$ in $t = 0$. This signal modulates the transverse wave amplitude moving toward the $z = z_1$ interface. Considering the carrier transverse wave for axial propagation in the center of the Bragg zone, electromagnetic field components of a time-dependent carrier transverse wave will be

$$\cos\left(\frac{2\pi t}{\lambda_0^{car}\sqrt{\varepsilon_0\mu_0}}+\phi\right)$$

or

$$\sin\left(\frac{2\pi t}{\lambda_0^{car}\sqrt{\varepsilon_0\mu_0}}+\phi\right)$$

where λ_0^{car} is the carrier wavelength, and φ is the initial phase. If the transverse wave carries RCP and STP is a right-handed helicoidal wave ($h = \pm 1$), as the condition will be optimized for the circular Bragg phenomenon, the selected pulse is offered with the following function:

$$g(t) = g_p \frac{t}{t_p} e^{-2\frac{t}{t_p}} \tag{6.29}$$

and the signal momentum Poynting vector is dependent from φ.

In Figure 6.4 the momentum scheme of an axial component (z) of the momentum Poynting vector is depicted in t = 30.1 fs. In the next times, the pulse is descended on an $z = z_1$ interface and reflection and refraction occur. In Figure 6.5, reflected pulses are shown for various media. It shows the Poynting vector component $P_z(z,t)$, at t = 132 fs, of the pulses reflected by a linear chiral

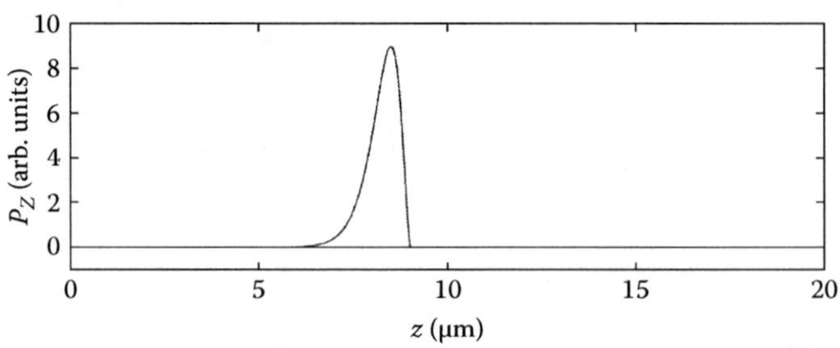

FIGURE 6.4
Plot of the Poynting vector $P_z(z,t)$ for the incident pulse at t = 30.1 fs. (Reprinted from Geddes Iii, J. B., and Lakhtakia, A., Effects of carrier phase on reflection of optical narrow-extent pulses from axially excited chiral sculptured thin films, *Optics Communications* 225 (1–3), 141–150, Copyright 2003, with permission from Elsevier.)

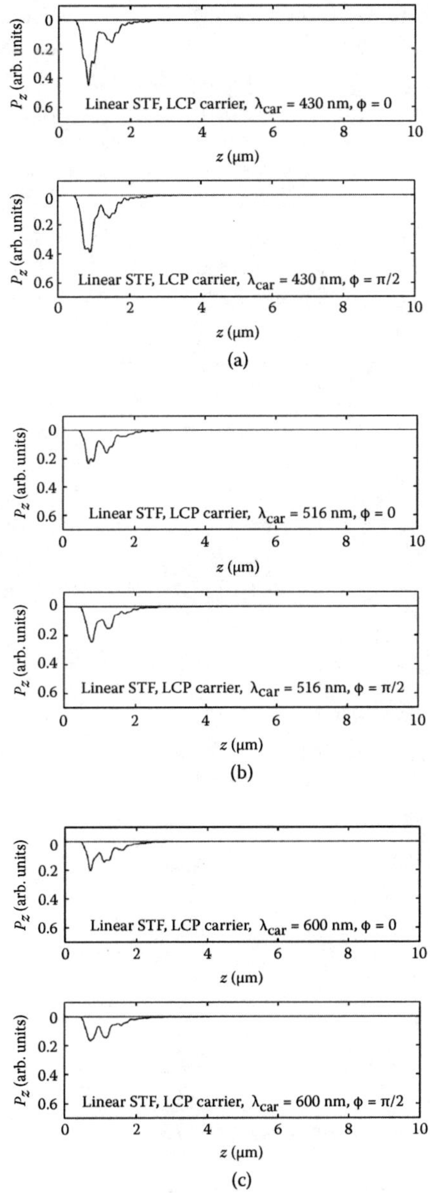

FIGURE 6.5
Plots of the Poynting vector component $P_z(z,t)$, at $t = 132$ fs, of the pulses reflected by a linear chiral sculptured thin film (STF). The carrier phase is either $\varphi = 0$ or $\varphi = \pi/2$; and the carrier wavelength is $\lambda_{car} = 430$ nm (a), $\lambda_{car} = 516$ nm (b), or $\lambda_{car} = 600$ nm (c). The carrier plane wave is left circularly polarized (LCP). The vacuum/STF boundary at $z_\ell = 20$ μm is not shown. (Reprinted from Geddes Iii, J. B., and Lakhtakia, A., Effects of carrier phase on reflection of optical narrow-extent pulses from axially excited chiral sculptured thin films, *Optics Communications* 225 (1–3), 141–150, Copyright 2003, with permission from Elsevier.)

STF. The carrier phase is either $\varphi = 0$ or $\varphi = \pi/2$; and the carrier wavelength is $\lambda_{car} = 430$ nm (a), $\lambda_{car} = 516$ nm (b), or $\lambda_{car} = 600$ nm (c). The carrier plane wave is LCP. When a refraction medium ($z \geq z_1$) is occupied with anisotropic substances, the momentum Poynting vector and the average of carrier reflective wave are not affected by the vertical transverse wave phase. It was shown in Figure 6.5 for $\varphi = 0, \pi/2$ clearly (both momentum schemes are similar). The circular Bragg phenomenon does not occur because of an homogenous medium. Momentum schemes are similar for the LCP as well. Because the medium is scattering, the reflected pulse is not the same as the vertical pulse.

If the refraction medium is anisotropic, the importance of the initial phase will be obvious in signals. If the refraction medium is anisotropic, the momentum Poynting vector will depend on the initial phase, but the average Poynting vector will be interdependent from the initial phase. Figure 6.6 shows that our refraction medium is a STF. Here, the importance of the initial phase is raised.

Phase, length, and time integration is indicated through the Goos–Hänchen effect. Newton intuitively found that if a finite beam is perpendicular to a medium with a diluter optical refractive index, total reflection phenomenon will occur, and the reflective beam will be replaced horizontally, which is known as the Goos–Hänchen effect. It is indicated schematically in Figure 6.7. Its importance was discovered in near-field optical microscopy in the recent decade. For studying on the Goos–Hänchen shift, incident and reflection media are presumed as anisotropic and nondispersive, the incident ray is Gaussian, and if θ_{inc} (is measured according to the perpendicular line to the interface) meets the total reflection then $\Phi(\rho)$ (ρ: reflective index of the transverse wave) will be determined. Thus, the Goos–Hänchen shift is calculated as

$$d = -\frac{d\phi(\rho)}{dk} \tag{6.30}$$

where $k = (\frac{2\pi}{\lambda_0})n_r \sin\theta_{inc}$ and n_r is the refractive index of incident and reflection media.

Figure 6.8 shows a measured value for this shift for waves with s and p polarizations. Figures 6.9 and 6.10 show measured values of $X = \pi/9$, $M_a = M_b = M_c = 50/\pi$, $\lambda_a = \lambda_c = 280$ nm, $\lambda_b = 290$ nm, $P_a = 0.4$, $P_b = 0.52$, and $P_c = 0.42$ for CTF and STF, respectively.

In small incident angles (in the case of total reflection), the Goos–Hänchen shift covers a noticeable part of the vacuum wavelength, but in larger incident

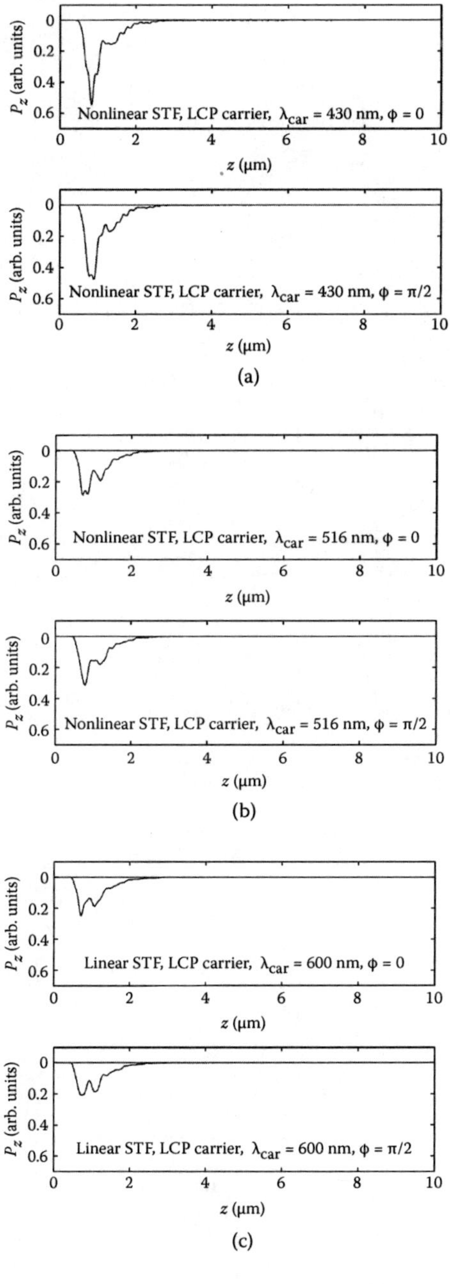

FIGURE 6.6
The same as in Figure 6.5, except that the chiral sculptured thin film (STF) is nonlinear. (Reprinted from Geddes Iii, J. B., and Lakhtakia, A., Effects of carrier phase on reflection of optical narrow-extent pulses from axially excited chiral sculptured thin films, *Optics Communications* 225 (1–3), 141–150, Copyright 2003, with permission from Elsevier.)

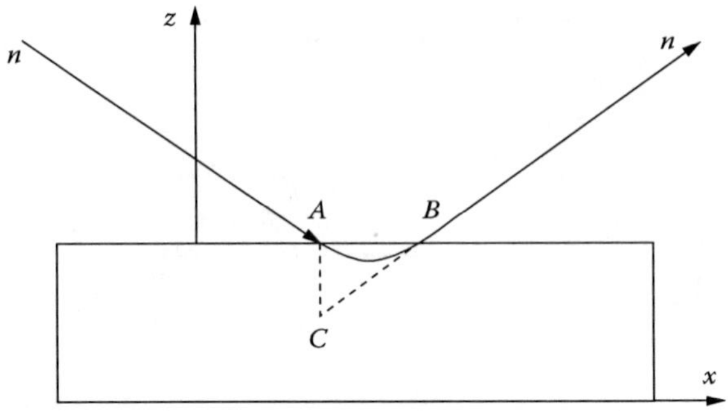

FIGURE 6.7
The Goos–Hänchen shift in total reflection. (Reprinted from Ignatovich, V. K., Neutron reflection from condensed matter, the Goos–Hänchen effect and coherence, *Physics Letters, Section A: General, Atomic and Solid State Physics* 322 (1–2), 36–46, Copyright 2004, with permission from Elsevier.)

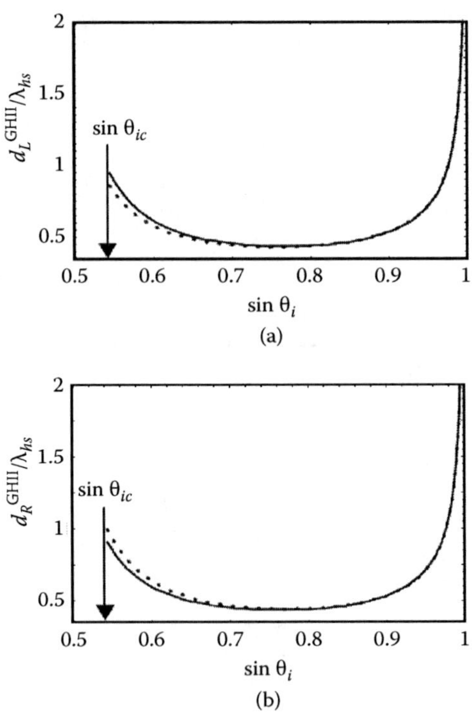

FIGURE 6.8
Normalized Goos–Hänchen shifts (a) $d_L^{GH\|}/\lambda_{hs}$ and (b) $d_R^{GH\|}/\lambda_{hs}$ as functions of $\theta_i \in [\sin\theta_{ic}, 1)$, calculated for $\alpha = 0°$ (dotted lines) and $\alpha = 15°$ (solid lines), when $\psi_i = 120°$, $n_{hs} = 4$, $\lambda_0 = 727$ nm, and $w_0 = 4\lambda_{hs}$. (Reprinted from Wang, F., and Lakhtakia, A., Lateral shifts of optical beams on reflection by slanted chiral sculptured thin films, *Optics Communications* 235 (1–3), 107–132, Copyright 2004, with permission from Elsevier.)

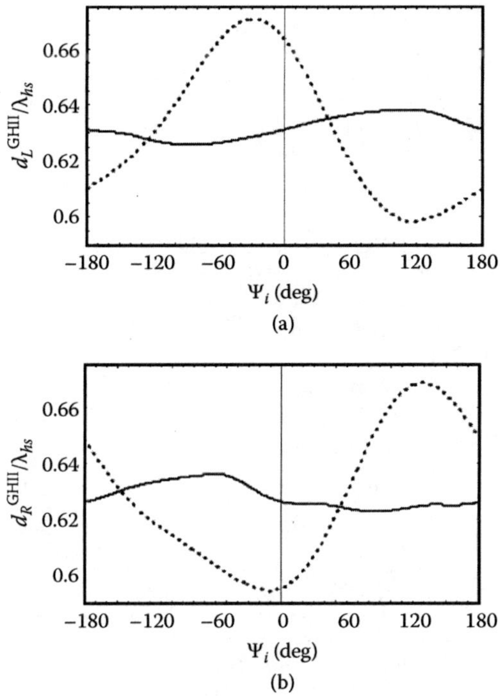

FIGURE 6.9
Same as Figure 6.8, except $d_L^{GH\|}/\lambda_{hs}$ and $d_R^{GH\|}/\lambda_{hs}$ are plotted against $\psi_i \in (-180°, 180°]$, for a fixed postcritical θ_i = 37.5°. (Reprinted from Wang, F., and Lakhtakia, A., Lateral shifts of optical beams on reflection by slanted chiral sculptured thin films, *Optics Communications* 235 (1–3), 107–132, Copyright 2004, with permission from Elsevier.)

angles (Figures 6.8, 6.9, and 6.10) it exceeds $v\lambda_0$ and the length role appears in the following two forms:

1. The first nanostructural effect of the refraction medium occurs during the Goos–Hänchen shift. Penetration depth of the ray that experiences this shift is estimated with $\frac{d}{2}\cot\theta_{inc}$. Thus, penetration in large incident angles is trivial; sensitivity of such a shift is larger in proportion than in nanostructural dimensions of the refraction medium in small θ_{inc}.
2. The second role of this effect emerges in the next application of the reflective beam. In three previous figures, the Goos–Hänchen shift exceeds 300 nm, which is a great distance in terms of nanoscales. For instance, the diameter of quantum dots is about 1 to 2 nm, the diameter of globular proteins is about 6 nm, and oxide films are (gate oxide films) approximately 2 nm wide. Thus, apart from such shifting, it applies serious deletions in nanoscales to problems.

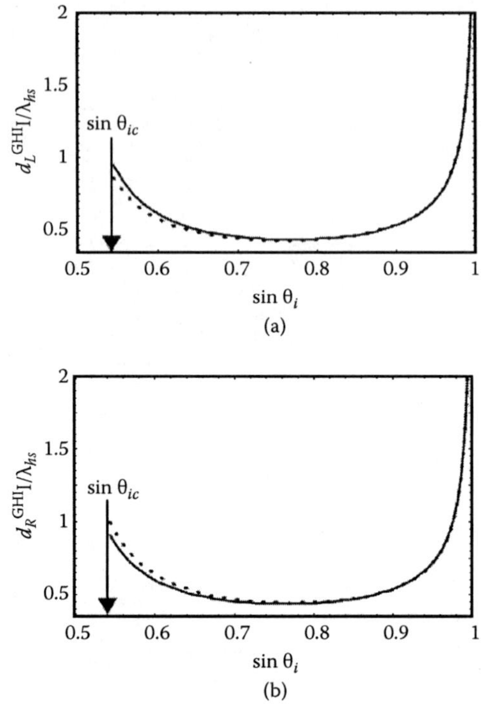

FIGURE 6.10
Normalized Goos–Hänchen shifts (a) $d_L^{GH\|}/\lambda_{hs}$ and (b) $d_R^{GH\|}/\lambda_{hs}$ as functions of $\theta_i \in [\sin\theta_{ic}, 1)$; calculated for $w_0 = 2\lambda_{hs}$ (dotted lines) and $w_0 = 4\lambda_{hs}$ (solid lines), when $\alpha = 15°$. Other parameters are the same as for Figure 6.8. (Reprinted from Wang, F., and Lakhtakia, A., Lateral shifts of optical beams on reflection by slanted chiral sculptured thin films, *Optics Communications* 235 (1–3), 107–132, Copyright 2004, with permission from Elsevier.)

The Goos–Hänchen shift discerns an extra distance that each photon must pass during the total reflection, and this shift brings about delays. Phase time delay is calculated as

$$t^{ph} = n_r \sqrt{\varepsilon_0 \mu_0} d \sin\theta_{inc} \tag{6.31}$$

For Figures 6.8 and 6.10, phase delay time of fs is 0.2, but group time delay is calculated by the following equation:

$$\tau = \frac{d\phi(\rho)}{d\omega} \tag{6.32}$$

The equation is related to optical pulse communications. This time delay in small θ_{inc} angles is larger than 0.2 fs, but in large oblique angles it is trivial. Delays as large as 0.2 fs are not significant in picosecond pulses, but modern optics tend to femtosecond and attosecond pulses and, apart from this delay, will result in more problems.

As a result, phase, length, and time will combine in any optical (macroscopic) system that is obvious to any optics researchers; introduction of a nanostructure with certain dimensions to thin films will integrate these three quantities in new methods and offers a new chain for optics, which is called PLT sandwich nanotechnology. S and P polarization beams fall on the transverse interface with θ_{inc} angle in proportion to the z axis and propagate in the xz plane.

6.6 A Model to Make a Relation between Features and Structures of Dielectric Helicoidal Sculptured Thin Films

STFs of any materials may be prepared through deposition, by physical evaporation, on any substrates in a reduced mobility condition of adsorbed atoms in the surface. Porosity of finite clusters is a natural state, and voids are as important as filled regions (deposited materials). The fact indicates usefulness of homogenization formalism to harmonic electromagnetic fields in a condition in which a STF nanostructure is replaced by a continuum in high wavelength regions. This equivalent continuum is called an homogenized composite medium (HCM), which must have unidirectional heterogeneity and alteration of STFs.

According to definitions, morphology of a TFHBM in any $z = z_1$ planes may coincide with the morphology of each $z = z_2$ plane by using a proper rotation, which the z axis shows heterogeneity direction. Because local morphology is spatially uniform in each xy plane, constitutive features of the source (location) appear. Therefore, universal constitutive features may be obtained from local constitutive features using rotational operators.

After homogenization, constitutive relations of a dielectric TFHBM are as follows:

$$\underline{D}(\underline{r}) = \varepsilon_0 \cdot \underline{\underline{\varepsilon}}(z) \cdot \underline{E}(\underline{r}) \tag{6.33}$$

$$\underline{B}(\underline{r}) = \mu_0 \, \underline{H}(\underline{r}) \tag{6.34}$$

Heterogeneous relative permittivity dyadics are presented in terms of an homogenous relative permittivity dyadic and two rotational dyadics:

$$\underline{\underline{\varepsilon}}(z) = \underline{\underline{S}}_{z}(z) \cdot \underline{\underline{S}}_{y}(\chi) \cdot \underline{\underline{\varepsilon}}^{\circ}_{ref} . \underline{\underline{S}}_{y}^{-1}(\chi) \cdot \underline{\underline{S}}_{z}^{-1}(z) \tag{6.35}$$

$$\underline{\underline{\varepsilon}}^{0}_{ref} = \varepsilon_a u_z u_z + \varepsilon_b u_x u_x + \varepsilon_c u_y u_y \tag{6.36}$$

$$\underline{\underline{S}}_{y}(\chi) = u_y u_y + (u_x u_x + u_z u_z)\cos\chi + (u_z u_x - u_x u_z)\sin\chi \tag{6.37}$$

$$\underline{\underline{S}}_{z}(z) = u_z u_z + (u_x u_x + u_y u_y)\cos\left(\frac{\pi z}{\Omega}\right) + (u_y u_x - u_x u_y)\left(\frac{\pi z}{\Omega}\right) \tag{6.38}$$

where 2Ω is constitutive duration, and X is the growth angle of helicoidal columns. Source permittivity dyadic ($\underline{\underline{\varepsilon}}_{ref} = \underline{\underline{S}}_y(\chi)\cdot\varepsilon^{\circ}\cdot \underline{\underline{S}}^{-1}_{ref=y}(\chi)$) is clear. Therefore, all universal permittivity dyadics of STF are attainable, and Equation (6.38) is the case for right-handed TFHBM. For left-handed TFHBM, Ω must convert to $-\Omega$.

If the relative permittivity dyadic of an homogenous medium is considered to be $\underline{\underline{\varepsilon}}_{HCM}$, in $\Omega \to \infty$, heterogeneous relative permittivity dyadic of a TFHBM ($\underline{\underline{\varepsilon}}_{ref}$) will be equal to $\underline{\underline{\varepsilon}}_{HCM}$, ($\lim_{\Omega\to\infty} \underline{\underline{\varepsilon}}_{TFHBM} = \underline{\underline{\varepsilon}}_{ref}$). The longest ellipsoid axis of this HCM is parallel with unit vector $\underline{u}'_x = \underline{\underline{S}}_y(\chi).\underline{u}_x$, and its shortest axis is parallel with unit vector $\underline{u}'_z = \underline{\underline{S}}_y(\chi)\cdot \underline{u}_z$. These ellipsoid forms are created by isotropic dielectric materials with scalar permittivity $\varepsilon_s(\omega)$, and permittivity of porous regions is equal to 1. The area of an ellipsoid form in proportion to its center in Cartesian coordinates is

$$z'^2 + \left(\frac{y'}{\gamma_2}\right)^2 + \left(\frac{x'}{\gamma_3}\right)^2 = \delta^2 \tag{6.39}$$

where δ is absolute size, $y_2 > 1$ (lateral transverse ratio), and $y_3 > 1$ (slenderness ratio). According to Bruggeman formalism,

$$f\,\underline{\underline{a}}_s + (1-f)\,\underline{\underline{a}}_v = \underline{\underline{0}} \tag{6.40}$$

where f ($0 < f < 1$) is the film volume occupied by an ellipse, and $\underline{\underline{0}}$ is the zero dyadic. Dyadic of ellipsoid polarization sinking in HCM in volume scale is measured as

$$\underline{\underline{a}}_s = \varepsilon_0(\varepsilon_s \underline{\underline{I}} - \underline{\underline{\varepsilon}}_{HCM})\cdot[\underline{\underline{I}} + i\omega\varepsilon_0 \underline{\underline{D}}_s \cdot (\varepsilon_s \underline{\underline{I}} - \underline{\underline{\varepsilon}}_{HCM})]^{-1} \tag{6.41}$$

And dyadic of ellipsoid depolarization in HCM is

$$\underline{\underline{D}}_s = \underline{\underline{S}}_y(\chi)\cdot[\frac{1}{i\omega\varepsilon_0}\frac{2}{\pi}\int_0^{\frac{\pi}{2}} d\phi \int_0^{\frac{\pi}{2}} d\theta \sin\theta \frac{\left(\frac{\cos^2\theta}{\gamma_3^2}\right)\underline{u}_x\underline{u}_x + \sin^2\theta(\varepsilon_a\cos^2\phi\,\underline{u}_z\underline{u}_z + \left(\frac{\sin^2\phi}{\gamma_2^2}\right)\underline{u}_y\underline{u}_y}{\varepsilon_b\left(\frac{\cos^2\theta}{\gamma_3^2}\right) + \sin^2\theta\left(\varepsilon_a\cos^2\phi + \varepsilon_c\left(\frac{\sin^2\phi}{\gamma_2^2}\right)\right)}\cdot \underline{\underline{S}}^1(\chi) \tag{6.42}$$

where a_v is similar to a_x and $\underline{\underline{D}}_v = \underline{\underline{D}}_s$, so

$$\underline{\underline{a}}_v = \varepsilon_0(\varepsilon_v \underline{\underline{I}} - \underline{\underline{\varepsilon}}_{HCM})\cdot[\underline{\underline{I}} + i\omega\varepsilon_0 \underline{\underline{D}}_v \cdot (\varepsilon_v \underline{\underline{I}} - \underline{\underline{\varepsilon}}_{HCM})]^{-1} \tag{6.43}$$

$\underline{\underline{\varepsilon}}_{HCM}$ is obtained after solving Equation (6.41), which is easily attainable through a repetitive method such as the Jacob method and successive iterations of $\underline{\underline{\varepsilon}}_{HCM}$ are obtained with this equation:

$$\underline{\underline{\varepsilon}}_{HCM}[n], \quad (n = 0,1,2,\ldots)$$

and

$$\underline{\underline{\varepsilon}}_{HCM}[n] = u(\underline{\underline{\varepsilon}}_{HCM}[n-1]) \tag{6.44}$$

We have $\underline{\underline{\varepsilon}}_{HCM}[0] = \underline{\underline{I}}(f\varepsilon_s + (1-f)\varepsilon_v]$, and u (an operator) is applied as the following:

$$u(\underline{\underline{\varepsilon}}_{HCM}) = \left\{ f\varepsilon_s[\underline{\underline{I}} + i\omega\varepsilon_0 \underline{\underline{D}}_s \cdot (\varepsilon_s \underline{\underline{I}} - \underline{\underline{\varepsilon}}_{HCM})]^{-1} + (1-f)\varepsilon_v[\underline{\underline{I}} + i\omega\varepsilon_0 \underline{\underline{D}}_v \cdot (\varepsilon_v \underline{\underline{I}} - \underline{\underline{\varepsilon}}_{HCM})]^{-1} \right\}$$
$$\cdot \left\{ f[\underline{\underline{I}} + i\omega\varepsilon_0 \underline{\underline{D}}_s \cdot (\varepsilon_s \underline{\underline{I}} - \underline{\underline{\varepsilon}}_{HCM})]^{-1} + (1-f)[\underline{\underline{I}} + i\omega\varepsilon_0 \underline{\underline{D}}_v \cdot (\varepsilon_v \underline{\underline{I}} - \underline{\underline{\varepsilon}}_{HCM})]^{-1} \right\} \tag{6.45}$$

If $0 \leq z \leq 1$ is occupied with a TFHBM and $z \geq 1$, and $z \leq 0$ is a vacuum and transverse wave with desired polarization, and wave number ($k = \omega\sqrt{\varepsilon_0\mu_0}$) falls perpendicularly to the TFHBM edge from $z \geq 0$, then fuzzy presentation of the electric field in $z \geq 0$ is as follows:

$$\begin{aligned} \underline{E}(\underline{r}, \lambda_0) &= (a_L \underline{u} + a_R \underline{u}_{-})e^{ik_0 z} + (r_L \underline{u}_{-} + r_R \underline{u}_{+})e^{-ik_0 z} \\ &= (a_s \underline{u}_y - a_p \underline{u}_x)e^{ik_0 z} + (r_s \underline{u}_y + r_p \underline{u}_x)e^{-ik_0 z}, \quad z \leq 0 \end{aligned} \tag{6.46}$$

where $\underline{u}_{\pm} = (\underline{u}_x \pm i\underline{u}_y)/\sqrt{2}$ is the fuzzy presentation of the transmission electric field and is calculated with

$$\underline{E}_{tr}(\underline{r}, \lambda_0) = (t_L \underline{u}_{+} + t_R \underline{u}_{-})e^{ik_0(z-1)} = (t_s \underline{u}_y - t_p \underline{u}_x)e^{ik_0(z-1)}, \quad z \leq 0 \tag{6.47}$$

where a_L and a_R, r_L, r_R, t_R, and t_L are amplitudes of incident, reflective, and transmissive transverse waves of LCP and RCP, respectively. Transmission and reflection amplitudes may be measured in terms of amplitudes of incident wave by using boundary value problems. Results are offered as matrix forms:

$$\begin{bmatrix} r_S \\ r_P \end{bmatrix} = \begin{bmatrix} r_{SS} & r_{SP} \\ r_{PS} & r_{PP} \end{bmatrix} \begin{bmatrix} a_L \\ a_R \end{bmatrix}, \begin{bmatrix} t_L \\ t_R \end{bmatrix} = \begin{bmatrix} t_{LL} & t_{LR} \\ t_{RL} & t_{RR} \end{bmatrix} \begin{bmatrix} a_L \\ a_R \end{bmatrix} \tag{6.48}$$

$$\begin{bmatrix} r_S \\ r_P \end{bmatrix} = \begin{bmatrix} r_{SS} & r_{SP} \\ r_{PS} & r_{PP} \end{bmatrix} \begin{bmatrix} a_S \\ a_P \end{bmatrix}, \begin{bmatrix} t_S \\ t_P \end{bmatrix} = \begin{bmatrix} t_{SS} & t_{SP} \\ t_{PS} & t_{PP} \end{bmatrix} \begin{bmatrix} a_S \\ a_P \end{bmatrix} \tag{6.49}$$

Similar and different entries of 2 × 2 matrixes point to copolarized and cross-polarized modes, respectively. Important observable optical features such as optical rotation, ellipticity, circular dichroism, linear dichroism, and apparent linear and circular dichroism may be obtained using measured entries. Circular dichroism is the difference in attracting LCP and RCP transverse waves:

$$CD = \frac{\sqrt{A_R} - \sqrt{A_L}}{\sqrt{A_R} + \sqrt{A_L}} \tag{6.50}$$

where $A_R = 1 - (|r_{RR}|^2 + |r_{LR}|^2 + |t_{RR}|^2 + |t_{LR}|^2)$ and $A_R = 1 - (|r_{LL}|^2 + |r_{RL}|^2 + |t_{LL}|^2 + |t_{RL}|^2)$ are absorption values.

Linear dichroism is measured for transverse waves with linear polarization:

$$LD = \frac{\sqrt{A_S} - \sqrt{A_P}}{\sqrt{A_S} + \sqrt{A_P}} \tag{6.51}$$

where A_s and A_p are measured as

$$A_S = 1 - \left(|r_{SS}|^2 + |r_{PS}|^2 + |t_{SS}|^2 + |t_{PS}|^2\right)$$

$$A_P = 1 - \left(|r_{PP}|^2 + |r_{SP}|^2 + |t_{PP}|^2 + |t_{SP}|^2\right)$$

And appeared circular dichroism is measured through

$$CD_{app} = \frac{\sqrt{T_R} - \sqrt{T_L}}{\sqrt{T_R} + \sqrt{T_L}} \tag{6.52}$$

where $T_L = |t_{LL}|^2 + |t_{RL}|^2$ and $T_R = |t_{RR}|^2 + |t_{LR}|^2$.

Appeared linear dichroism:

$$LD_{app} = \frac{\sqrt{T_S} - \sqrt{T_P}}{\sqrt{T_S} + \sqrt{T_P}} \tag{6.53}$$

where $T_P = |t_{PP}|^2 + |t_{SP}|^2$ and $T_S = |t_{SS}|^2 + |t_{PS}|^2$.

Incident and transmission ellipticities are measured through the following equations:

$$\phi_{inc} = -2\frac{\text{Im}[a_s a_p^*]}{\left(|a_s|^2 + |a_p|^2\right)} \text{ and } \phi_{tr} = -2\frac{\text{Im}[t_s t_p^*]}{\left(|t_s|^2 + |t_p|^2\right)} \tag{6.54}$$

In order to calculate optical rotation, which is an angular quantity and indicates rotation value of a transmission vibrational ellipsoid large axis in relation to the incident vibrational ellipsoid large axis, the following accessory vectors are used:

$$\underset{-}{F}_{inc} = [1 + (1 - \phi_{inc}^2)^{\frac{1}{2}}]\text{Re}[a_s \underset{-y}{u} - a_p \underset{-x}{u}] - \phi_{inc}\,\text{Im}[a_s \underset{-x}{u} + a_p \underset{-y}{u}] \tag{6.55}$$

$$\underset{-}{F}_{tr} = [1 + \left(1 - \phi_{tr}^2\right)^{\frac{1}{2}}]\text{Re}[t_s \underset{-y}{u} - t_p \underset{-x}{u}] - \phi_{tr}\,\text{Im}[t_s \underset{-x}{u} + t_p \underset{-y}{u}] \tag{6.56}$$

Oblique angles are measured as

$$\cos\tau_l = \frac{\underset{-l}{F} \cdot \underset{-y}{u}}{\left|\underset{-l}{F}\right|} \text{ and } \sin\tau_l = \frac{-\underset{-l}{F} \cdot \underset{-x}{u}}{\left|\underset{-l}{F}\right|} \text{ and } (l = inc, tr) \tag{6.57}$$

$$\phi_{tr} = \begin{cases} \tau_{tr} - \tau_{inc} + \pi & if \quad -\pi \le \tau_{tr} - \tau_{inc} \le \frac{-\pi}{2} \\ \tau_{tr} - \tau_{inc} & if \quad |\tau_{tr} - \tau_{inc}| \le \frac{\pi}{2} \\ \tau_{tr} - \tau_{inc} - \pi & if \quad \frac{\pi}{2} \le \tau_{tr} - \tau_{inc} \le \pi \end{cases} \tag{6.58}$$

And any change in the TFHBM constitutive direction will change the φ_{tr} mark.

Whenever TFHBM is stimulated with a transverse wave in vertical incidence, a circular Bragg phenomenon occurs. In narrow wavelength regions

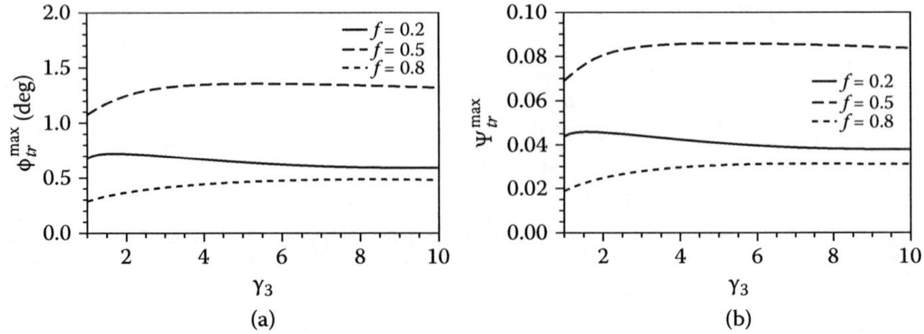

FIGURE 6.11
(a) Optical rotation maximum and (b) transmittance ellipticity maximum for p-polarized incidence, plotted as functions of γ_3. The maximums were determined over the spectral range $\lambda_0 \in [400,600]$nm, $\gamma_2 = 1$, $\Omega_2 = 200$ nm, L = 20 Ω, $f = 0.2, 0.5, 0.8$, and $\chi = 20°$. (Reprinted from Sherwin, J. A., and Lakhtakia, A., Nominal model for structure-property relations of chiral dielectric sculptured thin films, *Mathematical and Computer Modelling* 34 (12–13), 1499–1514, Copyright 2001, with permission from Elsevier.)

(the Bragg zone), if edge thickness is proper, TFHBM reflects RCP and passes LCP completely.

Spectra of optical rotation, transmission ellipiticity, linear and circular dichroisms, appeared linear and circular dichroisms may be measured in terms of functions of geometrical and constitutive parameters of ε_s, γ_2, γ_3, X, Ω, f, δ_x and vacuum wavelength ($\lambda_0 = \frac{2\pi}{k_0}$). Relative permittivity (ε_s) follows the resonance Lorentz model:

$$\varepsilon_s(\lambda_0) = 1 + \frac{p_s}{1 + (N_s^{-1} - i\lambda_s\lambda_0^{-1})^2} \tag{6.59}$$

In Figure 6.11, the slenderness ratio is calculated.

Figure 6.12 shows the transverse ratio effect. These diagrams indicate that there is a critical value for each χ by which optical activity orients to zero.

It has been proven that optical activity of a unidirectional TFHBM edge decreases through increasing X. It is observed in measuring optical activity of central wavelength (λ_0^{Br}) and total width in half maximum ($\Delta\lambda_0^{Br}$) of the Bragg zone. These parameters are estimated by the following equations:

$$\lambda_0^{Bragg} = \Omega\left[\sqrt{\varepsilon_c(\lambda_0^{Bragg})} + \sqrt{\tilde{\varepsilon}_d(\lambda_0^{Bragg})}\right] \tag{6.60}$$

$$(\Delta\lambda_0^{Bragg}) \approx 2\Omega\left[\sqrt{\varepsilon_c(\lambda_0^{Bragg})} - \sqrt{\tilde{\varepsilon}_d(\lambda_0^{Bragg})}\right] \tag{6.61}$$

$$\tilde{\varepsilon}_d = \frac{\varepsilon_a\varepsilon_b}{\varepsilon_a\cos^2\chi + \varepsilon_b\sin^2\chi} \tag{6.62}$$

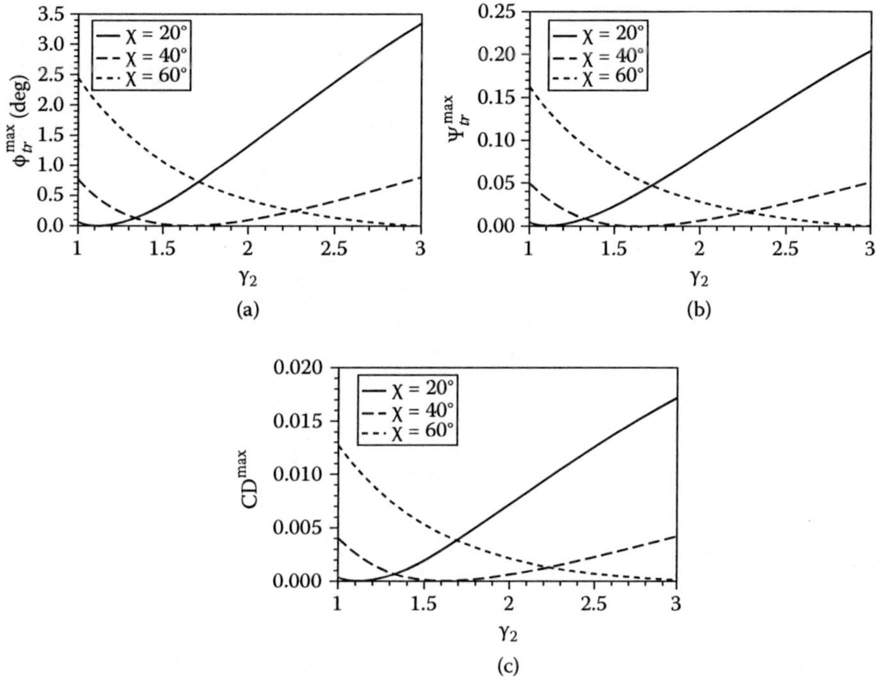

FIGURE 6.12
(a) Optical rotation maximum and (b) transmittance ellipticity maximum for p-polarized incidence, and (c) circular dichroism maximum plotted as functions of γ_2. The maximums were determined over the spectral range $\lambda_0 \in [400,600]$nm, $\gamma_3 = 20$, $\Omega = 200$ nm, $L = 20\ \Omega$, $f = 0.2, 0.5, 0.8$, and $\chi = 20°$. (Reprinted from Sherwin, J. A., and Lakhtakia, A., Nominal model for structure-property relations of chiral dielectric sculptured thin films, *Mathematical and Computer Modelling* 34 (12–13), 1499–1514, Copyright 2001, with permission from Elsevier.)

Figure 6.13 shows the same diagrams of Figure 6.12 when $\chi = 20°$.

Figure 6.14 shows effects due to the volume fraction of ellipsoids. These diagrams indicate that the maximum observed feature orients to zero ($f \to 0.1$). Because for $f = 0$ and 1, TFHBM is both isotropic and homogenous. And, for each γ_2 value there is a maximum value for each observable feature. f value is represented by f^0 in the maximum mode.

The role of Ω was analyzed by researchers and it was proven that if all the other parameters are presumed to be fixed, λ_0^{br} and $(\Delta\lambda_0^{br})$ will increase with Ω linearly.

In Bruggeman formalism, it is hypothesized that all helicoidal columns grow locally in the columnar angle (X) in proportion to the xy plane. Because orientation of deposition flux is not complete, the angular distribution must be considered for X. Therefore, because there is no quantitative study about this subject, it should be limited to a representative example. If $\chi \in [\chi_a - \delta_x, \chi_a + \delta_x]$, where X_a is the central value and δ_x is the

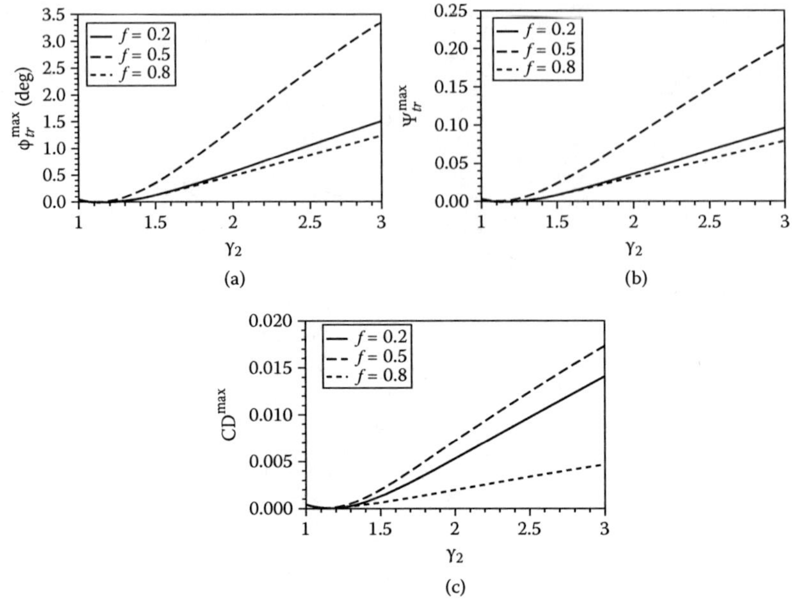

FIGURE 6.13
(a) Optical rotation maximum and (b) transmittance ellipticity maximum for p-polarized incidence, and (c) circular dichroism maximum plotted as functions of γ_2. The maximums were determined over the spectral range $\lambda_0 \in [400,600]$nm, $\gamma_3 = 20$, $\Omega = 200$ nm, $L = 20\,\Omega$, $f = 0.5$, and $\chi = 20°, 40°, 60°$. (Reprinted from Sherwin, J. A., and Lakhtakia, A., Nominal model for structure-property relations of chiral dielectric sculptured thin films, *Mathematical and Computer Modelling* 34 (12–13), 1499–1514, Copyright 2001, with permission from Elsevier.)

maximum declination, the average of orientation is measured through the following equation:

$$< \underset{=s,v}{a} >= \int_{-\delta_x}^{\delta_x} F(v) \underset{=s,v}{a} (\chi_a + v)dv \tag{6.63}$$

It brings about the average Bruggeman equation:

$$< f \underset{=s}{a} > +(1-f) < \underset{=v}{a} >= \underset{=}{0} \tag{6.64}$$

If $F(V)$ is symmetrical around $v = 0$, $\underset{=ref}{\varepsilon^{\circ}}$ and $< \underset{=s}{a} >, < \underset{=v}{a} >$ will be oblique in the initial coordination system and the distribution function will be as follows:

$$F(v) = \cos^m\left(\frac{\pi v}{2\delta_x}\right) \int_{-\delta_x}^{\delta_x} \cos^m\left(\frac{\pi v}{2\delta_x}\right) dv_0 \tag{6.65}$$

where m is the integer.

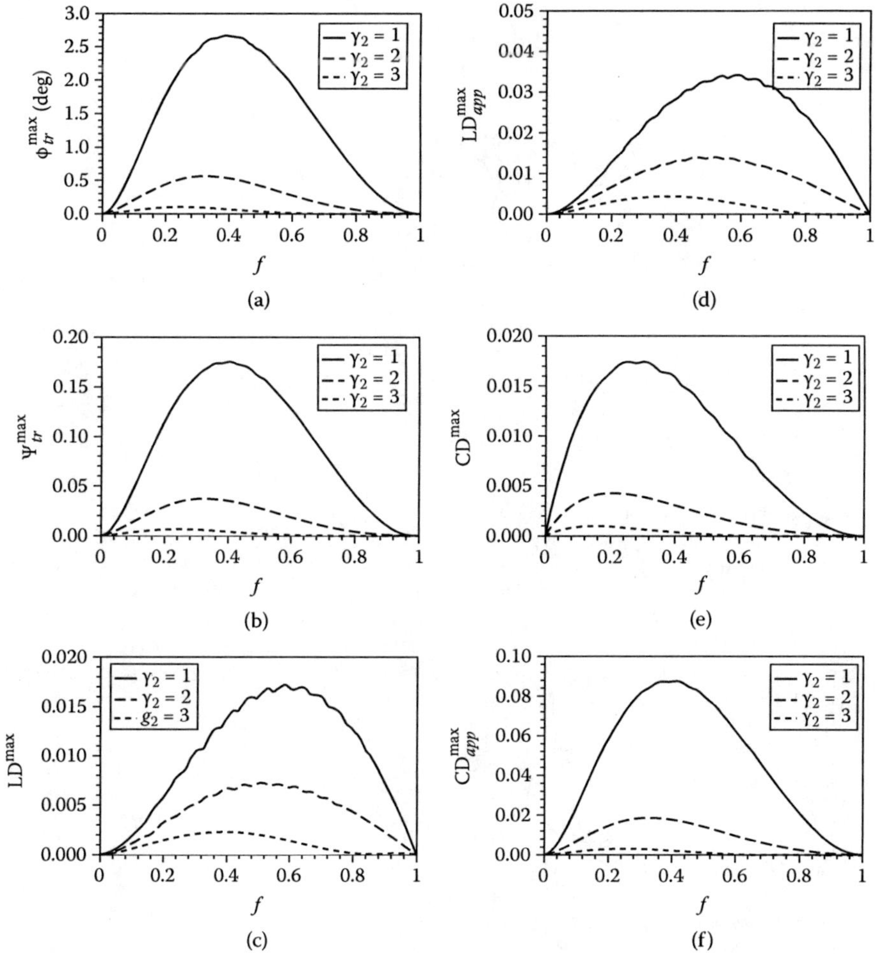

FIGURE 6.14
(a) Optical rotation maximum and (b) transmittance ellipticity maximum for p-polarized incidence, (c) linear dichroism maximum, (d) apparent linear dichroism maximum, (e) circular dichroism maximum, and (f) apparent circular dichroism maximum plotted as functions of f. The maximums were determined over the spectral range $\lambda_0 \in [400,600]$nm, $\gamma_3 = 20$, $\Omega = 200$ nm, $L = 20\Omega$, $\chi = 20°$, $\gamma_2 = 1, 2, 3$. (Reprinted from Sherwin, J. A., and Lakhtakia, A., Nominal model for structure-property relations of chiral dielectric sculptured thin films, *Mathematical and Computer Modelling* 34 (12–13), 1499–1514, Copyright 2001, with permission from Elsevier.)

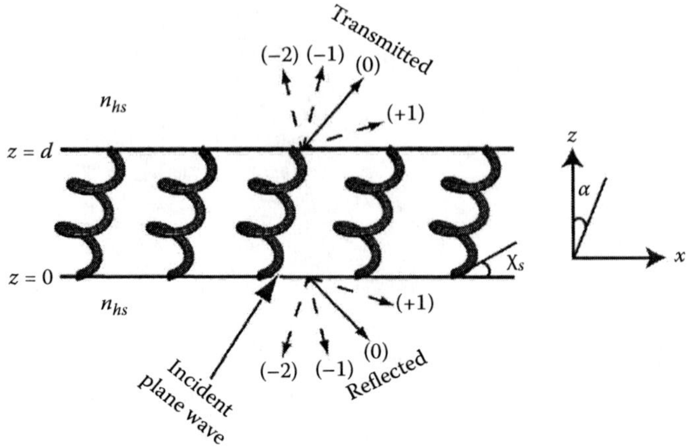

FIGURE 6.15
Boundary value problem involving a slanted chiral sculptured thin film (STF) of thickness d illuminated by an obliquely incident plane wave. The lower and the upper half-spaces are filled by an homogeneous, nondissipative, dielectric medium of refractive index n_{hs}. Both specular ($n = 0$) and nonspecular ($n \neq 0$) reflections and transmissions occur when $\alpha \neq 0$, but all nonspecular reflections/transmissions fold into the specular reflection/transmission when $\alpha = 0$. (Reprinted from Wang, F., and Lakhtakia, A., Lateral shifts of optical beams on reflection by slanted chiral sculptured thin films, *Optics Communications* 235 (1–3), 107–132, Copyright 2004, with permission from Elsevier.)

6.7 Analysis of Precise Couple Wave for the Incident Transverse Wave

It is hypothesized that region $0 < z < d$ is occupied with an oblique helicoidal STF and the half space of $z \geq d$ and $z \leq 0$ have been filled with an homogenous and isotropic dielectric material (r (refractive index) $= n_{hs}$) (Figure 6.15).

The permittivity dyadic of an oblique helicoidal STF is measured as

$$\underline{\underline{\varepsilon}}(r,\lambda_0) = \underline{\underline{S}}_y(-\alpha)\cdot\underline{\underline{S}}_z(r)\underline{\underline{S}}_y(\chi_s)\cdot\underline{\underline{\varepsilon}}_{ref}(\lambda_0)\underline{\underline{S}}_y^T(\chi_s)\underline{\underline{S}}_z^T(r)\underline{\underline{S}}_y^T(-\alpha), \quad 0 < z < d \tag{6.66}$$

Regarding local orthrombicity of STFs, the relative permittivity dyadic of the source is calculated as

$$\underline{\underline{\varepsilon}}_{ref}(\lambda_0) = \varepsilon_a(\lambda_0)u_z u_z + \varepsilon_b(\lambda_0)u_x u_x + \varepsilon_c(\lambda_0)u_y u_y \tag{6.67}$$

where $\varepsilon_{a,b,c}$ are measured through the single-resonance Lorentz model:

$$\varepsilon_{a,b,c}(\lambda_0) = 1 + \frac{p_{a,b,c}}{[1 + (N_{a,b,c}^{-1} - i\lambda_{a,b,c}\lambda_0^{-1})^2]} \tag{6.68}$$

where $P_{a,b,c}$ is oscillator power, $\lambda_{a,b,c}$ are resonance wavelengths, and $N_{a.b,c}$ are widths of absorption lines. Rotational heterogeneity of a thin film is measured through the following rotational dyadic:

$$\underset{=z}{S}(r) = u_z u_z + (u_x u_x + u_y u_y)\cos\left(\frac{\pi}{\Omega}(r, u_l) + h(u_y u_x - u_x u_y\right)\sin\left(\frac{\pi}{\Omega}(r, u_l)\right) \quad (6.69)$$

The helicoidal axis was considered as parallel with the unit vector $u_l = u_x \sin\alpha + u_z \cos\alpha$, 2Ω = constitutive duration, and $h = \pm 1$ show constitutive direction of the thin film (CW or CCW).

Dyadic:

$$\underset{=y}{S}(\sigma) = u_y u_y + (u_x u_x + u_z u_z)\cos\sigma + (u_x u_z - u_z u_x)\sin\sigma \quad (6.70)$$

It has two different roles: (1) $\underset{=y}{S}(\chi_s)$ introduces the role of the growth process, $\frac{\pi}{2} - \chi_s$ is gradient of the helicoidal axis; (2) $\underset{=y}{S}(-\alpha)$ indicates inclination of the helicoidal axis. Because nanowires of an oblique helicoidal STF have to grow in the half space over the substrate, $|\alpha| < \chi_s$ limitation is applied for α angle. α = 0, particularly, represents helicoidal STFs that decrease $\underset{=}{\varepsilon}(r, \lambda_0) \to \underset{=}{\varepsilon}(z, \lambda_0)$.

Now, reaction of oblique helicoidal STFs to transverse waves during the oblique incident is analyzed. If the transverse wave falls with the wave vector $k_+^{(0)} = k_x^{(0)} u_x + k_y^{(0)} u_y + k_z^{(0)} u_z$ from $z \le 0$ on the STF, incident, reflective, and transmissive electromagnetic fields are as the following in fuzzy presentation in terms of harmonic Floquet sets:

$$E_i = \sum_{n\in Z} (L_+^{(n)} a_L^{(n)} + R_+^{(n)} a_R^{(n)}) e^{iK_+^{(n)}.r}, \; z \le 0 \quad (6.71)$$

$$H_i = -\frac{i n_{hs}}{\eta_0} \sum_{n\in Z} (L_+^{(n)} a_L^{(n)} - R_+^{(n)} a_R^{(n)}) e^{iK_+^{(n)}.r}, \; z \le 0 \quad (6.72)$$

$$E_r = \sum_{n\in Z} (L_-^{(n)} r_L^{(n)} + R_-^{(n)} r_R^{(n)}) e^{iK_-^{(n)}.r}, \; z \le 0 \quad (6.73)$$

$$H_r = -\frac{i n_{hs}}{\eta_0} \sum_{n\in Z} (L_-^{(n)} r_L^{(n)} - R_-^{(n)} r_R^{(n)}) e^{iK_-^{(n)}.r}, \; z \le 0 \quad (6.74)$$

$$E_t = \sum_{n\in Z} (L_+^{(n)} t_L^{(n)} + R_+^{(n)} t_R^{(n)}) e^{iK_+^{(n)}.\tilde{r}} \; z \le 0 \quad (6.75)$$

$$H_t = -\frac{i n_{hs}}{\eta_0} \sum_{n\in Z} (L_+^{(n)} t_L^{(n)} - R_+^{(n)} t_R^{(n)}) e^{iK_+^{(n)}.\tilde{r}} \; z \le 0 \quad (6.76)$$

where $\eta_0 = \sqrt{\frac{\mu_0}{\varepsilon_0}}$ is innate impedance of vacuum; $\left\{r_L^{(n)}, r_R^{(n)}\right\}$, $\left\{a_L^{(n)}, a_R^{(n)}\right\}$, and $\left\{t_L^{(n)}, t_R^{(n)}\right\}$ are mixed amplitudes of nth rank of LCP and RCP waves; and $n = 0, \pm1, \pm2, \ldots$ and $\tilde{r} = r - du_z$.

Wave vectors ($K_{\pm}^{(n)}$) and polarization vectors ($L_{\pm}^{(n)}$ and $R_{\pm}^{(n)}$) are

$$K_{\pm}^{(n)} = K_x^{(n)} u_x + K_y^{(0)} u_y + K_z^{(n)} u_z \tag{6.77}$$

$$L_{\pm}^{(n)} = \pm(iS^{(n)} - P_{\pm}^{(n)})/\sqrt{2} \tag{6.78}$$

$$R_{\pm}^{(n)} = \mp(iS^{(n)} - P_{\pm}^{(n)})/\sqrt{2} \tag{6.79}$$

In the mentioned equations the following vectors are related to the s and p polarized fields:

$$\begin{cases} S^{(n)} = \dfrac{-k_y^{(0)}}{k_{xy}^{(n)}} u_x + \dfrac{k_x^{(n)}}{k_{xy}^{(n)}} u_y \\ P_{\pm}^{(n)} = \mp \dfrac{k_z^{(n)}}{k_0 n_{hs}} \left(\dfrac{k_x^{(n)}}{k_{xy}^{(n)}} u_x + \dfrac{k_y^{(0)}}{k_{xy}^{(n)}} u_y \right) + \dfrac{k_{xy}^{(n)}}{k_0 n_{hs}} u_z \end{cases} \tag{6.80}$$

And the following scalars depend on $STF\ \Lambda_x = \frac{2\Omega}{|\sin\alpha|}$ of the oblique helicoidal STF toward the X axis:

$$\begin{cases} k_x = \dfrac{\pi}{\Omega} |\sin\alpha| \\ k_x^{(n)} = k_x^{(0)} + nk_x \\ k_z^{(n)} = +\sqrt{k_0^2 n_{hs}^2 - (k_{xy}^{(n)})^2} \\ k_{xy}^{(n)} = +\sqrt{(k_x^{(n)})^2 - (k_y^{(0)})^2} \end{cases} \tag{6.81}$$

Wavelength of the vacuum space is $k_0 = \frac{2\pi}{\lambda_0} = \omega\sqrt{\varepsilon_0 \mu_0}$.

If the Floquet harmonic incident transverse wave is zero (i.e., for any $n \neq 0$, $a_L = a_R = 0$), mixed amplitudes $\{t_L^{(n)}, t_R^{(n)}\}$, $\{r_L^{(n)}, r_R^{(n)}\}$ may be determined based on known amplitudes $\{a_L^{(n)}, a_R^{(n)}\}$.

Alternative spatial changes $\underline{\underline{\varepsilon}}(r)$ are measured through extension of the February set from Equation (6.66):

$$\underline{\underline{\varepsilon}}(r) = \sum_{n \in Z} \underline{\underline{\varepsilon}}^{(n)} e^{in(k_x x + k_z z)}, \; 0 < z < d \tag{6.82}$$

where

$$\underset{=}{\varepsilon}^{(n)} = \sum_{\sigma,\sigma'} \varepsilon_{\sigma\sigma'}^{(n)} u_\sigma u_{\sigma'} \text{ , } \sigma,\sigma' = x,y,z \tag{6.83}$$

are fixed value dyadics. $k_z = \frac{\pi}{\Omega}\cos\alpha$ and $STF\ \Lambda_x = \frac{2\Omega}{|\sin\alpha|}$ determine the oblique helicoidal STF perpendicular to the substrate (z axis).

Electromagnetic fields in fuzzy presentation are separated as

$$E(r) = \sum_{n\in Z} [E_x^{(n)}(z)u_x + E_y^{(n)}(z)u_y + E_z^{(n)}(z)u_z]e^{i(k_x^{(n)}x + k_y^{(0)}y)} \tag{6.84}$$

$$H(r) = \sum_{n\in Z} [H_x^{(n)}(z)u_x + H_y^{(n)}(z)u_y + H_z^{(n)}(z)u_z]e^{i(k_x^{(n)}x + k_y^{(0)}y)} \tag{6.85}$$

where $E_{x,y,x}{}^{(\mathrm{r})}$ and $H_{x,y,x}{}^{(\mathrm{n})}$ are unknown functions from $z \in (0,d)$.

Following Chateau and Hugonin, we have

$$\begin{cases} \tilde{E}_\sigma^{(n)}(z) = E_\sigma^{(n)}(z)e^{-ink_z z} \\ \qquad\qquad\qquad\qquad , \quad \sigma = x\cdot y\cdot z \\ \tilde{H}_\sigma^{(n)}(z) = H_\sigma^{(n)}(z)e^{-ink_z z} \end{cases} \tag{6.86}$$

By replacing Equation (6.82) with Equation (6.86) in Curl–Maxwell:

$$\begin{cases} \nabla\times E(r) = i\omega\mu_0 H(r) \\ \qquad\qquad\qquad\qquad , \quad 0<z<d \\ \nabla\times H(r) = -i\omega\varepsilon_0\,\underset{=}{\varepsilon}\,E(r) \end{cases} \tag{6.87}$$

and by applying $e^{ik_\pm^{(n),r}}$ in the fixed plane of z, the following coupled wave equations are obtainable in $z \in (0,d)$:

$$\frac{d}{dz}\tilde{E}_x^{(n)}(z) + ink_z\tilde{E}_x^{(n)}(z) - ink_x^{(n)}\tilde{E}_x^{(n)}(z) = ik_0\eta_0\tilde{H}_y^{(n)}(z) \tag{6.88}$$

$$\frac{d}{dz}\tilde{E}_y^{(n)}(z) + ink_z\tilde{E}_y^{(n)}(z) - ink_y^{(0)}\tilde{E}_z^{(n)}(z) = -ik_0\eta_0\tilde{H}_x^{(n)}(z) \tag{6.89}$$

$$k_y^{(0)}\tilde{E}_x^{(n)}(z) - k_x^{(n)}\tilde{E}_y^{(n)}(z) = -k_0\eta_0\tilde{H}_z^{(n)}(z) \tag{6.90}$$

$$\frac{d}{dz}\tilde{H}_x^{(n)}(z)+ink_z\tilde{H}_x^{(n)}(z)-ik_x^{(n)}\tilde{H}_z^{(n)}(z)$$

$$=\frac{-ik_0}{\eta_0}\sum_{n'\in z}[\varepsilon_{yx}^{(n-n')}\tilde{E}_x^{(n')}(z)+\varepsilon_{yy}^{(n-n')}\tilde{E}_y^{(n')}(z)+\varepsilon_{yz}^{(n-n')}\tilde{E}_z^{(n')}(z)] \tag{6.91}$$

$$\frac{d}{dz}\tilde{H}_y^{(n)}(z)+ink_z\tilde{H}_y^{(n)}(z)-ik_y^{(0)}\tilde{H}_z^{(n)}(z)$$

$$=\frac{ik_0}{\eta_0}\sum_{n'\in z}[\varepsilon_{xx}^{(n-n')}\tilde{E}_x^{(n')}(z)+\varepsilon_{xy}^{(n-n')}\tilde{E}_y^{(n')}(z)+\varepsilon_{xz}^{(n-n')}\tilde{E}_z^{(n')}(z)] \tag{6.92}$$

$$k_y^{(0)}\tilde{H}_x^{(n)}(z)-k_x^{(n)}\tilde{H}_y^{(n)}(z)=\frac{k_0}{\eta_0}\sum\varepsilon_{zx}^{(n-n')}\tilde{E}_x^{(n')}(z)+\varepsilon_{zy}^{(n-n')}\tilde{E}_y^{(n')}(z)+\varepsilon_{zz}^{(n-n')}\tilde{E}_z^{(n')}(z) \tag{6.93}$$

The mentioned equations are infinite systems of the first-degree equation ($n \in z$). For digital solution $|n| \le N_t$ may be applied.

The four next columnar vectors $(2N_t + 1)$ are introduced:

$$\begin{cases}[\tilde{E}_{-\sigma}(z)]=[\tilde{E}_\sigma^{(n)}(z)], \quad [E_{-\sigma}(z)]=[E_\sigma^{(n)}(z)] \\ [\tilde{H}_{-\sigma}(z)]=[\tilde{H}_\sigma^{(n)}(z)], \quad [H_{-\sigma}(z)]=[H_\sigma^{(n)}(z)]\end{cases} \tag{6.94}$$

Likewise oblique matrixes of $(2N_t + 1) \times (N_t + 1)$ are as follows:

$$\begin{cases}[K_{=x}]=[k_x^{(n)}\delta_{n,n'}] \\ \qquad\qquad\qquad\qquad , \quad n,n' \in [-N_l, N_t] \\ [k_{=z}]=k_z[n\delta_{n,n'}]\end{cases} \tag{6.95}$$

where $\delta_{n,n}$ is the Kronecker delta. Toeplitz matrixes:

$$[\varepsilon_{=\sigma,\sigma'}]=[\varepsilon_{\sigma,\sigma'}^{(n-n')}], \sigma=x,y,z; n,n' \in [-N_l,N_t] \tag{6.96}$$

By replacing Equations (6.90) and (6.93) in (6.88) to (6.96), omitting vertical components of electromagnetic fields ($\tilde{H}_z^{(n)}, \tilde{E}_z^{(n)}$) and some mathematical operations, the first-degree ordinary columnar equation is achieved:

$$\frac{d}{dz}[\tilde{f}_-(z)]=i[\tilde{P}_=][\tilde{f}_-(z)], \quad 0<z<d \tag{6.97}$$

The kernel matrix, which is independent from $[\underline{\underline{\tilde{P}}}]z$ and next $4(2N_t+1)\times 4(2N_t+1)$, are written as

$$[\underline{\underline{\tilde{P}}}]=\begin{bmatrix}[\underline{\underline{\tilde{P}}}_{11}][\underline{\underline{\tilde{P}}}_{12}][\underline{\underline{\tilde{P}}}_{13}][\underline{\underline{\tilde{P}}}_{14}]\\ [\underline{\underline{\tilde{P}}}_{21}][\underline{\underline{\tilde{P}}}_{22}][\underline{\underline{\tilde{P}}}_{23}][\underline{\underline{\tilde{P}}}_{24}]\\ [\underline{\underline{\tilde{P}}}_{31}][\underline{\underline{\tilde{P}}}_{32}][\underline{\underline{\tilde{P}}}_{33}][\underline{\underline{\tilde{P}}}_{34}]\\ [\underline{\underline{\tilde{P}}}_{41}][\underline{\underline{\tilde{P}}}_{42}][\underline{\underline{\tilde{P}}}_{43}][\underline{\underline{\tilde{P}}}_{44}]\end{bmatrix} \tag{6.98}$$

And also the 16 following matrixes will be obtained:

$$\begin{cases}[\underline{\underline{\tilde{P}}}_{11}]=-[\underline{\underline{k}}_z]-[\underline{\underline{k}}_x][\underline{\underline{\varepsilon}}_{zz}]^{-1}[\underline{\underline{\varepsilon}}_{zx}]\\ [\underline{\underline{\tilde{P}}}_{12}]=-[\underline{\underline{k}}_x][\underline{\underline{\varepsilon}}_{zz}]^{-1}[\underline{\underline{\varepsilon}}_{zy}]\\ [\underline{\underline{\tilde{P}}}_{13}]=\dfrac{k_y^{(0)}}{k_0}[\underline{\underline{k}}_x][\underline{\underline{\varepsilon}}_{zz}]^{-1}\\ [\underline{\underline{\tilde{P}}}_{14}]=k_0[\underline{\underline{I}}]-\dfrac{1}{k_0}[\underline{\underline{k}}_x][\underline{\underline{\varepsilon}}_{zz}]^{-1}[\underline{\underline{k}}_x]\end{cases} \tag{6.99}$$

$$\begin{cases}[\underline{\underline{\tilde{P}}}_{21}]=-[\underline{\underline{k}}_y^{(0)}][\underline{\underline{\varepsilon}}_{zz}]^{-1}[\underline{\underline{\varepsilon}}_{zx}]\\ [\underline{\underline{\tilde{P}}}_{22}]=-[\underline{\underline{k}}_z]-k_y^{(0)}[\underline{\underline{\varepsilon}}_{zz}]^{-1}[\underline{\underline{\varepsilon}}_{zy}]\\ [\underline{\underline{\tilde{P}}}_{23}]=-k_0[\underline{\underline{I}}]+\dfrac{(k_y^{(0)})^2}{k_0}[\underline{\underline{\varepsilon}}_{zz}]^{-1}\\ [\underline{\underline{\tilde{P}}}_{24}]=-\dfrac{k_y^{(0)}}{k_0}[\underline{\underline{\varepsilon}}_{zz}]^{-1}[\underline{\underline{k}}_x]\end{cases} \tag{6.100}$$

$$\begin{cases}[\underline{\underline{\tilde{P}}}_{31}]=-\dfrac{k_y^{(0)}}{k_0}[\underline{\underline{k}}_x]+k_0([\underline{\underline{\varepsilon}}_{yz}][\underline{\underline{\varepsilon}}_{zz}]^{-1}[\underline{\underline{\varepsilon}}_{zx}]-[\underline{\underline{\varepsilon}}_{yx}])\\ [\underline{\underline{\tilde{P}}}_{32}]=\dfrac{1}{k_0}[\underline{\underline{k}}_x][\underline{\underline{k}}_x]+k_0([\underline{\underline{\varepsilon}}_{yz}][\underline{\underline{\varepsilon}}_{zz}]^{-1}[\underline{\underline{\varepsilon}}_{zy}]-[\underline{\underline{\varepsilon}}_{yy}])\\ [\underline{\underline{\tilde{P}}}_{33}]=-[\underline{\underline{k}}_z]-k_y^{(0)}[\underline{\underline{\varepsilon}}_{yz}][\underline{\underline{\varepsilon}}_{zz}]^{-1}\\ [\underline{\underline{\tilde{P}}}_{34}]=[\underline{\underline{\varepsilon}}_{yz}][\underline{\underline{\varepsilon}}_{zz}]^{-1}[\underline{\underline{k}}_x]\end{cases} \tag{6.101}$$

$$\begin{cases} [\underset{=41}{\tilde{P}}] = -\dfrac{(k_y^{(0)})^2}{k_0}[\underset{=}{I}] - k_0([\underset{=xz}{\varepsilon}][\underset{=zz}{\varepsilon}]^{-1}[\underset{=zx}{\varepsilon}] - [\underset{=xx}{\varepsilon}]) \\ [\underset{=42}{\tilde{P}}] = \dfrac{k_y^{(0)}}{k_0}[\underset{=x}{k}] - k_0([\underset{=xz}{\varepsilon}][\underset{=zz}{\varepsilon}]^{-1}[\underset{=zy}{\varepsilon}] - [\underset{=xy}{\varepsilon}]) \\ [\underset{=43}{\tilde{P}}] = k_y^{(0)}[\underset{=xz}{\varepsilon}][\underset{=zz}{\varepsilon}]^{-1} \\ [\underset{=44}{\tilde{P}}] = -[\underset{=z}{k}] - [\underset{=xz}{\varepsilon}][\underset{=zz}{\varepsilon}]^{-1}[\underset{=x}{k}] \end{cases} \tag{6.102}$$

The following columnar vector includes $4(2N_t + 1)$ components:

$$[\underset{-}{\tilde{f}}(z)] = [[\underset{-x}{\tilde{E}}(z)]^T, [\underset{-y}{\tilde{E}}(z)]^T, \eta_0[\underset{-x}{\tilde{H}}(z)]^T, \eta_0[\underset{-y}{\tilde{H}}(z)]^T]^T \tag{6.103}$$

The result of the matrix of the ordinary differential equation is

$$[\underset{-}{\tilde{f}}(z_2)] = [\underset{=}{\tilde{G}}]e^{i(z_2 - z_1)[\underset{-}{\tilde{D}}]}[\underset{=}{\tilde{G}}]^{-1}[\underset{-}{\tilde{f}}(z_1)] \tag{6.104}$$

Columns of the square matrix $[\underset{=}{\tilde{G}}]$ and the oblique matrix $[\underset{=}{\tilde{D}}]$ are composed of consecutive vectors $[\underset{=}{\tilde{P}}]$ and corresponding values $[\underset{=}{\tilde{P}}]$, respectively. It is supposed that $[\underset{=}{\tilde{P}}]$ may be converted to the independent linear oblique vector by using $4(2N_t + 1)$. For solving the boundary value problem of the vector, Equation (6.106) is determined instead of $[\underset{-}{\tilde{f}}(z)]$:

$$[\underset{-}{\tilde{f}}(z)] = [[\underset{-x}{\tilde{E}}(z)]^T, [\underset{-y}{\tilde{E}}(z)]^T, \eta_0[\underset{-x}{\tilde{H}}(z)]^T, \eta_0[\underset{-y}{\tilde{H}}(z)]^T]^T \tag{6.105}$$

These two columnar vectors will be related to each other through the following vector:

$$[\underset{-}{f}(z)] = [\underset{=}{c}(z)][\underset{-}{\tilde{f}}(z)] \tag{6.106}$$

Oblique matrix:

$$[\underset{=}{c}(z)] = [e^{ink_z z}\delta_{n,n'}],\ n, n' \in [1, 4(2N_t + 1] \tag{6.107}$$

And $Mod[n,n']$, $\tilde{n} = Mod[n-1, 2N_t+1] - N_t$ is the result of dividing n by n'. n and n' are positive integers. Regarding Equations (6.104) and (6.105), the following equation is obtainable:

$$[\underline{f}(z_2)] = [\underline{\underline{G}}(z_2)]e^{i(z_2-z_1)[\underline{\underline{\tilde{D}}}]}[\underline{\underline{G}}(z_1)]^{-1}[\underline{f}(z_1)] \tag{6.108}$$

Matrix (6.109) is an alternative function of z:

$$[\underline{\underline{G}}(z)] = [\underline{\underline{c}}(z)][\underline{\underline{\tilde{G}}}] \tag{6.109}$$

Thus,

$$[\underline{f}(d)] = [\underline{\underline{G}}(d)]e^{i(d_1)[\underline{\underline{\tilde{D}}}]}[\underline{\underline{G}}(0)]^{-1}[\underline{f}(0)] \tag{6.110}$$

Continuity of tangential components of electromagnetic fields in fuzzy presentation in boundaries $z = 0, d$ should be applied for each harmonic Floquet of nth rank. Therefore, we have

$$[\underline{f}(0)] = \begin{bmatrix} [\underline{\underline{Y}}_e^+] & [\underline{\underline{Y}}_e^-] \\ [\underline{\underline{Y}}_h^+] & [\underline{\underline{Y}}_h^-] \end{bmatrix} \begin{bmatrix} [\underline{A}] \\ [\underline{B}] \end{bmatrix}; \quad [\underline{f}(d)] = \begin{bmatrix} [\underline{\underline{Y}}_e^+] & [\underline{\underline{0}}] \\ [\underline{\underline{Y}}_h^+] & [\underline{\underline{0}}] \end{bmatrix} \begin{bmatrix} [\underline{T}] \\ [\underline{0}] \end{bmatrix} \tag{6.111}$$

$$[\underline{A}] = \begin{bmatrix} a_L^{(n)} \\ a_R^{(n)} \end{bmatrix}; \quad [\underline{R}] = \begin{bmatrix} r_L^{(n)} \\ r_R^{(n)} \end{bmatrix}; \quad [\underline{T}] = \begin{bmatrix} t_L^{(n)} \\ t_R^{(n)} \end{bmatrix} \tag{6.112}$$

Equation (6.112) shows the next $4(2N_t + 1)$ matrixes. Square matrixes $[\underline{\underline{Y}}_h^{\pm}]$ and $[\underline{\underline{Y}}_e^{\pm}]$ with sizes of $4(2N_t + 2) \times 4(2N_t + 2)$ are rare. Their nonzero sentences are

$$\begin{cases} [\underline{\underline{Y}}_e^{\pm}] = \left(\dfrac{-i}{n_{hs}}\right)[\underline{\underline{Y}}_h^{\pm}]_{nn'} = L_{\pm}^{(n)} \cdot u_x \quad if \quad n = n' \in [1, (2N_t+1)] \\ [\underline{\underline{Y}}_e^{\pm}] = \left(\dfrac{-i}{n_{hs}}\right)[\underline{\underline{Y}}_h^{\pm}]_{nn'} = L_{\pm}^{(n)} \cdot u_y \quad if \quad n = n' + 2N_t + 1 \\ [\underline{\underline{Y}}_e^{\pm}] = \left(\dfrac{-i}{n_{hs}}\right)[\underline{\underline{Y}}_h^{\pm}]_{nn'} = R_{\pm}^{(n)} \cdot u_x \quad if \quad n = n' - 2N_t - 1 \\ [\underline{\underline{Y}}_e^{\pm}] = \left(\dfrac{i}{n_{hs}}\right)[\underline{\underline{Y}}_h^{\pm}]_{nn'} = R_{\pm}^{(n)} \cdot u_y \quad if \quad n = n' \in [(2N_t+2), (4N_t+2)] \end{cases} \tag{6.113}$$

By replacing (6.112) in (6.113), we will have

$$\begin{bmatrix}[\underset{=}{U}_T]\\[\underset{=}{V}_T]\end{bmatrix}\left[\underset{-}{T}\right]+\begin{bmatrix}e^{id[\tilde{\underset{=}{D}}_1]} & [\underset{-}{0}]\\ [\underset{-}{0}] & e^{id[\tilde{\underset{=}{D}}_2]}\end{bmatrix}\begin{bmatrix}[\underset{=}{U}_R]\\[\underset{=}{V}_R]\end{bmatrix}[\underset{-}{R}]=\begin{bmatrix}e^{id[\tilde{\underset{=}{D}}_1]} & [\underset{-}{0}]\\ [\underset{-}{0}] & e^{id[\tilde{\underset{=}{D}}_2]}\end{bmatrix}\begin{bmatrix}[\underset{=}{U}_A]\\[\underset{=}{V}_A]\end{bmatrix}[\underset{-}{A}] \quad (6.114)$$

$[\tilde{\underset{=}{D}}_2]$ and $[\tilde{\underset{=}{D}}_1]$ are oblique submatrixes of up and down of $[\tilde{\underset{=}{D}}]$ and

$$\begin{bmatrix}[\underset{=}{U}_T]\\[\underset{=}{V}_T]\end{bmatrix}=[\underset{=}{G}(d)]^{-1}\begin{bmatrix}[\underset{=}{Y}^+_e]\\[\underset{=}{V}^+_h]\end{bmatrix} \quad (6.115)$$

$$\begin{bmatrix}[\underset{=}{U}_R]\\[\underset{=}{V}_R]\end{bmatrix}=-[\underset{=}{G}(0)]^{-1}\begin{bmatrix}[\underset{=}{Y}^-_e]\\[\underset{=}{V}^-_h]\end{bmatrix} \quad (6.116)$$

$$\begin{bmatrix}[\underset{=}{U}_A]\\[\underset{=}{V}_A]\end{bmatrix}=[\underset{=}{G}(0)]^{-1}\begin{bmatrix}[\underset{=}{Y}^+_e]\\[\underset{=}{V}^+_h]\end{bmatrix} \quad (6.117)$$

The following algebraic relation may be achieved through Equation (6.114):

$$\begin{bmatrix}e^{-id[\tilde{\underset{=}{D}}_1]}[\underset{=}{U}_T] & [\underset{=}{U}_R]\\ [\underset{=}{V}_T] & e^{id[\tilde{\underset{=}{D}}_2]}[\underset{=}{U}_R]\end{bmatrix}\begin{bmatrix}[\underset{-}{T}]\\[\underset{-}{V}]\end{bmatrix}=\begin{bmatrix}[\underset{=}{U}_A]\\ e^{id[\tilde{\underset{=}{D}}_2]}[\underset{=}{V}_A]\end{bmatrix}[\underset{-}{A}] \quad (6.118)$$

$[\underset{-}{R}]$ and $[\underset{-}{T}]$ of any d and N_t may be calculated through using the reversal matrix operator and standard techniques. $\{t_L^{(n)}, t_R^{(n)}\}$ and $\{r_L^{(n)}, r_R^{(n)}\}$ for all ranks of n may be calculated by using rigorous coupled-wave analysis (RCWA) for transmission and reflection coefficients of the n rank in terms of incident transverse wave vector of $k_+^{(0)}$.

6.8 Physical Principles and Applications of Different Fabrication Methods

6.8.1 Excimer Laser

As previously mentioned, the excimer laser is a powerful type of chemical laser that produces nanopulses with a wavelength within UV waves. To produce such a laser beam, a gas mixture, mainly composed of a noble gas such as Argon,

Xenon, Krypton, with a halogen, such as fluorine or chlorine, is pumped into a high-voltage electrical discharge with short pulses (nanosecond) or is exposed to an electronic beam. This leads to the production of some quasi-molecules called *excimer*. These complexes are only present in the high content of energy or in an excited state and are not stable in the basic electronic state. Due to spontaneous transfer between these two energy levels, a kind of laser beam with a wavelength in the UV range and pulses of nanoseconds will be generated.

Because a dimer is representative of a molecule made of two identical parts, an excimer is a vague concept and is not a correct term to use. Instead, it is recommended to use the terms *rare gas halogen laser* or *exciplex laser*. Various types of the excimer laser produce wavelengths between 157 and 351 nm. A common excimer laser emits pulses with frequencies of 100 Hz and has average output power of several to several thousand watts, which makes them most powerful in the range of UV, especially wavelengths of lower than 300 nm. The output power is varied between 0.2% and 2%.

A radiated UV beam from an excimer laser can be efficiently absorbed by biomaterials and organic materials. It is worth noting that lower than a threshold value, which depends on the material, the laser has no effect and it is just above this value that the laser beam is effective. Once a beam of the excimer laser hits a material surface, it results in molecular bonds breakdown, instead of their cutting or burning, and leads to a very controlled removal of the particles from the surface and their releasing into the air, without any burning. The other characteristic of this laser is its capability of removing very thin layers of the material from the surface, without any temperature change in substrate and periphery of removed particles. This makes an excimer laser an efficient tool for micromachining organic materials such as some polymers and plastics or some delicate surgeries such as LASIK (laser-assisted in situ keratomileusis).

Using excimer lasers has created different applications such as

1. Production of very delicate patterns in lithography methods (e.g., in production of semiconductors)
2. Laser resource for laser ablation
3. Use as a laser resource coating by pulse laser (PLD)
4. Eye surgery with ArF with a 193 nm wavelength laser

Once a laser has been used, safety tips such as working with high voltages, transporting toxic gases, as well as skin cancer threat or eye damages due to UV light radiation, are of great importance.[87–92]

6.8.2 Molecular Beam Epitaxy (MBE)

In MBE technology, elements of a crystal in a molecular beam form, such as a semiconductor, are put on a heated crystalline substrate and then very thin layers of epitaxy start to develop. Sources for creating this beam are different and are mostly made of pure elements sublimed with the heat. However, there are some other sources such as gas sources MBE and group III metallic-organic precursors that can be used in some of the applications. This technique is performed in very high amounts of vacuum (10^{-8} Pa) that contribute to its least rate of pollution. The most important characteristic of the MBE is its low speed of coating layer, enabling the epitaxial growth of the deposited layer. Figure 6.16 illustrates a view of a MBE system.

Through the MBE method with a solid element source, very pure elements such as gallium and arsenic are heated in a separate diffusion cell and allowed to be slowly sublimed. Then, gas elements descend on a heated wafer and sometimes react with each other. For instance, in gallium and arsenic the monocrystal of gallium-arsenide would be developed. Here, the term *beam* implies that the evaporated atoms do not react with each other until they reach the wafer surface, which induces long average free movement of the atoms.

There is a shutter in front of each cell which is controlled by a computer system. This enables us to control the exact thickness of the layer in the atomic scale. Also, it is possible to decrease the growth rate of the crystals up to several angstroms per second. Once it is necessary for the substrate to remain cold, the vacuum environment of the container can be cooled using cryopanels, cryopumps, and liquid or gas nitrogen.

The substrate is installed on a manipulation system that facilitates sample movement within the container. Heating the underlying samples is also

FIGURE 6.16
Image of a molecular beam epitaxy (MBE) system. (Reprinted from Shimizu, S., Tsukakoshi, O., Komiya, S., and Makita, Y., Molecular and ion-beam epitaxy system for the growth of III-V compound semiconductors using a mass-separated, low-energy group-V ion beam, *Japanese Journal of Applied Physics, Part 1: Regular Papers and Short Notes* 24 (9), 1130–1140, Copyright 1985, with permission from the Japan Society of Applied Physics.)

done in this part. In order to show the in situ growth of the crystalline layers within the process, the reflected high energy electrons diffraction (RHEED) method can be applied. Using this system, the surface morphology, surface temperature, atoms arrangement, and many other surface parameters can be controlled and even regulated. RHEED guns diffuse the 10 KeV electrons, then the electrons hit the surface and are received by a phosphoric fluorescent plate and the surface information is interpreted; a camera takes the photos from the surface, and a barometer measures the container pressure and the beam.

MBE technology has made it possible to make multicomponent and multilayer structures from different materials. Using the MBE mechanism we can produce structures in which the electrons can be limited in the space, which leads to creation of quantum dots or points. Such multilayer structures are crucial in so many semiconductor panels such as semiconductor lasers. MBE technology is also used in the fabrication of some semiconductor organic lasers. Through this method, molecules, instead of atoms, are evaporated and deposited on the wafer.

The MBE is generally used to produce layers with very pure composition, precisely controlled thickness, and a very sharp interface. Accurate control of the combination and doping of the products are among noticeable characteristics of this method. These properties have made MBE an outstanding technique for making sophisticated electronic and optic-electric tools.[94–102]

6.8.3 Ion Implantation

In general, each ion implantation system is composed of an ion source, an accelerator (which electrostatically speeds up the ions to higher energy levels), and a target container where the ions are shut on the target material's surface. Figures 6.17 and 6.18 show real and schematic figures of ion implantation rigs.

The number of ions implanted on the target is known as ion dose, which depends on the current of produced ions in the machine. Typically, the produced ionic current in the implantation machine is very low (in the microampere range); thus, the amount of implanted ions number in a period span is not very much. This method will be useful whenever there is no need for wide chemical modifications. In this method, the typical energy of the ions is between 10 and 500 KeV, though the energy of 1 to 10 KeV is also available, which results in material diffusion in the nanometer or lower dimensions. The lower energy level of ions minimizes the destruction of the target material.

It is worth mentioning that in some accelerators the ion energy reaches up to 5 MeV, which is involved structural destructions of the material, as well as amorphism of the combination. Ion diffusion depth rests upon the material composition, as well as the ion's energy. The average of diffusion depth is known as "range of ions." Considering some required measures, the range of ions would be 10 nm to 1 μm.

FIGURE 6.17
Ultra high vacuum (UHV) low-energy ion implantation system. (Reprinted from Shimizu, S., Tsukakoshi, O., Komiya, S., and Makita, Y., Molecular and ion-beam epitaxy system for the growth of III-V compound semiconductors using a mass-separated, low-energy group-V ion beam, *Japanese Journal of Applied Physics, Part 1: Regular Papers and Short Notes* 24 (9), 1130–1140, Copyright 1985, with permission from the Japan Society of Applied Physics.)

The equipment of ion implantation involves dramatic advances. Manufacturers of these machines have optimized the factors such as ion number, energy content of the ions, and impact angle control. The current challenging issue in this technology is enhancing the ionic current in energy contents lower than 10 eV, which reinforces application of the machine in nanotechnology.

Furthermore, the pollution induced from the machine and the equipment has dramatically decreased in this method. These impurities have variant sources, and the manufacturers come out with different solutions for them.[103–115]

As previously mentioned, the ion implantation method cannot be used to create physical, chemical, and structural modifications in the sample's surface. Some of the main applications of this method are discussed in the next sections.

6.8.3.1 Doping

Doping is the most common application of ion implantation. Elements such as Ar, P (phosphorous), B (Boron), In (Indium), Sb (Antimony), Ge (Germanium), Si, N, H, and He (helium) can be used for this purpose. The dominant doping method has been thermal diffusion of the deposited additives. Because the method is more accurate, reliable, and repeatable, solving some of its problems can quickly make it a good alternative for previous methods.

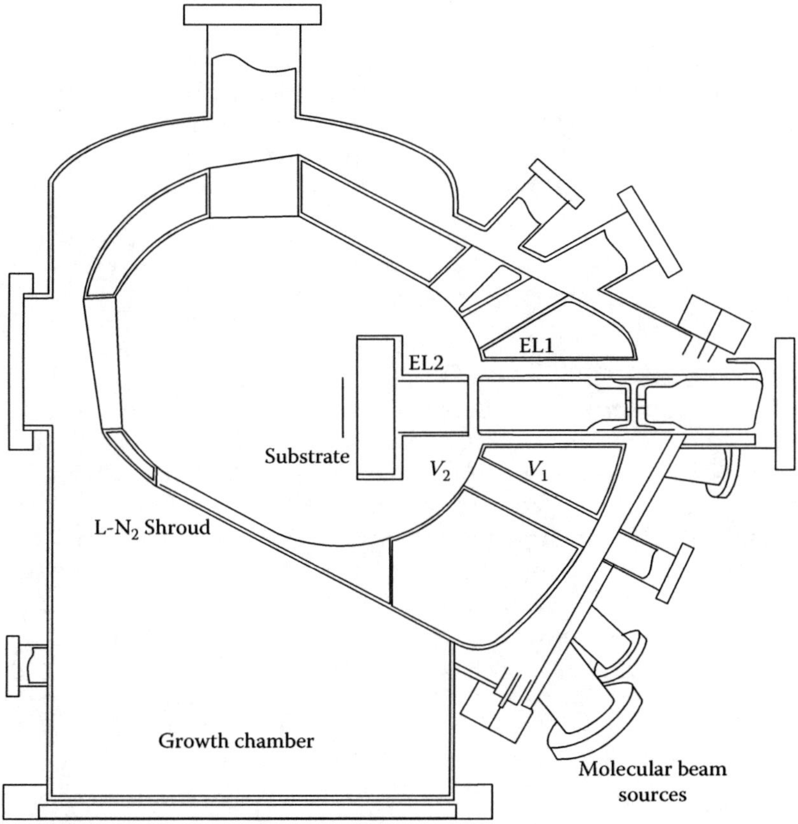

FIGURE 6.18
Arrangement of substrate, effusion cells, and deceleration electrodes in growth chambers. (Reprinted from Shimizu, S., Tsukakoshi, O., Komiya, S., and Makita, Y., Molecular and ion-beam epitaxy system for the growth of III-V compound semiconductors using a mass-separated, low-energy group-V ion beam, *Japanese Journal of Applied Physics, Part 1: Regular Papers and Short Notes* 24 (9), 1130–1140, Copyright 1985, with permission from the Japan Society of Applied Physics.)

6.8.3.2 Mesotaxy

This process is somehow similar to epitaxy, regarding that it develops under the wafer surface or substrate. Here, considering the required measures during the selection of the materials for adaption of both crystalline networks is of great importance.

6.8.3.3 Surface Operations

Surface ion implantation is applied to increase the chemical strength or resistance against corrosion in surfaces of different pieces, including metals and ceramics, as well as enhance the resistance against the development of surface crack.

6.8.3.4 Amorphous Fabrication

The impact of ions with high energy content to the surface can lead to the development of an amorphous film. Through the accurate adjusting of the effective parameters in the process, it is feasible to create an amorphous nanometer film. As noticed before, while the energy content of the ions is in the range of 10 KeV, the application of this method will be strongly concentrated in nanotechnology to prohibit deep diffusion.

6.8.4 Focused Ion Beam (FIB) in Nanotechnology

A focused beam of focused ions that are mainly gallium ions is used in the focused ion beam (FIB) method. Applying a very strong electrical field (as a liquid metal ion source, LMIS) leads to generation of the ions that will be concentrated by some electrostatic lenses. LMIS development is the starting point of FIB technology. Bombardment of the sample surface by ion current results in generation of secondary electrons, secondary ions, and scattered materials that are among the most important cases of FIB's variant applications. In the FIB method, generation of the secondary electron occurs from higher levels of the material (about 10 nm), though in the SEM method these electrons develop from levels lower than 100 nm. In the ion bombardment process there would be no backscattered electrons (BSEs). Releasing specific gases near the surface during the ion bombardment defines special applications, such as deposition of the materials, or promoted coat removing, for the FIB. In this method, different materials such as SiO_2 and Pt can be inserted into the coating layer. Another characteristic of the FIB system is its capability of scanning from the material surface as well as precise surface coating.[116–118]

In total, three main actions are made by FIB systems:

1. Imaging
2. Ion milling
3. Deposition

Based on potentials of the FIB system, its major applications are (it must be mentioned that there is some additional equipment, as well as main ones, for each)

1. Ion lithography
2. Producing and manufacturing semiconductors
3. Repairing the lithographic masks
4. Solving the structural defects
5. Preparing the circuit patterns
6. Sample preparation for transmission electron microscopy (TEM)

NOTE: The advantage of the FIB in sample preparation of the TEM, in contrast with Argon Ion Milling, is its higher speed and better performance in multilayer structures; however, its main disadvantage is the possibility of damage to the sample surface.

1. Imaging
2. Defining the grain size

NOTE: Due to differences in ion diffusion depths in different ions, contrast of the grains can be used to define the grain size.

1. Nanomachining

There are also some advantages for the FIB system:

1. Capability of performing simultaneous scanning and coating or coat removing
2. Spot sizes lower than 10 nm make it possible to generate components with a thickness of 100 nm
3. Flexibility of the method: capability of the coat removing from different materials with different shapes and coating of different materials
4. Uniqueness of the method in abrasion and milling of the oblique cuts
5. Absence of reflective electrons—FIB creates high-density lithography in comparison with the electron-beam lithography (EBL) method

6.8.5 FIB-SEM Dual Beam System

Applications of this method are

1. Analysis of defects and fractures in the substrate of nanostructure materials
2. Creation of cross section and analysis of internal and deep defects
3. Accurate cuts, where the sample cannot bear stresses induced by mechanical shear
4. Imaging equal to quality and resolution of the SEM
5. Sample preparation of the TEM with high speed and accuracy

6.8.6 Langmuir–Blodgett Film (LBF)

To create stratified structures (from 1 to 100 layers), the LBF is a suitable technique that enables us to accurately control the thickness and preserve the particles during the coating. This method is mainly used to create stratified structures from organic and polymer compositions. However, it has been proved in recent years that the metallic and semiconductor nanometer particles coated by ligand membrane can also be deposited as a layer on substrates in this way. In molecular and optical electronic science, development of the above-mentioned multilayer structures can be performed by techniques such as rotational coating, deposit coating, and simple floating. In all of these methods, the exact control of the thickness is really difficult, but the LBF technique solves this problem.

Another advantage of this method is the possibility of preserving particle orientation during the deposition, which is induced from polarization of the particles and can have particular applications. It is just feasible by controlling the surface pressure, temperature, and molecular density to bind the layers to substrates and layers to layers. There is no need for any other adhesive. The numeric value of these parameters, depending on layer and substrate material, is varied and can be empirically defined for each case, but the surface pressure is 10 to 40 MN/m^2.

The substrate can be made of any material. The only difference caused in the process by the material type is that if the substrate is hydrophilic (e.g., glass or SiO_2) the deposition of the first layer is performed by "heaving" the substrate from the underlying phase (liquid) toward the surface monolayer, but once the material of the substrate is hydrophobic (e.g., highly ordered pyrolitic graphite, HOPG), the first layer deposits by "lowering" the substrate into the underlying phase across the monolayer.

Regarding all discussed points about these films, it can be concluded that using the LBF technique it is possible to produce stratified structures with one to several hundred layers and a very controlled thickness and orientation. Lack of need for any adhesive to bind the layers to substrate and layers to layers is another significant advantage of this method. However, this method has some drawbacks, too. To date, the method is mostly applied in polymers and organic materials because the particles' polarity is an obligation for performing this method. But there are some studies to solve these problems, and it is expanding to other under-processing compositions.[119–129]

6.8.7 Electrospinning

Most common applications of the electrospinning method are as follows:

1. Production of catalysts from organic polymers to maintain the enzymes
2. Texture engineering in medicine and biology
3. Production of nanofiber ceramics

The unique properties of the ceramics such as high chemical resistance, high thermal strength, efficient mechanical strength, and favorable catalyst characteristics have created very interesting applications for them. During the last few years, there have been so many attempts to produce nanofibers made of TiO_2 and Al_2O_3 with the electrospinning method. However, their main application is production of nanocomposites with very favorable properties.

Among the most important advantages of this method, one can name its easy performance, low cost of equipment, and high operation speed. However, its dependence on fibers made of particular materials is its disadvantage. But there is a wide attempt to develop the range of electrospinning applications for other types of materials.[130–137]

References

1. Hodgkinson, I., Wu, Q. H., Knight, B., Lakhtakia, A., and Robbie, K., Vacuum deposition of chiral sculptured thin films with high optical activity, *Applied Optics* 39 (4), 642–649, 2000.
2. Hodgkinson, I. J., Wu, Q. H., Thorn, K. E., Lakhtakia, A., and McCall, M. W., Spacerless circular-polarization spectral-hole filters using chiral sculptured thin films: Theory and experiment, *Optics Communications* 184 (1), 57–66, 2000.
3. Lakhtakia, A., Sculptured thin films: Accomplishments and emerging uses, *Materials Science and Engineering* C 19 (1–2), 427–434, 2002.
4. Messier, R., Sculptured thin films—II. Experiments and applications, *Materials Research Innovations* 2 (4), 217–222, 1999.
5. Messier, R., Gehrke, T., Frankel, C., Venugopal, V. C., Otaño, W., and Lakhtakia, A., Engineered sculptured nematic thin films, *Journal of Vacuum Science and Technology A: Vacuum, Surfaces and Films* 15 (4), 2148–2152, 1997.
6. Messier, R., Venugopal, V. C., and Sunal, P. D., Origin and evolution of sculptured thin films, *Journal of Vacuum Science and Technology A: Vacuum, Surfaces and Films* 18 (4 II), 1538–1545, 2000.
7. Robbie, K., and Brett, M. J., Sculptured thin films and glancing angle deposition: Growth mechanics and applications, *Journal of Vacuum Science and Technology A: Vacuum, Surfaces and Films* 15 (3), 1460–1465, 1997.
8. Robbie, K., Brett, M. J., and Lakhtakia, A., Chiral sculptured thin films, *Nature* 384 (6610), 616, 1996.
9. Suzuki, M., Ito, T., and Taga, Y., Photocatalysis of sculptured thin films of TiO_2, *Applied Physics Letters* 78 (25), 3968–3970, 2001.
10. Wu, Q., Hodgkinson, I. J., and Lakhtakia, A., Circular polarization filters made of chiral sculptured thin films: Experimental and simulation results, *Optical Engineering* 39 (7), 1863–1868, 2000.
11. Hodgkinson, I. J., Wu, Q. H., Lakhtakia, A., and McCall, M. W., Spectral-hole filter fabricated using sculptured thin-film technology, *Optics Communications* 177 (1), 79–84, 2000.

12. Lakhtakia, A., McCall, M. W., Sherwin, J. A., Wu, Q. H., and Hodgkinson, I. J., Sculptured-thin-film spectral holes for optical sensing of fluids, *Optics Communications* 194 (1–3), 33–46, 2001.
13. Lakhtakia, A., Sculptured thin films—I. Concepts, *Materials Research Innovations* 1 (3), 145–148, 1997.
14. Horn, M. W., Pickett, M. D., Messier, R., and Lakhtakia, A., Blending of nanoscale and microscale in uniform large-area sculptured thin-film architectures, *Nanotechnology* 15 (3), 303–310, 2004.
15. Lakhtakia, A., and McCall, M., Sculptured thin films as ultranarrow-bandpass circular-polarization filters, *Optics Communications* 168 (5), 457–465, 1999.
16. Suzuki, M., and Taga, Y., Integrated sculptured thin films, *Japanese Journal of Applied Physics, Part 2: Letters* 40 (4 A), L358–L359, 2001.
17. Demirel, M. C., So, E., Ritty, T. M., Naidu, S. H., and Lakhtakia, A., Fibroblast cell attachment and growth on nanoengineered sculptured thin films, *Journal of Biomedical Materials Research—Part B Applied Biomaterials* 81 (1), 219–223, 2007.
18. Hodgkinson, I. J., Lakhtakia, A., and Wu, Q. H., Experimental realization of sculptured-thin-film polarization-discriminatory light-handedness inverters, *Optical Engineering* 39 (10), 2831–2834, 2000.
19. Lakhtakia, A., Venugopal, V. C., and McCall, M. W., Spectral holes in Bragg reflection from chiral sculptured thin films: Circular polarization filters, *Optics Communications* 177 (1), 57–68, 2000.
20. Yu, D. P., Lee, C. S., Bello, I., Sun, X. S., Tang, Y. H., Zhou, G. W., Bai, Z. G., Zhang, Z., and Feng, S. Q., Synthesis of nano-scale silicon wires by excimer laser ablation at high temperature, *Solid State Communications* 105 (6), 403–407, 1998.
21. Georgiev, D. G., Baird, R. J., Avrutsky, I., Auner, G., and Newaz, G., Controllable excimer-laser fabrication of conical nano-tips on silicon thin films, *Applied Physics Letters* 84 (24), 4881–4883, 2004.
22. Adikaari, A. A. D. T., and Silva, S. R. P., Thickness dependence of properties of excimer laser crystallized nano-polycrystalline silicon, *Journal of Applied Physics* 97 (11), 1–7, 2005.
23. Radhakrishnan, G., Adams, P. M., and Bernstein, L. S., Plasma characterization and room temperature growth of carbon nanotubes and nano-onions by excimer laser ablation, *Applied Surface Science* 253 (19), 7651–7655, 2007.
24. Crespo-Sosa, A., Schaaf, P., Reyes-Esqueda, J. A., Seman-Harutinian, J. A., and Oliver, A., Excimer laser absorption by metallic nano-particles embedded in silica, *Journal of Physics D: Applied Physics* 40 (7), 1890–1895, 2007.
25. Huang, H. W., Kao, C. C., Chu, J. T., Wang, W. C., Lu, T. C., Kuo, H. C., Wang, S. C., Yu, C. C., and Kuo, S. Y., Investigation of InGaN/GaN light emitting diodes with nano-roughened surface by excimer laser etching method, *Materials Science and Engineering B: Solid-State Materials for Advanced Technology* 136 (2–3), 182–186, 2007.
26. Kajiyama, S., Harada, K., Fukusaki, E., and Kobayashi, A., Single cell-based analysis of torenia petal pigments by a combination of ArF excimer laser micro sampling and nano-high performance liquid chromatography (HPLC)-mass spectrometry, *Journal of Bioscience and Bioengineering* 102 (6), 575–578, 2006.

27. Yoshizawa, M., Kikuchi, A., Fujita, N., Kushi, K., Sasamoto, H., and Kishino, K., Self-organization of GaN/$Al_{0.18}Ga_{0.82}N$ multi-layer nano-columns on (0 0 0 1) Al_2O_3 by RF molecular beam epitaxy for fabricating GaN quantum disks, *Journal of Crystal Growth* 189–190, 138–141, 1998.
28. Kusakabe, K., Kikuchi, A., and Kishino, K., Characterization of overgrown GaN layers on nano-columns grown by RF-molecular beam epitaxy, *Japanese Journal of Applied Physics, Part 2: Letters* 40 (3 A), L192–L194, 2001.
29. Kusakabe, K., Kikuchi, A., and Kishino, K., Overgrowth of GaN layer on GaN nano-columns by RF-molecular beam epitaxy, *Journal of Crystal Growth* 237–239 (1–4 II), 988–992, 2002.
30. Nomura, Y., Morishita, Y., Goto, S., and Katayama, Y., Lateral growth of GaAs on patterned {-1-1-1}B substrates for the fabrication of nano wires using metalorganic molecular beam epitaxy, *Journal of Electronic Materials* 23 (2), 97–100, 1994.
31. Lubyshev, D. I., Rossi, J. C., Gusev, G. M., and Basmaji, P., Nano-scale wires of GaAs on porous Si grown by molecular beam epitaxy, *Journal of Crystal Growth* 132 (3–4), 533–537, 1993.
32. Liu, H. F., Xiang, N., and Chua, S. J., Growth of InAs on micro- and nano-scale patterned GaAs(001) substrates by molecular beam epitaxy, *Nanotechnology* 17 (20), 5278–5281, 2006.
33. Pretorius, A., Yamaguchi, T., Kübel, C., Kröger, R., Hommel, D., and Rosenauer, A., Structural investigation of growth and dissolution of $In_xGa_{1-x}N$ nano-islands grown by molecular beam epitaxy, *Journal of Crystal Growth* 310 (4), 748–756, 2008.
34. Springholz, G., and Bauer, G., Molecular beam epitaxy of IV-VI semiconductor hetero- and nano-structures, *Physica Status Solidi (B) Basic Research* 244 (8), 2752–2767, 2007.
35. Cirlin, G. E., Egorov, V. A., Volovik, B. V., Tsatsul'nikov, A. F., Ustinov, V. M., Ledentsov, N. N., Zakharov, N. D., Werner, P., and Gösele, U., Optical and structural properties of Ge submonolayer nano-inclusions in a Si matrix grown by molecular beam epitaxy, *Nanotechnology* 12 (4), 417–420, 2001.
36. Shinada, T., Koyama, H., Hinoshita, C., Imamura, K., and Ohdomari, I., Improvement of focused ion-beam optics in single-ion implantation for higher aiming precision of one-by-one doping of impurity atoms into nano-scale semiconductor devices, *Japanese Journal of Applied Physics, Part 2: Letters* 41 (3 A), L287–L290, 2002.
37. Wołowski, J., Badziak, J., Czarnecka, A., Parys, P., Pisarek, M., Rosiski, M., Turan, R., and Yerci, S., Application of pulsed laser deposition and laser-induced ion implantation for formation of semiconductor nano-crystallites, *Laser and Particle Beams* 25 (1), 65–69, 2007.
38. Ila, D., Zimmerman, R. L., Muntele, C. I., Thevenard, P., Orucevic, F., Santamaria, C. L., Guichard, P. S., Schiestel, S., Carosella, C. A., Hubler, G. K., Poker, D. B., and Hensley, D. K., Nano-cluster engineering: A combined ion implantation/co-deposition and ionizing radiation, *Nuclear Instruments and Methods in Physics Research, Section B: Beam Interactions with Materials and Atoms* 191 (1–4), 416–421, 2002.
39. Fukumi, K., Chayahara, A., Kageyama, H., Kadono, K., Akai, T., Kitamura, N., Mizoguchi, H., Horino, Y., Makihara, M., Fujii, K., and Hayakawa, J., Formation process of CuCl nano-particles in silica glass by ion implantation, *Journal of Non-Crystalline Solids* 259 (1–3), 93–99, 1999.

40. Hayashi, N., Sakamoto, I., Toriyama, T., Wakabayashi, H., Okada, T., and Kuriyama, K., Embedded iron nano-clusters prepared by Fe ion implantation into MgO crystals, *Surface and Coatings Technology* 169–170, 540–543, 2003.
41. Kögler, R., Eichhorn, F., Kaschny, J. R., Mücklich, A., Reuther, H., Skorupa, W., Serre, C., and Perez-Rodriguez, A., Synthesis of nano-sized SiC precipitates in Si by simultaneous dual-beam implantation of C+ and Si+ ions, *Applied Physics A: Materials Science and Processing* 76 (5), 827–835, 2003.
42. Yoo, J. H., Yoon, K. S., Kim, J. S., and Won, T., Atomistic simulation for a nano-CMOS process: From ion implantation to diffusion, *Journal of the Korean Physical Society* 49 (3), 1260–1265, 2006.
43. Zhu, S., Xiang, X., Zu, X. T., and Wang, L. M., Magnetic nano-particles of Ni in MgO single crystals by ion implantation, *Nuclear Instruments and Methods in Physics Research, Section B: Beam Interactions with Materials and Atoms* 242 (1–2), 114–117, 2006.
44. Ho, H. P., Lo, K. C., Fu, K. Y., Chu, P. K., Li, K. F., and Cheah, K. W., Synthesis of beta gallium oxide nano-ribbons from gallium arsenide by plasma immersion ion implantation and rapid thermal annealing, *Chemical Physics Letters* 382 (5–6), 573–577, 2003.
45. San, J., Liu, J., Zhu, B., Mo, Z., Zhang, Q., and Dong, C., Metal-ion implantation effects on nano-hardness and tribological properties of Nylon6, *Surface and Coatings Technology* 161 (1), 1–10, 2002.
46. Morita, T., Kometani, R., Watanabe, K., Kanda, K., Haruyama, Y., Hoshino, T., Kondo, K., Kaito, T., Ichihashi, T., Fujita, J. I., Ishida, M., Ochiai, Y., Tajima, T., and Matsui, S., Free-space-wiring fabrication in nano-space by focused-ion-beam chemical vapor deposition, *Journal of Vacuum Science and Technology B: Microelectronics and Nanometer Structures* 21 (6), 2737–2741, 2003.
47. Hao, L., MacFarlane, J. C., Gallop, J. C., Cox, D., Beyer, J., Drung, D., and Schurig, T., Measurement and noise performance of nano-superconducting-quantum-interference devices fabricated by focused ion beam, *Applied Physics Letters* 92 (19), art no. 192507, 2008.
48. Youn, S. W., Takahashi, M., Goto, H., and Maeda, R., Microstructuring of glassy carbon mold for glass embossing—Comparison of focused ion beam, nano/femtosecond-pulsed laser and mechanical machining, *Microelectronic Engineering* 83 (11–12), 2482–2492, 2006.
49. Nagase, T., Gamo, K., Kubota, T., and Mashiko, S., Direct fabrication of nano-gap electrodes by focused ion beam etching, *Thin Solid Films* 499 (1–2), 279–284, 2006.
50. Nagase, M., Takahashi, H., Shirakawabe, Y., and Namatsu, H., Nano-four-point probes on microcantilever system fabricated by focused ion beam, *Japanese Journal of Applied Physics, Part 1: Regular Papers and Short Notes and Review Papers* 42 (7 B), 4856–4860, 2003.
51. Lehrer, C., Frey, L., Petersen, S., Sulzbach, T., Ohlsson, O., Dziomba, T., Danzebrink, H. U., and Ryssel, H., Fabrication of silicon aperture probes for scanning near-field optical microscopy by focused ion beam nano machining, *Microelectronic Engineering* 57–58, 721–728, 2001.
52. Nagase, T., Kubota, T., and Mashiko, S., Fabrication of nano-gap electrodes for measuring electrical properties of organic molecules using a focused ion beam, *Thin Solid Films* 438–439, 374–377, 2003.

53. Kometani, R., Hoshino, T., Kanda, K., Haruyama, Y., Kaito, T., Fujita, J. I., Ishida, M., Ochiai, Y., and Matsui, S., Three-dimensional high-performance nanotools fabricated using focused-ion-beam chemical-vapor-deposition, *Nuclear Instruments and Methods in Physics Research, Section B: Beam Interactions with Materials and Atoms* 232 (1–4), 362–366, 2005.
54. Ren, H. X., Huang, X. J., Kim, J. H., Choi, Y. K., and Gu, N., Pt/Au bimetallic hierarchical structure with micro/nano-array via photolithography and electrochemical synthesis: From design to GOT and GPT biosensors, *Talanta* 78 (4–5), 1371–1377, 2009.
55. Van Meerbeek, B., Conn Jr., L. J., Steven Duke, E., Schraub, D., and Ghafghaichi, F., Demonstration of a focused ion-beam cross-sectioning technique for ultrastructural examination of resin-dentin interfaces, *Dental Materials* 11 (2), 87–92, 1995.
56. Li, W., Lalev, G., Dimov, S., Zhao, H., and Pham, D. T., A study of fused silica micro/nano patterning by focused-ion-beam, *Applied Surface Science* 253 (7), 3608–3614, 2007.
57. Li, W., Dimov, S., and Lalev, G., Focused-ion-beam direct structuring of fused silica for fabrication of nano-imprinting templates, *Microelectronic Engineering* 84 (5–8), 829–832, 2007.
58. Singh, D. R. P., Chawla, N., and Shen, Y. L., Focused Ion Beam (FIB) tomography of nanoindentation damage in nanoscale metal/ceramic multilayers, *Materials Characterization* 61 (4), 481–488, 2010.
59. Krause, K. M., Vick, D. W., Malac, M., and Brett, M. J., Taking a little off the top: Nanorod array morphology and growth studied by focused ion beam tomography, *Langmuir* 26 (22), 17558–17567, 2010.
60. Chun, S., Han, K. S., Shin, J. H., Lee, H., and Kim, D., Fabrication and characterization of CdTe nano pattern on flexible substrates by nano imprinting and electrodeposition, *Microelectronic Engineering* 87 (11), 2097–2102, 2010.
61. Ariga, K., Nakanishi, T., and Michinobu, T., Immobilization of biomaterials to nano-assembled films (self-assembled monolayers, Langmuir-Blodgett films, and layer-by-layer assemblies) and their related functions, *Journal of Nanoscience and Nanotechnology* 6 (8), 2278–2301, 2006.
62. Iwamoto, M., and Itoh, E., Nano-electrostatic phenomena in Langmuir-Blodgett films, *Thin Solid Films* 331 (1–2), 15–24, 1998.
63. Elliot, D. J., Furlong, D. N., and Grieser, F., Fabrication of nano-sized particles of metallic copper and copper sulfide in Langmuir-Blodgett films, *Colloids and Surfaces A: Physicochemical and Engineering Aspects* 141 (1), 9–17, 1998.
64. Ohnuki, H., Saiki, T., Kusakari, A., Ichihara, M., and Izumi, M., Immobilization of glucose oxidase in Langmuir-Blodgett films containing Prussian blue nanoclusters, *Thin Solid Films* 516 (24), 8860–8864, 2008.
65. Sarkar, J., Pal, P., and Talapatra, G. B., Self-assembly of silver nano-particles on stearic acid Langmuir-Blodgett film: Evidence of fractal growth, *Chemical Physics Letters* 401 (4–6), 400–404, 2005.
66. Garnaes, J., Schwartz, D. K., Viswanathan, R., and Zasadzinski, J. A. N., Nano scale defects in Langmuir-Blodgett film observed by atomic force microscopy, *Synthetic Metals* 57 (1), 3795–3800, 1993.
67. Morandi, S., Puggelli, M., and Caminati, G., Antibiotic association with phospholipid nano-assemblies: A comparison between Langmuir-Blodgett films and

supported lipid bilayers, *Colloids and Surfaces A: Physicochemical and Engineering Aspects* 321 (1–3), 125–130, 2008.

68. Del Caño, T., Goulet, P. J. G., Pieczonka, N. P. W., Aroca, R. F., and De Saja, J. A., Nano-structured Langmuir-Blodgett mixed films of titanyl(IV) phthalocyanine and bis(neopentylimido)perylene: Unique degree of miscibility, *Synthetic Metals* 148 (1), 31–35, 2005.
69. Linh, N. T. B., Lee, K. H., and Lee, B. T., Fabrication of photocatalytic PVA-TiO_2 nano-fibrous hybrid membrane using the electro-spinning method, *Journal of Materials Science* 46 (17), 5615–5620, 2011.
70. Liang, S., Li, Q., Tang, X. F., Feng, Y., and He, D. Q., Cytocompatibility of electrospinning nano-fibrous scaffolds for skeletal muscle tissue engineering, *Journal of Clinical Rehabilitative Tissue Engineering Research* 15 (12), 2171–2174, 2011.
71. Zeugolis, D. I., Khew, S. T., Yew, E. S. Y., Ekaputra, A. K., Tong, Y. W., Yung, L. Y. L., Hutmacher, D. W., Sheppard, C., and Raghunath, M., Electro-spinning of pure collagen nano-fibres—Just an expensive way to make gelatin?, *Biomaterials* 29 (15), 2293–2305, 2008.
72. Ali, A. A., New generation of super absorber nano-fibroses hybrid fabric by electro-spinning, *Journal of Materials Processing Technology* 199 (1), 193–198, 2008.
73. Wang, F., and Lakhtakia, A., Lateral shifts of optical beams on reflection by slanted chiral sculptured thin films, *Optics Communications* 235 (1–3), 107–132, 2004.
74. Sherwin, J. A., Lakhtakia, A., and Hodgkinson, I. J., On calibration of a nominal structure-property relationship model for chiral sculptured thin films by axial transmittance measurements, *Optics Communications* 209 (4–6), 369–375, 2002.
75. Lakhtakia, A., and McCall, M., Simple expressions for Bragg reflection from axially excited chiral sculptured thin films, *Journal of Modern Optics* 49 (9), 1525–1535, 2002.
76. Lakhtakia, A., Pseudo-isotropic and maximum-bandwidth points for axially excited chiral sculptured thin films, *Microwave and Optical Technology Letters* 34 (5), 367–371, 2002.
77. Venugopal, V. C., Lakhtakia, A., Messier, R., and Kucera, J. P., Low-permittivity nanocomposite materials using sculptured thin film technology, *Journal of Vacuum Science and Technology B: Microelectronics and Nanometer Structures* 18 (1), 32–36, 2000.
78. Lakhtakia, A., On bioluminescent emission from chiral sculptured thin films, *Optics Communications* 188 (5–6), 313–320, 2001.
79. Wang, F., and Lakhtakia, A., Specular and nonspecular, thickness-dependent, spectral holes in a slanted chiral sculptured thin film with a central twist defect, *Optics Communications* 215 (1–3), 79–92, 2002.
80. Lakhtakia, A., Dielectric sculptured thin films for polarization-discriminatory handedness-inversion of circularly polarized light, *Optical Engineering* 38 (9), 1596–1602, 1999.
81. Lakhtakia, A., Dielectric sculptured thin films as Šolc filters, *Optical Engineering* 37 (6), 1870–1875, 1998.
82. Patzig, C., and Rauschenbach, B., Temperature effect on the glancing angle deposition of Si sculptured thin films, *Journal of Vacuum Science and Technology A: Vacuum, Surfaces and Films* 26 (4), 881–886, 2008.
83. Buzea, C., Kaminska, K., Beydaghyan, G., Brown, T., Elliott, C., Dean, C., and Robbie, K., Thickness and density evaluation for nanostructured thin films

by glancing angle deposition, *Journal of Vacuum Science and Technology B: Microelectronics and Nanometer Structures* 23 (6), 2545–2552, 2005.
84. Geddes III, J. B., and Lakhtakia, A., Effects of carrier phase on reflection of optical narrow-extent pulses from axially excited chiral sculptured thin films, *Optics Communications* 225 (1–3), 141–150, 2003.
85. Ignatovich, V. K., Neutron reflection from condensed matter, the Goos-Hänchen effect and coherence, *Physics Letters, Section A: General, Atomic and Solid State Physics* 322 (1–2), 36–46, 2004.
86. Sherwin, J. A., and Lakhtakia, A., Nominal model for structure-property relations of chiral dielectric sculptured thin films, *Mathematical and Computer Modelling* 34 (12–13), 1499–1514, 2001.
87. Bittl, J. A., Physical aspects of excimer laser angioplasty for undilatable lesions, *Catheterization and Cardiovascular Interventions* 71 (6), 808–809, 2008.
88. Zvorykin, V. D., Ionin, A. A., Losev, V. F., Mikheev, L. D., Konyashenko, A. V., Kovalchuk, B. M., Krokhin, O. N., Mesyats, G. A., Molchanov, A. G., Starodub, A. N., Tarasenko, V. F., and Yakovlenko, S. I., Petawatt excimer laser project at Lebedev Physical Institute, in *AIP Conference Proceedings*, 176–181, 2006.
89. Tsuda, N., and Yamada, J., Physical properties of dense plasma produced by XeCl excimer laser in high-pressure argon gases, *Japanese Journal of Applied Physics, Part 1: Regular Papers and Short Notes and Review Papers* 38 (6 A), 3712–3715, 1999.
90. Dostálová, T., Jelínek, M., Himmlová, L., Pešáová, V., and Adam, M., Physical and biological evaluation of hydroxylapatite films formed on Ti_6Al_4V substrates by excimer laser ablation, *Cells and Materials* 6 (1–3), 117–126, 1996.
91. Watanabe, H., Takata, T., and Tsuge, M., Polymer surface modification due to excimer laser radiation—Chemical and physical changes in the surface structure of poly(ethylene terephthalate), *Polymer International* 31 (3), 247–254, 1993.
92. Neev, J., Raney, D. V., Whalen, W. E., Fujishige, J. T., Ho, P. D., McGrann, J. V., and Berns, M. W., Dentin ablation with two excimer lasers: A comparative study of physical characteristics, *Lasers in the Life Sciences* 5 (1–2), 129–153, 1992.
93. Shimizu, S., Tsukakoshi, O., Komiya, S., and Makita, Y., Molecular and ion-beam epitaxy system for the growth of III-V compound semiconductors using a mass-separated, low-energy group-V ion beam, *Japanese Journal of Applied Physics, Part 1: Regular Papers and Short Notes* 24 (9), 1130–1140, 1985.
94. Yu, Z., Overgaard, C. D., Droopad, R., Passlack, M., and Abrokwah, J. K., Growth and physical properties of Ga_2O_3 thin films on GaAs(001) substrate by molecular-beam epitaxy, *Applied Physics Letters* 82 (18), 2978–2980, 2003.
95. Wang, S. Z., Yoon, S. F., He, L., and Shen, X. C., Physical mechanisms of photoluminescence of chlorine-doped ZnSe epilayers grown by molecular beam epitaxy, *Journal of Applied Physics* 90 (5), 2314–2320, 2001.
96. Dimakis, E., Iliopoulos, E., Tsagaraki, K., and Georgakilas, A., Physical model of InN growth on Ga-face GaN (0001) by molecular-beam epitaxy, *Applied Physics Letters* 86 (13), 1–3, 2005.
97. Baraldi, A., Colonna, F., Ghezzi, C., Magnanini, R., Parisini, A., Tarricone, L., Bosacchi, A., and Franchi, S., Electron mobility and physical magnetoresistance in n-type GaSb layers grown by molecular beam epitaxy, *Semiconductor Science and Technology* 11 (11), 1656–1667, 1996.

98. Ozasa, K., Yuri, M., Tanaka, S., and Matsunami, H., Effect of misfit strain on physical properties of InGaP grown by metalorganic molecular-beam epitaxy, *Journal of Applied Physics* 68 (1), 107–111, 1990.
99. Buell, A. A., Pham, L. T., Newton, M. D., Venzor, G. M., Norton, E. M., Smith, E. P., Varesi, J. B., Harper, V. B., Johnson, S. M., Coussa, R. A., de Lyon, T., Roth, J. A., and Jensen, J. E., Physical structure of molecular-beam epitaxy growth defects in HgCdTe and their impact on two-color detector performance, *Journal of Electronic Materials* 33 (6), 662–666, 2004.
100. Zhang, D. H., Shi, W., Zheng, H. Q., Yoon, S. F., Kam, C. H., and Wang, X. Z., Physical properties of InGaAsP/InP grown by molecular beam epitaxy with valve phosphorous cracker cell, *Journal of Crystal Growth* 211 (1), 384–388, 2000.
101. Herman, M. A., Physical problems concerning effusion processes of semiconductors in molecular beam epitaxy, *Vacuum* 32 (9), 555–565, 1982.
102. Zhou, J. J., Li, Y., Thompson, P., Sato, D. L., Lee, H. P., and Kuo, J. M., Physical origins of temperature variation and background radiation associated with pyrometric interferometry measurement during III-V molecular-beam-epitaxy growth, *Applied Physics Letters* 69 (18), 2683–2685, 1996.
103. Magruder III, R. H., Haglund Jr., R. F., Yang, L., Wittig, J. E., and Zuhr, R. A., Physical and optical properties of Cu nanoclusters fabricated by ion implantation in fused silica, *Journal of Applied Physics* 76 (2), 708–715, 1994.
104. Armelao, L., Bertoncello, R., Cattaruzza, E., Gialanella, S., Gross, S., Mattei, G., Mazzoldi, P., and Tondello, E., Chemical and physical routes for composite materials synthesis: Ag and Ag2S nanoparticles in silica glass by sol-gel and ion implantation techniques, *Journal of Materials Chemistry* 12 (8), 2401–2407, 2002.
105. Ensinger, W., Klein, J., Usedom, P., and Rauschenbach, B., Characteristic features of an apparatus for plasma immersion ion implantation and physical vapour deposition, *Surface and Coatings Technology* 93 (2–3), 175–180, 1997.
106. Colwell, J. M., Wentrup-Byrne, E., Bell, J. M., and Wielunski, L. S., A study of the chemical and physical effects of ion implantation of micro-porous and nonporous PTFE, *Surface and Coatings Technology* 168 (2–3), 216–222, 2003.
107. Wei, Z., Xie, H., Han, G., and Li, W., Physical mechanisms of mutation induced by low energy ion implantation, *Nuclear Inst. and Methods in Physics Research, B* 95 (3), 371–378, 1995.
108. Zhu, H. N., and Liu, B. X., $CrSi_2$ films synthesized by high current Cr ion implantation and their physical properties, *Applied Surface Science* 161 (1), 240–248, 2000.
109. Battaglin, G., Bertoncello, R., Casarin, M., Cattaruzza, E., Mattei, G., Mazzoldi, P., Trivillin, F., and Urbani, M., Pd ion implantation in silica and alumina: Chemical and physical interactions, *Journal of Non-Crystalline Solids* 253 (1–3), 251–260, 1999.
110. Yang, D., Zhang, X., Xue, Q., Ding, X., and Lin, W., Surface physical and chemical changes of pure iron after molybdenum ion implantation and their effects on the tribological behaviour II: Tribological behaviour, *Thin Solid Films* 250 (1–2), 126–131, 1994.
111. Yang, D., Zhang, X., Xue, Q., Ding, X., and Lin, W., Surface physical and chemical changes of pure iron after molybdenum ion implantation and their effects on the tribological behaviour I: Physical and chemical analysis, *Thin Solid Films* 240 (1–2), 92–96, 1994.

112. Miotello, A., and Mazzoldi, P., Sputtering process during ion implantation in glasses: Mathematical and physical analysis, *Journal of Physics C: Solid State Physics* 16 (1), 221–228, 1983.
113. Pranevichus, L., and Tamulevichus, S., The physical properties of thin Ag films formed under the simultaneous ion implantation in the substrate, *Nuclear Instruments and Methods in Physics Research* 209–210 (Part 1), 179–184, 1983.
114. Buller, J. F., Brown, L., Cheek, J., Neal, T., Ballast, L., and Stallings, P., Physical and electrical examination of CMOS transistors with dual gate oxide process formed by ion implantation, *Journal of the Electrochemical Society* 150 (2), G103–G106, 2003.
115. Szekeres, A., Alexandrova, S., and Paneva, A., Effect of hydrogen ion implantation on the physical properties of SiO_2/Si system, *Vacuum* 58 (2), 166–173, 2000.
116. Kaito, T., Oba, H., Sugiyama, Y., Yasaka, A., Fujita, J. I., Suzuki, T., Kanda, K., and Matsui, S., Deposition yield and physical properties of carbon films deposited by focused-ion-beam chemical vapor deposition, *Japanese Journal of Applied Physics* 49 (6 Part 2), 06GH081–06GH084, 2010.
117. Tripathi, S. K., Shukla, N., and Kulkarni, V. N., Correlation between ion beam parameters and physical characteristics of nanostructures fabricated by focused ion beam, *Nuclear Instruments and Methods in Physics Research, Section B: Beam Interactions with Materials and Atoms* 266 (8), 1468–1474, 2008.
118. Fu, Y., and Bryan, N. K. A., Investigation of physical properties of quartz after focused ion beam bombardment, *Applied Physics B: Lasers and Optics* 80 (4–5), 581–585, 2005.
119. Ding, H., Erokhin, V., Ram, M. K., Paddeu, S., Valkova, L., and Nicolini, C., Physical insight into the gas-sensing properties of copper (II) tetra-(tert-butyl)-5,10,15,20-tetraazaporphyrin Langmuir-Blodgett films, *Thin Solid Films* 379 (1–2), 279–286, 2000.
120. Dourthe, C., Izumi, M., Garrigou-Lagrange, C., Buffeteau, T., Desbat, B., and Delhaes, P., Physical properties of mixed Langmuir-Blodgett conducting films based on a tetrathiafulvalene derivative, *Journal of Physical Chemistry* 96 (7), 2812–2820, 1992.
121. Ram, M. K., Carrara, S., Paddeu, S., and Nicolini, C., Effect of annealing on physical properties of conducting poly(ortho-anisidine) Langmuir-Blodgett films, *Thin Solid Films* 302 (1–2), 89–97, 1997.
122. Nakamura, T., Tachibana, H., Yumura, M., Matsumoto, M., and Tagaki, W., Structure and physical properties of Langmuir-Blodgett films of C60 with amphiphilic matrix molecules, *Synthetic Metals* 56 (2–3), 3131–3136, 1993.
123. Colbrook, R., and Roberts, G. G., Physical mechanisms of pyroelectricity in Langmuir-Blodgett films, *Thin Solid Films* 179 (1 –2 pt 2), 335–341, 1989.
124. Xiao, Y., Yao, Z., and Jin, D., Physical properties of a mixed conducting Langmuir-Blodgett film based on tetrathiafulvalene derivative with or without iodine oxidation, *Thin Solid Films* 249 (2), 210–214, 1994.
125. Dourthe-Lalanne, C., Izumi, M., Dupart, E., Flandrois, S., Delhaes, P., Buffeteau, T., Desbat, B., and Morand, J. P., Physical properties of mixed Langmuir-Blodgett conducting films based on a TTF derivative, *Synthetic Metals* 42 (1–2), 1451–1455, 1991.
126. Xiao, Y., Ye, Y., and Yao, Z., Preparation and physical properties of TMT-TTF-C60Br6 Langmuir-Blodgett film, *Thin Solid Films* 261 (1–2), 120–123, 1995.

127. Tieke, B., Wegmann, A., Fischer, W., Hilti, B., Mayer, C. W., and Pfeiffer, J., Novel conducting Langmuir-Blodgett films based on 2-n-octyloxy-5,6,11,12-tetrathiotetracene—preparation methods and physical properties, *Thin Solid Films* 179 (1 –2 pt 2), 233–238, 1989.
128. Sadagopan, K., Sawant, S. N., Kulshreshtha, S. K., and Jarori, G. K., Physical and chemical characterization of enolase immobilized polydiacetylene Langmuir-Blodgett film, *Sensors and Actuators, B: Chemical* 115 (1), 526–533, 2006.
129. Nakamura, T., Matsumoto, M., Tachibana, H., Tanaka, M., Manda, E., and Kawabata, Y., The structure and physical properties of N-docosylpyridinium-bistetracyanoquinodimethane Langmuir-Blodgett films, *Thin Solid Films* 178 (1–2), 413–419, 1989.
130. Yeh, F. B., Wei, P. S., and Chiu, S. H., Distinct property effects on rapid solidification of a thin liquid layer on a substrate subject to self-consistent melting, *Journal of Crystal Growth* 247 (3–4), 563–575, 2003.
131. Lee, K. S., Lee, B. S., Park, Y. H., Park, Y. C., Kim, Y. M., Jeong, S. H., and Kim, S. D., Dyeing properties of nylon 66 nano fiber with high molecular mass acid dyes, *Fibers and Polymers* 6 (1), 35–41, 2005.
132. Puppi, D., Piras, A. M., Chiellini, F., Chiellini, E., Martins, A., Leonor, I. B., Neves, N., and Reis, R., Optimized electro- and wet-spinning techniques for the production of polymeric fibrous scaffolds loaded with bisphosphonate and hydroxyapatite, *Journal of Tissue Engineering and Regenerative Medicine* 5 (4), 253–263, 2011.
133. Singh, G., Rana, D., Matsuura, T., Ramakrishna, S., Narbaitz, R. M., and Tabe, S., Removal of disinfection byproducts from water by carbonized electrospun nanofibrous membranes, *Separation and Purification Technology* 74 (2), 202–212, 2010.
134. Ceacutecile, C., and Hsieh, Y. L., Organic and aqueous compatible polystyrene—Maleic anhydride copolymer ultra-fine fibrous membranes, *Journal of Applied Polymer Science* 114 (2), 784–793, 2009.
135. Samani, F., Kokabi, M., and Valojerdi, R. M., Optimising the electrospinning process conditions to produce polyvinyl alcohol nanofibres, *International Journal of Nanotechnology* 6 (10–11), 1031–1040, 2009.
136. Changsarn, S., Mendez, J. D., Weder, C., and Supaphol, P., Morphology and photophysical properties of electrospun light-emitting polystyrene/poly-(p-phenylene ethynylene) fibers, *Macromolecular Materials and Engineering* 293 (12), 952–963, 2008.
137. Eldering, C. A., Kowel, S. T., Knoesen, A., Anderson, B. L., and Higgins, B. G., Characterization of modulated spin-coated and Langmuir-Blodgett thin film etalons, *Thin Solid Films* 179 (1–2), 535–542, 1989.

Index

D

S